Plane Trigonometry, Second Edition

Author: **Joseph Lloyd Harris**
             Professor of Mathematics
             Gulf Coast State College

Editor: **Sharon Hudson**
             Professor of Mathematics
             Gulf Coast State College

Special thank you to:

Gulf Coast State College District Board of Trustees

Dr. John Holdnak, President

Dr. Holly Kuehner, Vice-President of Academic Affairs

Angelia Reynolds, Chair of Mathematics

Scott Spencer, Senior Administrative Assistant, Mathematics

Mathematics Division

Copyright 2017, Gulf Coast State College

Plane Trigonometry

Table of Contents

Section 1: Angle Measure: Degrees and Radians, Coterminal Angles

Learning Outcome: The student will correctly memorize and apply trigonometric formulas, definitions, identities, and properties.

Objectives: At the conclusion of this lesson you should be able to:

1. Determine the quadrant an angle lies in.
2. Convert between decimal degrees and degrees-minutes-seconds.
3. Perform computations with degrees-minutes-seconds.
4. Convert between degrees and radians.
5. Find the complement and/or supplement of an angle.
6. Find coterminal angles.

An angle is formed by two lines or rays diverging from a common point called the vertex. The starting point of the angle is called the initial side and the position after the rotation is called the terminal side as shown in **Figure 1.1**. The amount of rotation from the initial side to the terminal side of an angle is the measure of the angle. The two ways to measure an angle are degrees and radians.

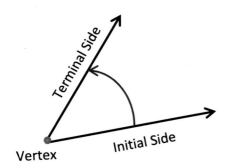

Figure 1.1

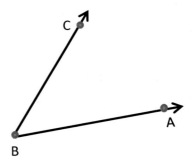

Figure 1.2

Angles are labeled with upper case letters such as A, B, and C and with Greek letters α (alpha), β (beta), γ (gamma), and θ (theta). In **Figure 1.2** the angle can be labeled three different ways: $\angle ABC$, $\angle CBA$, and $\angle B$.

An angle is in standard position when the vertex is at the origin and the initial side is on the x-axis as shown in **Figure 1.3**.

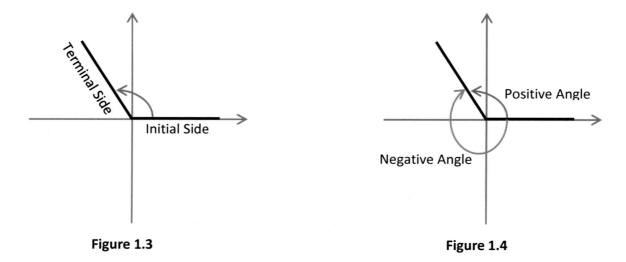

Figure 1.3   Figure 1.4

A positive angle is measured counterclockwise and a negative angle is measured clockwise as shown in **Figure 1.4**. Two angles that share the same initial side and the same terminal side are called coterminal angles. The two angles in **Figure 1.4** are coterminal.

Degrees are denoted by the symbol °. A complete counterclockwise rotation is 360°, a half rotation is 180°, one-fourth rotation is 90° as shown in **Figure 1.5**. A rotation of $\frac{1}{360}$ is a measure of 1°. Negative angles are measured clockwise as shown in **Figure 1.6**.

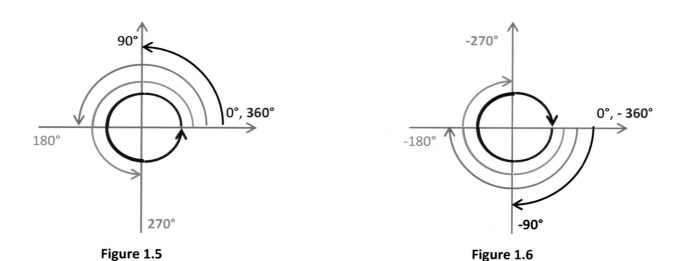

Figure 1.5   Figure 1.6

► EXAMPLE 1

Determine the quadrant each angle lies in.

a.) $60°$ b.) $-85°$

c.) $235°$ d.) $-275°$

e.) $-310°$ f.) $310°$

►► Solution

a.)   Quadrant I   b.)   Quadrant IV

c.)   Quadrant III   d.)   Quadrant I

e.)   Quadrant I   f.)   Quadrant IV

Degrees can be expressed as degrees($°$), minutes ($'$) and seconds($''$). A degree is divided into 60 minutes and a minute is divided into 60 seconds. These minutes and seconds have nothing to do with time. Minutes are very small and are used with measuring latitude and longitude.

$$60' = 1° \ or \ 1' = \frac{1}{60}°, \quad 60'' = 1' \ or \ 1'' = \frac{1'}{60}, \quad 3600'' = 1° \ or \ 1'' = \frac{1}{3600}°$$

► EXAMPLE 2

a.)   Convert $83°14'26''$ to decimal degrees rounded to the nearest ten-thousandth.

b.)   Convert $48.2685°$ to degrees, minutes, and seconds.

►► Solution

a.) $83°14'26'' = \left(83 + \dfrac{14}{60} + \dfrac{26}{3600}\right)° \approx 83.2406°$

b.) $48.2685° = 48°\left(0.2685 \times 60\right)'$

$$= 48° \ 16.11'$$
$$= 48° \ 16' \left(0.11 \times 60\right)''$$
$$= 48° \ 16' \ 6.6''$$
$$= 48° \ 16' \ 7''$$

▶EXAMPLE 3

Perform the indicated operations. Express answers in degrees, minutes, and seconds.

a.) $53°34'46" + 28°41'37"$ b.) $93°16'21" - 58°51'27"$

▶▶Solution

a.)   Add the degrees, minutes, and seconds:

$$53°34'46"$$
$$+ 28°41'37"$$
___
$$81°75'83"$$

Since $83" = 1' + 23"$ we have $81°76'23"$.

Since $76' = 1° + 16'$ we have $82°16'23"$

Therefore $53°34'46" + 28°41'37" = 82°16'23"$.

b.)

$$93°16'21"$$
$$- 58°51'27"$$

We cannot subtract 27 from 21. We know that $1' = 60"$ so we have to borrow 1 minute from the 16' to give us $93°15'81"$.

$$92°75'81"$$
$$- 58°51'27"$$
___
$$34°24'54"$$

We cannot subtract 51 from 15. We know $1° = 60'$ so we have to borrow 1 degree from the 93° to give us $92°75'81"$.

Therefore $93°16'21" - 58°51'27" = 34°24'54"$.

Recall that the standard form of the equation of a circle is $(x - h)^2 + (y - k)^2 = r^2$. When the center of the circle is at the origin we have what is called a central circle $x^2 + y^2 = r^2$ as shown in **Figure 1.7**. When the radius of a central circle is one, we have $x^2 + y^2 = 1$ which is called a unit circle as shown in **Figure 1.8**.

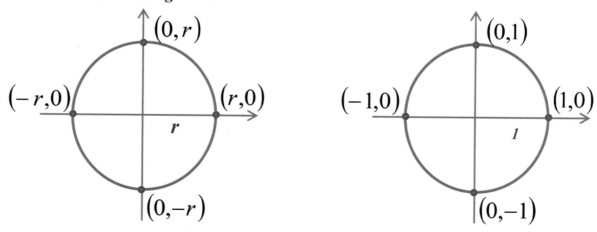

Figure 1.7   **Figure 1.8**

A central angle of a circle is an angle whose vertex is at the center of the circle. A central angle that intercepts an arc on the circle equal in length to the radius of the circle has a measure of one radian abbreviated as 1 rad. See **Figure 1.9**.

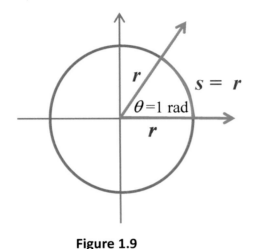

Arc Length

If θ is a central angle on a circle of radius *r*, then the length of the subtended arc *s* is given by $s = r\theta$ where θ is measured in radians.

Radians

If central angle θ is subtended by an arc that is equal to the radius, then $\theta = 1$ radian.

Figure 1.9

Since the circumference of a circle is $C = 2\pi r$ units, a central angle that makes one full revolution counterclockwise would have an arc length of $s = 2\pi r$. The units of measure for both *s* and *r* are the same which means the ratio of *s* to *r* has no units (they would cancel out). By the definition of a radian, the arc length *s* is equal to the radius *r*. Therefore one full revolution counterclockwise would be 2π radians. One-half revolution counterclockwise would be π radians and one-fourth revolution counterclockwise would be $\frac{\pi}{2}$ radians as shown in **Figure 1.10**. Clockwise yields negative angles as shown in **Figure 1.11**.

$$\pi \approx 3.14159 \qquad 2\pi \approx 6.28319$$

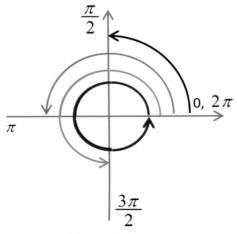

Figure 1.10

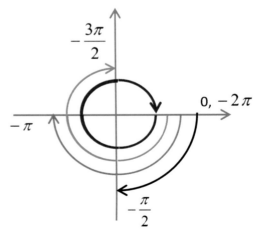

Figure 1.11

►EXAMPLE 4

Determine the quadrant each angle lies in.

a.) $\dfrac{2\pi}{3}$

b.) $-\dfrac{2\pi}{3}$

c.) $\dfrac{9\pi}{7}$

d.) -3.815

e.) $-\dfrac{11\pi}{6}$

f.) $\dfrac{7\pi}{4}$

►►Solution

a.) Quadrant II

b.) Quadrant III

c.) Quadrant III

d.) Quadrant II

e.) Quadrant I

f.) Quadrant IV

One revolution counterclockwise in degrees is 360° and in radians it is 2π rad which means $360°=2\pi$ and $180°=\pi$. We us the relationship $180°=\pi$ rad to develop a method for converting degrees and radians.

$$1° = \frac{\pi}{180}\,radian \qquad 1\,radian = \frac{180°}{\pi}$$

Converting from Radians to Degrees	**Converting from Degrees to Radians**
Multiply a radian measure by $\dfrac{180°}{\pi}$ and simplify to convert to degrees.	Multiply a degree measure by $\dfrac{\pi}{180}$ radian and simplify to convert to radians.

►EXAMPLE 5

Convert each degree measure to radians.

a.) 60°

b.) -225°

c.) 113.6° Round to the nearest ten-thousandth.

▶▶**Solution**

a.) $60° = 60\left(\dfrac{\pi}{180} \, radian\right) = \dfrac{\pi}{3} \, radians$

b.) $-225° = -225\left(\dfrac{\pi}{180} \, radian\right) = -\dfrac{5\pi}{4} \, radians$

c.) $113.6° = 113.6\left(\dfrac{\pi}{180} \, radian\right) \approx 1.9827 \, radians$

▶ **EXAMPLE 6**

Convert each radian measure to degrees.

a.) $\dfrac{5\pi}{6}$ b.) $-\dfrac{4\pi}{3}$ c.) 3.785

▶▶**Solution**

a.) $\dfrac{5\pi}{6} = \dfrac{5\pi}{6}\left(\dfrac{180°}{\pi}\right) = 150°$

b.) $-\dfrac{4\pi}{3} = -\dfrac{4\pi}{3}\left(\dfrac{180°}{\pi}\right) = -240°$

c.) $3.785 = 3.785\left(\dfrac{180°}{\pi}\right) = 216.8645°$ *or* $216°51'52''$

A right angle measures $90°$ or $\frac{\pi}{2}$ rad and is denoted by a box as shown in **Figure 1.12**. A straight angle measures $180°$ or π rad shown in **Figure 1.13**.

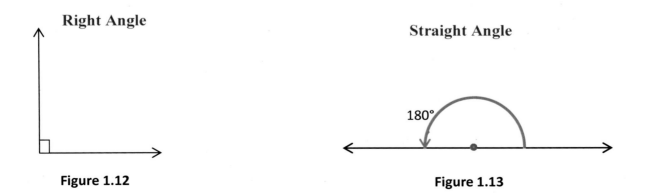

Figure 1.12   **Figure 1.13**

Two angles that add to 90° or $\frac{\pi}{2}$ rad are called complementary angles as shown in **Figure 1.14**.
Two angles that add to 180° or π rad are called supplementary angles as shown in **Figure 1.15**.

Complementary Angles   **Supplementary Angles**

$$\alpha + \beta = 90°$$

$$\alpha + \beta = 180°$$

$$\alpha + \beta = \frac{\pi}{2}$$

$$\alpha + \beta = \pi$$

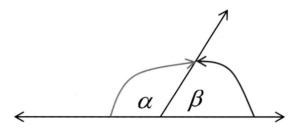

Figure 1.14   **Figure 1.15**

▶ EXAMPLE 7

Determine the measure of each angle described.

a.)   The complement of 62°.   b.)   The supplement of 112°.

c.)   The complement of $\dfrac{\pi}{3}$.   d.)   The supplement of $\dfrac{3\pi}{4}$.

e.)   The complement of 28°45'16".   f.)   The supplement of 88°29'41".

▶▶ Solution

a.) $90° - 62° = 28°$ b.) $180° - 112° = 68°$

c.) $\dfrac{\pi}{2} - \dfrac{\pi}{3} = \dfrac{3\pi}{6} - \dfrac{2\pi}{6} = \dfrac{\pi}{6}$ d.) $\pi - \dfrac{3\pi}{4} = \dfrac{4\pi}{4} - \dfrac{3\pi}{4} = \dfrac{\pi}{4}$

e.) $90° - 28°45'16" = 89°59'60" - 28°45'16" = 61°14'44"$

f.) $180° - 88°29'41" = 179°59'60" - 88°29'41" = 91°30'19"$

Coterminal angles are angles that share the same initial side and the same terminal side as shown in **Figure 1.16**.

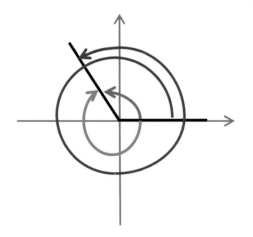

Figure 1.16

> ### Coterminal Angles
>
> - Angles that share the same initial side and the same terminal side.
> - Coterminal Angles will always differ by multiples of $360°$ or 2π .

> ### Formula for Finding Coterminal Angles
>
> Angle $+ 360°k$
>
> Angle $+ 2\pi k$
>
> Where k is an integer.

►EXAMPLE 8

Find two positive and two negative coterminal angles.

a.) $215°$ b.) $\dfrac{11\pi}{6}$

►►Solution

a.) $k = 1$ $215° + 360° \cdot 1 = 215° + 360° = 575°$

     $k = 2$ $215° + 360° \cdot 2 = 215° + 720° = 935°$

     $k = -1$ $215° + 360° \cdot (-1) = 215° - 360° = -145°$

     $k = -2$ $215° + 360° \cdot (-2) = 215° - 720° = -505°$

b.) $k = 1$ $\dfrac{11\pi}{6} + 2\pi \cdot 1 = \dfrac{11\pi}{6} + 2\pi = \dfrac{11\pi}{6} + \dfrac{12\pi}{6} = \dfrac{23\pi}{6}$

     $k = 2$ $\dfrac{11\pi}{6} + 2\pi \cdot 2 = \dfrac{11\pi}{6} + 4\pi = \dfrac{11\pi}{6} + \dfrac{24\pi}{6} = \dfrac{35\pi}{6}$

     $k = -1$ $\dfrac{11\pi}{6} + 2\pi \cdot (-1) = \dfrac{11\pi}{6} - 2\pi = \dfrac{11\pi}{6} - \dfrac{12\pi}{6} = -\dfrac{\pi}{6}$

     $k = -2$ $\dfrac{11\pi}{6} + 2\pi \cdot (-2) = \dfrac{11\pi}{6} - 4\pi = \dfrac{11\pi}{6} - \dfrac{24\pi}{6} = -\dfrac{13\pi}{6}$

HOMEWORK

Objective 1

Determine the quadrant each angle lies in.

See example 1 and example 4.

1.   65°

2. $\dfrac{11\pi}{6}$

3. $-95°$

4.   245°

5. $\dfrac{7\pi}{9}$

6. $-\dfrac{7\pi}{6}$

7.   2.25

8. $-225°$

9.   315°

Objective 2

Convert each angle measure to degree, minutes, and seconds.

See example 2b.

10.   126.76°

11.   31.4296°

12.   174.255°

13.   59.0854°

Objective 2

Convert each angle measure to decimal degrees. If applicable, round to the nearest ten-thousandth.

See example 2a.

14.   91°35'54"

15.   111°51'24"

16.   27°37'44"

17.   251°6'28"

Objective 3

Perform the indicated operations. Express answers in degrees, minutes, and seconds.
See example 3.

18. $16°\,23'\,41'' + 44°\,43'\,39''$

19. $34°\,39'\,12'' - 9°\,49'\,18''$

20. $7°\,55'\,42'' + 8°\,22'\,28''$

21. $68°\,19'\,22'' - 12°\,45'\,58''$

Objective 4

Convert each degree measure to radians. Round to the nearest ten-thousandth for #25.
See example 5.

22. $30°$ 23. $240°$ 24. $-210°$ 25. $75.8°$

Objective 4

Convert each radian measure to degrees. Round to the nearest ten-thousandth for #29.
See example 6.

26. $\dfrac{5\pi}{3}$ 27. $-\dfrac{5\pi}{12}$ 28. $\dfrac{5\pi}{4}$ 29. 2.87

Objective 5

Find the complement of each angle.

See example 7.

30. $\dfrac{\pi}{4}$ 31. $71°$ 32. $51°16'39''$ 33. $\dfrac{\pi}{6}$

Objective 5

Find the supplement of each angle.

See example 7.

34. $\dfrac{5\pi}{6}$ 35. $136°$ 36. $88°29'41''$ 37. $\dfrac{7\pi}{9}$

Objective 6

Find two positive and two negative coterminal angles for each angle.

See example 8.

38. $165°$ 39. $-215°$ 40. $\dfrac{5\pi}{8}$ 41. $-\dfrac{7\pi}{6}$

ANSWERS

1.	Quadrant 1	2.	Quadrant IV	3.	Quadrant III
4.	Quadrant III	5.	Quadrant II	6.	Quadrant II
7.	Quadrant II	8.	Quadrant II	9.	Quadrant IV

10. 126°45'36" 11. 31°25'47" 12. 174°15'18" 13. 59°5'7"

14. 91.5983° 15. 111.8567° 16. 27.6289° 17. 251.1078°

18. 61°7'20" 19. 24°49'54" 20. 16°18'10" 21. 55°33'24"

22. $\dfrac{\pi}{6}$ 23. $\dfrac{4\pi}{3}$ 24. $-\dfrac{7\pi}{6}$ 25. 1.3230

26. 300° 27. -75° 28. 225° 29. 164.4389°

30. $\dfrac{\pi}{4}$ 31. 19° 32. 38°43'21" 33. $\dfrac{\pi}{3}$

34. $\dfrac{\pi}{6}$ 35. 44° 36. 91°30'19" 37. $\dfrac{2\pi}{9}$

38. 525°, 885°, -195°, -555° 39. 145°, 505°, -575°, -935°

40. $\dfrac{21\pi}{8}, \dfrac{37\pi}{8}, -\dfrac{11\pi}{8}, -\dfrac{27\pi}{8}$ 41. $\dfrac{5\pi}{6}, \dfrac{17\pi}{6}, -\dfrac{19\pi}{6}, -\dfrac{31\pi}{6}$

Section 2: Arc Length, Angular Velocity, Linear Velocity, and Area of a Sector

Learning Outcomes:

- The student will correctly memorize and apply trigonometric formulas, definitions, identities, and properties.
- The student will correctly use mathematical symbols and mathematical structure to examine and solve real world applications.

Objectives: At the conclusion of this lesson you should be able to:

1. Compute arc length.
2. Solve applications involving arc length.
3. Solve applications involving angular velocity and linear velocity.
4. Compute the area of a sector.
5. Solve applications involving the area of a sector.

An arc is a part of the circumference of a circle. Given the central circle with radius r in **Figure 2.1**, the central angle θ intercepts an arc of length s given by $s = r\theta$ where θ is measured in radians.

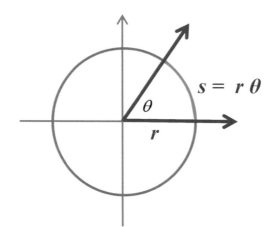

$$s = r\theta$$

Arc Length

If θ is a central angle on a circle of radius r, then the length of the subtended arc s is given by $s = r\theta$ where θ is measured in radians.

Figure 2.1

► EXAMPLE 1

Use the formula for arc length to find the value of the remaining unknown: $s = r\theta$. Round answers to the nearest tenth when necessary.

a.) $r = 2.6\,m, \theta = 112°$ b.) $r = 5.8\,cm, s = 29\,cm$ c.) $\theta = 64°, s = 89.7\,in$

►►**Solution**

a.) $s = r\theta$

$$s = (2.6\,cm)\left(112° \cdot \frac{\pi}{180°}\right)$$

$$s = 5.082398782\,cm$$

$$s = 5.1\,cm$$

b.) $s = r\theta$

$$29\,cm = (5.8\,cm)\theta$$

$$\frac{29\,cm}{5.8\,cm} = \theta$$

$$5\,rad = \theta$$

c.) $s = r\theta$

$$89.7\,in = r\left(64° \cdot \frac{\pi}{180°}\right)$$

$$\frac{89.7\,in}{\dfrac{64\pi}{180}} = r$$

$$80.3036\,in = r$$

$$80.3\,in = r$$

Latitude

The latitude of a fixed point on the Earth's surface tells how many degrees north or south of the equator the point is, as measured from the center of the Earth.

Longitude

The longitude of a fixed point on the Earth's surface tells how many degrees east or west of the Greenwich Meridian (located in England) the point is, as measured along the equator to the north/south line going through the point.

►**EXAMPLE 2**

a.) Pittsburgh, Pennsylvania and West Palm Beach, Florida are on the same longitude (one city is due north of the other). Pittsburgh is at $40°18'$ north latitude, while West Palm Beach is at $26°24'$ north latitude. Assuming the radius of the earth is 6378 kilometers (km), how far apart are the cities? Round to the nearest whole number.

b.) The rotation of the smaller wheel in the figure causes the larger wheel to rotate. Through how many degrees will the larger wheel rotate if the smaller one rotates through 54°? Round to the nearest tenth of a degree.

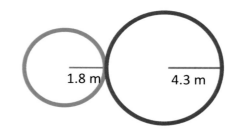

1.8 m   4.3 m

c.) Johannesburg, South Africa and Jerusalem, Israel are on the same longitude. Johannesburg is $26°10'$ south latitude and Jerusalem is $31°47'$ north latitude. Assuming the radius of the earth is 6378 km, how far apart are the cities. Round to the nearest whole number.

►►**Solution**

a.)

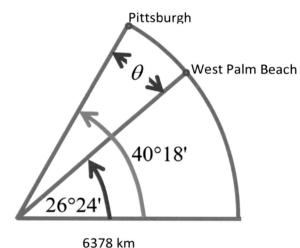

Pittsburgh

West Palm Beach

$40°18'$

$26°24'$

6378 km

$$\theta = 40°18' - 26°24' = 13°54'$$

$$s = r\theta$$

$$s = (6378\,km)\left(13°54' \cdot \frac{\pi}{180°}\right)$$

$$s = 1547.307686\,km$$

$$s = 1547\,km$$

b.)   The arc length of the larger wheel will be the same as the smaller wheel.

Smaller Wheel	Larger Wheel	Convert radians to degrees.

Smaller Wheel

$$s = r\theta$$

$$s = (1.8\,m)\left(54° \cdot \frac{\pi}{180°}\right)$$

$$s = \frac{27\pi}{50}\,m$$

Larger Wheel

$$s = r\theta$$

$$\frac{27\pi}{50}\,m = (4.3\,m)\theta$$

$$\frac{\dfrac{27\pi}{50}}{4.3} = \theta$$

Convert radians to degrees.

$$\frac{\dfrac{27\pi}{50}}{4.3} \cdot \frac{180°}{\pi} = \theta$$

$$22.6047° = \theta$$

$$22.6° = \theta$$

c.)

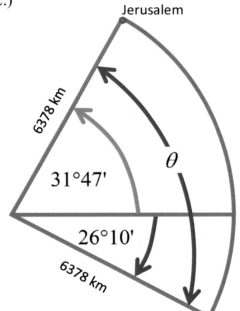

Jerusalem

6378 km

$31°47'$

$26°10'$

6378 km

Johannesburg

$$\theta = 31°47' + 26°10' = 57°57'$$

$$s = r\theta$$

$$s = (6378\,km)\left(57°57' \cdot \frac{\pi}{180°}\right)$$

$$s = 6450.825927\,km$$

$$s = 6451\,km$$

Linear velocity also called linear speed is the distance traveled per unit of time. Consider a particle **P** moving at a constant speed along a circular arc of radius r through an angle of θ radians. If s is the length of the arc traveled in time t, then the linear velocity v of the particle **P** is

$v = \dfrac{s}{t}$. **See Figure 2.2.** Similarly, angular velocity is the amount of rotation per unit of time.

Linear velocity is the arc length over time $v = \dfrac{s}{t}$ and angular velocity is the angle θ in radians

over time $\omega = \dfrac{\theta}{t}$ (ω is the lower case Greek letter omega).

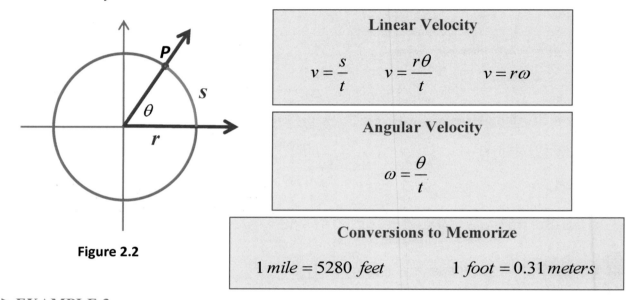

Figure 2.2

Linear Velocity

$$v = \frac{s}{t} \qquad v = \frac{r\theta}{t} \qquad v = r\omega$$

Angular Velocity

$$\omega = \frac{\theta}{t}$$

Conversions to Memorize

$1\ mile = 5280\ feet \qquad 1\ foot = 0.31\ meters$

► EXAMPLE 3

a.)   The wheels of a bicycle have radius 14 inches and are turning at the rate of 225 revolutions per minute. How fast is the bicycle traveling in miles per hour (mph)? Round to the nearest tenth.

b.)   A Ferris wheel with a 31 meter diameter makes 2 revolutions per minute. What is the angular velocity in radians per minute? What is the linear velocity in miles per hour? Round answers to the nearest tenth.

c.)   The planet Neptune orbits the Sun at a distance of 4500 million kilometers and completes one revolution every 165 years. Find the linear velocity in kilometers/hour as it orbits the Sun. Round to the nearest tenth.

d.)   The two pulleys in the figure have radii of 12 cm and 5 cm respectively and are connected by a belt. The larger pulley rotates 50 times in 45 seconds. Find the angular velocity of the smaller pulley in radians per seconds. Round to the nearest tenth.

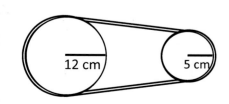

▶▶**Solution**

a.) $v = r\omega$

$$v = \frac{14\ in}{1} \cdot \frac{225\,rev}{1\,min} \cdot \frac{2\pi}{1\,rev} \cdot \frac{1\,ft}{12\,in} \cdot \frac{1\,mile}{5280\,ft} \cdot \frac{60\,min}{1\,hour}$$

$\underbrace{\qquad}_{\text{radius}}$ $\underbrace{\qquad\qquad}_{\text{Angular velocity}}$ $\underbrace{\qquad\qquad\qquad}_{\text{Conversion to mph}}$

$v = 18.7\,mph$

b.)   Angular Velocity

$$\omega = \frac{\theta}{t}$$

$$\omega = \frac{2\,rev}{1\,min} \cdot \frac{2\pi}{1\,rev}$$

$\omega = 4\pi\ rad\,/\,min$

$\omega = 12.6\,rad\,/\,min$

$radius = \dfrac{1}{2}\,diameter$

$r = \dfrac{1}{2}\left(31m\right)$ $r = 15.5\,m$

Linear Velocity

$v = r\omega$

$$v = \frac{15.5\,m}{1} \cdot \frac{4\pi}{1\,min} \cdot \frac{1\,ft}{0.31m} \cdot \frac{1\,mile}{5280\,ft} \cdot \frac{60\,min}{1\,hour}$$

$v = 7.1\,mph$

c.)   Linear Velocity

$v = r\omega$

$$v = \frac{4500\,km}{1} \cdot \frac{1\,rev}{165\,yr} \cdot \frac{2\pi}{1\,rev} \cdot \frac{1\,yr}{365\,days} \cdot \frac{1\,day}{24\,hr}$$

$v = 0.0195615981\ million\ \ km/\,hr$

$v = 19,561.6\ km/\,hr$

d.)   Both pulleys are connected by a belt and are turning at the same time which means both pulleys have the same linear velocity. Find the linear velocity of the larger pulley and use it to find the angular velocity of the smaller pulley.

Linear Velocity of Larger Pulley

$v = r\omega$

$$v = \frac{12\,cm}{1} \cdot \frac{50\,rot}{45\,sec} \cdot \frac{2\pi}{1\,rot}$$

$$v = \frac{80\pi}{3}\,cm\,/\,sec$$

Angular Velocity of the Smaller Pulley

$v = r\omega$ $\dfrac{v}{r} = \omega$

$$\frac{\dfrac{80\pi\ cm}{3\ sec}}{5\ cm} = \omega$$

$16.8\ rad\,/\,sec = \omega$

A sector of a circle is the region bounded by two radii of the circle and their intercepted arc. The area of a circle is $A = \pi r^2$ for a full rotation of 2π. If we let A equal the area of the sector and θ represent the central angle, we get the proportion: $\dfrac{Area\ of\ Sector\,(A)}{Area\ of\ Circle\left(\pi r^2\right)} = \dfrac{Central\ Angle\,(\theta)}{2\pi}$.

If we solve for A in the proportion $\dfrac{A}{\pi r^2} = \dfrac{\theta}{2\pi}$ we cross multiply to get $2\pi A = \theta \pi r^2$ and then

divide by 2π which gives us the area of a sector: $A = \dfrac{1}{2}\theta r^2$. The area of the sector shown in

Figure 2.3 with radius r and central angle θ measured in radians is given by $A = \dfrac{1}{2}r^2\theta$.

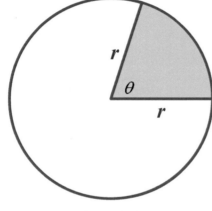

Figure 2.3

Area of a Sector
$A = \dfrac{1}{2}r^2\theta$
θ is measured in radians.

▶EXAMPLE 4

a.)   Find the area of a sector with radius 6 inches and central angle of 4π . Round to the nearest tenth.

b.)   A lawn sprinkler is set to spray water over a distance of 70 feet and rotates through an angle of 120°. Find the area of the lawn watered by the sprinkler. Round to the nearest tenth.

c.)   Find the area of the sector. Round to the nearest tenth.

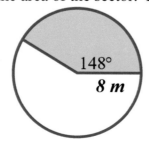

▶▶ **Solution**

a.) $A = \dfrac{1}{2}r^2\theta$

$A = \dfrac{1}{2}(6\,in)^2(4\pi)$

$A = 226.194671\,in^2$

$A = 226.2\,in^2$

b.) $A = \dfrac{1}{2}r^2\theta$

$A = \dfrac{1}{2}(70\,ft)^2\left(120° \cdot \dfrac{\pi}{180°}\right)$

$A = 5131.268\,ft^2$

$A = 5131.3\,ft^2$

c.) $A = \dfrac{1}{2}r^2\theta$

$A = \dfrac{1}{2}(8\,m)^2\left(148° \cdot \dfrac{\pi}{180°}\right)$

$A = 82.65879337\,m^2$

$A = 82.7\,m^2$

HOMEWORK

Objective 1

Use the formula for arc length to find the value of the remaining unknown. Round answers to the nearest tenth when necessary.

See example 1.

1. $r = 9.8m, \theta = 75°$

2. $r = 8cm, s = 12cm$

3. $r = 12.8m, s = 59m$

4. $\theta = \dfrac{4\pi}{3}, s = 24in$

5. $\theta = 42°, s = 16.2m$

6. $r = 9cm, \theta = \dfrac{7\pi}{6}$

7. $r = 10.8\,miles, \theta = 115°$

8. $r = 84cm, s = 175cm$

Objective 2

Solve. *See example 2.*

9.   Dallas, Texas and Omaha Nebraska are on the same longitude. Dallas is at $32°47'$ north latitude, while Omaha is at $41°16'$ north latitude. Assuming the radius of the earth is 6378 km, how far apart are the cities? Round to the nearest whole number.

10.   Miami, Florida and Erie, Pennsylvania are on the same longitude. Miami is at $25°46'37''$ north latitude, while Erie is at $42°7'15''$ north latitude. Assuming the radius of the earth is 3960 miles (mi), how far apart are the cities? Round to the nearest whole number.

11.   Find the distance between each pair of cities assuming they lie on the same north-south line. The radius of the earth is 6378 km. Round to the nearest whole number.

    a.)   Panama City, Panama $9°N$

            Pittsburgh, Pennsylvania $40°N$

    b.)   Halifax, Nova Scotia $45°N$

            Buenos Aires, Argentina $34°S$

    c.)   New York City, New York $41°N$

            Lima, Peru $12°S$

    d.)   San Francisco, California $37°48'N$

            Seattle, Washington $47°38'N$

12. How many inches to the nearest tenth will the weight in the figure rise if the pulley is rotated through an angle of 28°16'? Through what angle, to the nearest minute, must the pulley be rotated to raise the weight 25 inches?

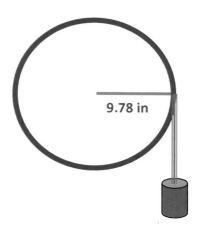

13.

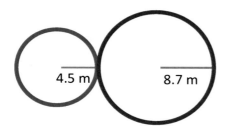

The rotation of the smaller wheel in the figure causes the larger wheel to rotate. Through how many degrees will the larger wheel rotate if the smaller one rotates through 96°? Round to the nearest tenth of a degree.

14. Find the radius of the larger wheel in the figure if the smaller wheel rotates 88° when the larger wheel rotates 48°. Round to the nearest tenth.

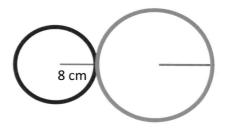

15.

The wagon wheel has a diameter of 28 inches and an angle of 30° between each spoke. What is the length of the arc (to the nearest tenth of an inch) between two adjacent spokes?

16. A hoist is being used to lift a beam. The diameter of the drum on the hoist is 13 inches and the beam must be raised 4 feet. Find the number of degrees through which the drum must rotate. Round to the nearest tenth of a degree.

Objective 3

Solve. *See example 3.*

17. A round-a-bout on the playground has a radius of 48 inches. If the round-a-bout is turning at $\frac{5}{6}$ revolutions per second: What is the angular velocity? What is the linear velocity in miles per hour? Round answers to the nearest tenth.

18. A Ferris Wheel has a 30 meter diameter making three revolutions per minute. What is the angular velocity of the Ferris Wheel? What is the linear velocity of the Ferris Wheel in miles per hour? Round answers to the nearest tenth.

19. What is linear velocity in miles per hour of a bicycle with 26 inch diameter tires making 100 revolutions per minute? Round to the nearest tenth.

20. A 12 inch diameter blade on a table saw rotates at 4000 revolutions per minute. How fast is the blade striking a piece wood in feet per second?

21. The diameter of a DVD is approximately 12 centimeters and makes 200 revolutions per minute. What is the angular velocity of the DVD?

22. Find the velocity v in miles per hour of a point on the edge of a flywheel of radius 18 inches rotating 400 times per minute.   Round to the nearest tenth.

23. Find the angular velocity of a pulley in radians per second if the radius of a pulley is 5 inches and the pulley is turning at 58 rpm (revolutions per minute). Round to the nearest tenth.

24. A water wheel has 12-ft radius. The wheel makes 15 revolutions per minute. What is the speed of the river in miles per hour?

25. The rear wheel on a John Deere farm tractor has a 23-in radius. If the wheel makes 20 rotations in 12 seconds, how fast is the tractor traveling in miles per hour?

26. A merry-go- round competes 60 revolutions per minute. Find the angular velocity in radians per minute. If the radius of the merry-go-round is 6 feet, what is the linear velocity in miles per hour?   Round to the nearest hundredth.

Objective 4and 5

Solve. *See example 4.*

27.   A sprinkler on a golf course is set to spray water over a distance of 100 feet and rotates through an angle of 120°. Find the area of the fairway watered by the sprinkler. Round to the nearest hundredth.

28.   What is the area of a slice of pizza to the nearest tenth if the pizza has a radius of 10 inches and the slice of pizza has a central angle of 25.1°?

29.   A weather radar system scans a circular area of radius 40 miles. In one second it scans a sector with central angle of 80°. What area to the nearest square mile is scanned at that time?

30.   Find the area of the sector. Round to the nearest tenth of a meter.

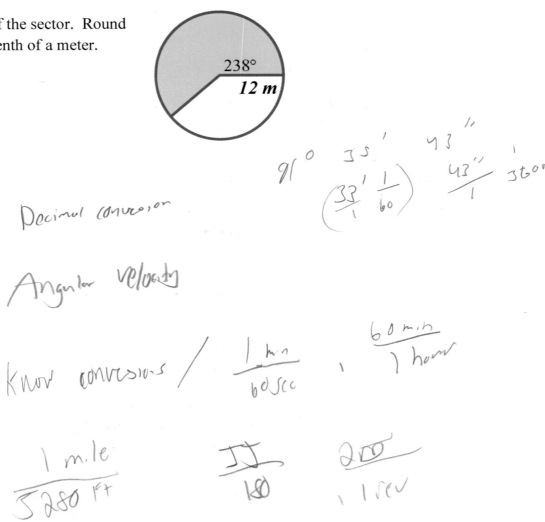

238°

12 m

$91°$ $35'$ $43''$

$\left(33' \dfrac{1}{60}\right)$ $\dfrac{43''}{1}$ 3600

Decimal conversion

Angular velocity

Know conversions / $\dfrac{1 \, min}{60 \, sec}$, $\dfrac{60 \, min}{1 \, hour}$

$\dfrac{1 \, mile}{5280 \, ft}$ $\dfrac{II}{180}$ $\dfrac{2\pi}{1 \, rev}$

ex. $48°26'85°$

48 (.2685 × 60)
4.44   (44 × 60)

ANSWERS

1.	$s = 12.8\,m$	2.	$\theta = 1.5\,rad$	3.	$\theta = 4.6\,rad$
4.	$r = 5.7\,in$	5.	$r = 22.1\,m$	6.	$s = 33.0\,cm$
7.	$s = 21.7\,miles$	8.	$\theta = 2.1\,rad$	9.	$944\,km$
10.	$1130\,miles$	11a.	$3451\,km$	11b.	$8794\,km$
11c.	$5900\,km$	11d.	$1095\,km$	12.	$4.8\,in,\ 146°28'$
13.	$49.7°$	14.	$14.7\,cm$	15.	$7.3\,in$
16.	$423.1°$	17.	5.2 rad/sec, 14.3 mph	18.	18.8 rad/min, 10.3 mph
19.	7.7 mph	20.	209.4 ft/sec	21.	1256.6 rad/min
22.	42.8 mph	23.	6.1 rad/sec	24.	12.9 mph
25.	13.7 mph	26.	376.99 rad/min, 25.70 mph		
27.	$10{,}471.98\,ft^2$	28.	$21.9\,in^2$		
29.	$1117.0\,mi^2$	30.	$299.1\,m^2$		

Section 3: Triangles: Similar, Right, and Special

Learning Outcomes:

- The student will correctly memorize and apply trigonometric formulas, definitions, identities, and properties.
- The student will correctly use mathematical symbols and mathematical structure to examine and solve real world applications.

Objectives: At the conclusion of this lesson you should be able to:

1. Find the angles of a triangle.
2. Solve applications involving similar triangles.
3. Use the Pythagorean Theorem to find sides of a right triangle.
4. Solve applications involving special triangles.

A triangle is a closed plane figure with three angles and three straight sides. The angles of a triangle are labeled with upper case letters and the corresponding sides with lower case letters as shown in **Figure 3.1.** The measures of the three angles of a triangle will always add to equal 180°. If we add the measure of any two sides of a triangle the sum will exceed the measure of the remaining side. The largest angle of a triangle is always opposite the largest side of the triangle and the smallest angle is always opposite the smallest side. These properties of a triangle are important and need to be remembered.

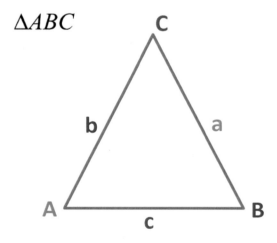

ΔABC

Figure 3.1

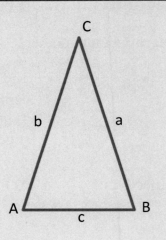

Properties of a Triangle

I. The sum of the angles of a triangle is 180°:
$A + B + C = 180°$

II. The combined length of any two sides exceeds that of the third side: $a + b > c$, $a + c > b$, and $b + c > a$

III. The largest angle is opposite the largest side: $B > A$, then $b > a$.

►EXAMPLE 1

1.) Find the measure of each angle for $\triangle ABC$.

A = _____   B= _____   C= _____

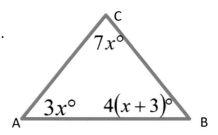

2.) Find the measure of the indicated angles.

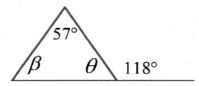

$\theta =$ _____

$\beta =$ _____

►►Solution

1.)
$$A + B + C = 180°$$
$$3x + 4(x + 3) + 7x = 180$$
$$3x + 4x + 12 + 7x = 180$$
$$14x + 12 = 180$$
$$14x = 168$$
$$x = 12$$

$$A = 3(12) = 36°$$
$$B = 4(12 + 3) = 4(15) = 60°$$
$$C = 7(12) = 84°$$

2.) θ and 118° are supplemental

$$\theta = 180° - 118°$$
$$\theta = 62°$$

$$\beta = 180° - (57° + 62°)$$
$$\beta = 180° - 119°$$
$$\beta = 61°$$

Similar triangles are triangles that have corresponding equal angles and corresponding sides that are proportional. We use the ~ symbol to represent similar. Two triangles are similar if they have at least two equal angles. In **Figure 3.2** $\triangle ABC$ and $\triangle DEF$ are similar triangles and we write this as $\triangle ABC \sim \triangle DEF$. Angle A corresponds to angle D, angle B corresponds to angle E, and angle C corresponds to angle F. Since the triangles are similar, the corresponding angles are equal and the ratios of their corresponding sides must be equal: $\dfrac{a}{d} = \dfrac{b}{e} = \dfrac{c}{f}$ or

$\dfrac{d}{a} = \dfrac{e}{b} = \dfrac{f}{c}$. We can also have proportions such as: $\dfrac{a}{b} = \dfrac{d}{e}$, $\dfrac{b}{c} = \dfrac{e}{f}$, and $\dfrac{a}{c} = \dfrac{d}{f}$. In

Figure 3.3 $\triangle XYZ \sim \triangle WVZ$ which means angle X corresponds to angle W, angle Y corresponds to angle V, and angle Z is the same in both triangles.

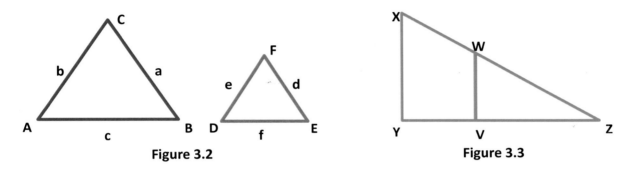

Figure 3.2   **Figure 3.3**

Similar triangles are useful for measuring inaccessible objects. For example, to measure the height of a building, we can use the following procedure: measure the shadow of the building, then hold a stick vertically nearby, and measure the shadow that the stick casts. The height of the building is a vertical measurement and the stick is vertical, so we have a pair of right angles. The sun is a fixed point in the sky and so very far away that we can assume the rays are parallel; the sun's rays create equal angles at the top of the building and the top of the stick. See **Figure 3.4**. Therefore the two triangles are similar and we can write a proportion and find the height of the building.

$$\frac{height\ of\ building}{height\ of\ stick} = \frac{shadow\ of\ building}{shadow\ of\ stick}$$

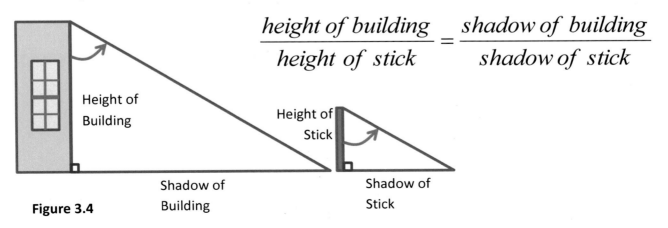

Figure 3.4

Properties of Similar Triangles

- Corresponding angles are equal

- Corresponding sides are proportional

►EXAMPLE 2

1.) $\triangle ABC \sim \triangle DEF$ complete the following:

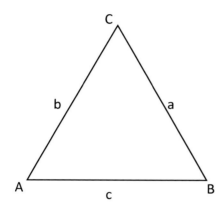
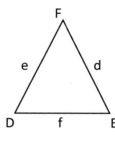

A = _____   D = _____

B = _____   E = 30°

C = 72°   F = _____

a = _____   d =   48 ft.

b =   12 ft.   e =   36 ft.

c =   15 ft.   f = _____

►►Solution

$\triangle ABC \sim \triangle DEF$: Angle B corresponds to angle E. Since angle E equals 30°, angle B will equal 30°. Angle C corresponds with angle F, so angle F will equal 72°. Adding angles B and C and subtracting from 180° gives us angle A which is 78°. Angle A corresponds with angle D so angle D is 78°.   The sides are proportional:

$$\frac{a}{12} = \frac{48}{36}$$
$$36a = 576$$
$$a = 16$$

$$\frac{12}{36} = \frac{15}{f}$$
$$12f = 540$$
$$a = 45$$

A = 78°   D = 78°

B = 30°   E = 30°

C = 72°   F = 72°

a =   16 ft.   d =   48 ft.

b =   12 ft.   e =   36 ft.

c =   15 ft.   f =   45 ft.

► EXAMPLE 3

If Amy is 5 ft tall and casts a
shadow of 8 ft, find the height of
the statue to the nearest foot if the
statue casts a shadow of 23 ft.

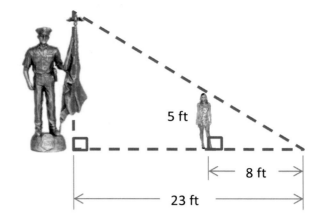

5 ft

8 ft

23 ft

►► Solution

$$\frac{x \ ft}{5 \ ft} = \frac{23 \ ft}{8 \ ft}$$

$$8x = 115 \ ft$$

$$x = 14.375 \ ft$$

$$x = 14 \ ft$$

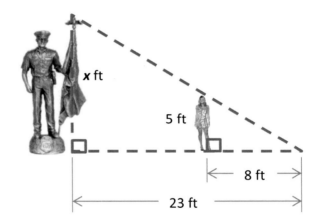

x ft

5 ft

8 ft

23 ft

► EXAMPLE 4

A house is 25 ft. tall and casts a shadow of 60 ft at the same time the shadow of a
nearby building is 300 ft.   Find the height of the building to the nearest foot.

►► Solution

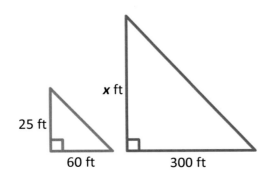

x ft

25 ft

60 ft   300 ft

$$\frac{25 \ ft}{60 \ ft} = \frac{x \ ft}{300 \ ft}$$

$$60x = 7500$$

$$x = 125 \ ft$$

In a right triangle, the two sides that form the 90° are called legs and the side opposite the 90° is called the hypotenuse as shown in **Figure 3.5**. The sum of the squares of the lengths of the legs is equal to the square of the length of the hypotenuse. This is called the Pythagorean Theorem.

If we let a and b represent the legs and c represent the hypotenuse we have: $a^2 + b^2 = c^2$.

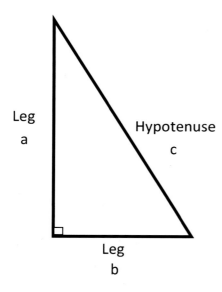

Leg
a

Hypotenuse
c

Leg
b

Figure 3.5

Pythagorean Theorem

$$(leg)^2 + (leg)^2 = (hypotenuse)^2$$

$$a^2 + b^2 = c^2$$

► EXAMPLE 5

Find the measure of the missing side. Round to the nearest tenth of a meter.

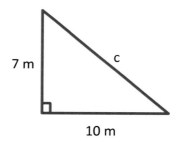

7 m

c

10 m

►►Solution

$$a^2 + b^2 = c^2$$

$$(7)^2 + (10)^2 = c^2$$

$$49 + 100 = c^2$$

$$149 = c^2$$

$$\sqrt{149} = \sqrt{c^2}$$

$$12.2\,m = c$$

▶ EXAMPLE 6

A 20 foot ladder is leaning against the wall of a house. If the ladder is 4 feet from the wall, how far up the wall is the ladder? Round to the nearest tenth of a foot.

▶▶ Solution

$$a^2 + b^2 = c^2$$

$$(4)^2 + b^2 = (20)^2$$

$$16 + b^2 = 400$$

$$b^2 = 384$$

$$\sqrt{b^2} = \sqrt{384}$$

$$b = 19.6 \; ft$$

A triangle that has angles of 45°, 45°, and 90° is called a **45°-45°-90° triangle**. Recall that an isosceles triangle has two equal sides and the angles opposite the two equal sides are equal. A **45°-45°-90° triangle** is an isosceles right triangle where the two legs are equal and the angles opposite the equal legs are 45° as shown in **Figure 3.6**. If we apply the Pythagorean Theorem we have: $a^2 + a^2 = c^2$. If we solve for **a** (see **Figure 3.7**) we get $a = \dfrac{c}{\sqrt{2}}$ which tells us

that a leg is equal to the hypotenuse divided by the $\sqrt{2}$. If we solve for **c** (see **Figure 3.8**) we get $c = a\sqrt{2}$ which tells us that the hypotenuse is equal to the $\sqrt{2}$ times a leg.

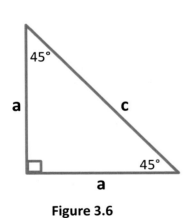

Figure 3.6

$$a^2 + a^2 = c^2$$

$$2a^2 = c^2$$

$$a^2 = \frac{c^2}{2}$$

$$\sqrt{a^2} = \sqrt{\frac{c^2}{2}}$$

$$a = \frac{c}{\sqrt{2}}$$

Figure 3.7

$$a^2 + a^2 = c^2$$

$$2a^2 = c^2$$

$$\sqrt{2a^2} = \sqrt{c^2}$$

$$a\sqrt{2} = c$$

Figure 3.8

$$45°\text{-}45°\text{-}90° \text{ triangle}$$

$$leg = \frac{hypotenuse}{\sqrt{2}}$$

$$hypotenuse = leg\sqrt{2}$$

►EXAMPLE 7

a.)   Find the measure of the missing side.

b.)   Find the measure of the missing side.

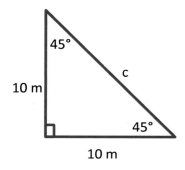

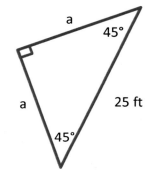

►►Solution

a.) $hypotenuse = leg\sqrt{2}$

$c = 10\sqrt{2} \ m$

b.) $leg = \dfrac{hypotenuse}{\sqrt{2}}$

$a = \dfrac{25}{\sqrt{2}} \ ft$

$a = \dfrac{25}{\sqrt{2}} \cdot \dfrac{\sqrt{2}}{\sqrt{2}} \ ft$

$a = \dfrac{25\sqrt{2}}{2} \ ft$

An equilateral triangle is a triangle with three equal sides and three equal angles measuring 60° each. If we draw an altitude from one angle to the opposite side, the altitude bisects both the side and angle. It also forms two triangles with angles of 30°, 60°, and 90° as shown in **Figure 3.9**. A triangle with angles of 30°, 60°, and 90° is called a **30°-60°-90° triangle**. Notice that the side opposite the 30° angle is half the measure of the hypotenuse and it is called the short leg. Using the Pythagorean Theorem (shown in **Figure 3.**10) we find the side opposite the 60° angle to equal $\sqrt{3}$ times the short leg and this side is called the long leg as shown in **Figure 3.11**.

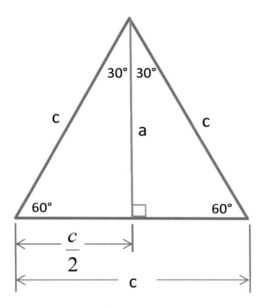

Figure 3.9

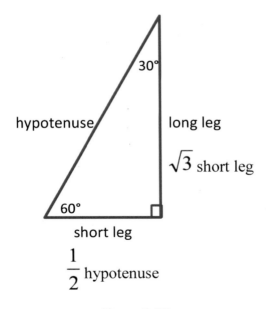

Figure 3.11

$$a^2 + b^2 = c^2$$

$$a^2 + \left(\frac{c}{2}\right)^2 = c^2$$

$$a^2 + \frac{c^2}{4} = c^2$$

$$a^2 = c^2 - \frac{c^2}{4}$$

$$a^2 = \frac{4c^2 - c^2}{4}$$

$$a^2 = \frac{3c^2}{4}$$

$$\sqrt{a^2} = \sqrt{\frac{3c^2}{4}}$$

$$a = \frac{\sqrt{3c^2}}{\sqrt{4}}$$

$$a = \frac{c\sqrt{3}}{2}$$

$$a = \left(\frac{c}{2}\right)\sqrt{3}$$

Figure 3.10

30°-60°-90° triangle

$$short\ leg = \frac{1}{2} \cdot hypotenuse$$

$$long\ leg = \sqrt{3} \cdot short\ leg$$

$$hypotenuse = 2 \cdot short\ leg$$

►EXAMPLE 8

a.) Find the measure of the missing side.

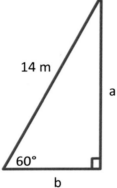

b.) Find the measure of the missing side.

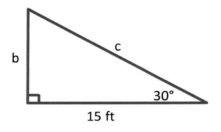

►►Solution

a.) $b = short\ leg$

$$b = \frac{1}{2} \cdot hypotenuse$$

$$b = \frac{1}{2} \cdot 14\,m$$

$$b = 7\,m$$

$$a = long\ leg$$

$$a = \sqrt{3} \cdot short \cdot leg$$

$$a = \sqrt{3}(7\,m)$$

$$a = 7\sqrt{3}\ m$$

b.) $b = short\ leg$

$$b = \frac{1}{\sqrt{3}} \cdot long\ leg$$

$$b = \frac{1}{\sqrt{3}} \cdot 15\,ft$$

$$b = \frac{15}{\sqrt{3}} \cdot \frac{\sqrt{3}}{\sqrt{3}}\ ft$$

$$b = \frac{15\sqrt{3}}{3}\ ft$$

$$b = 5\sqrt{3}\ ft$$

$c = hypotenuse$

$$c = 2 \cdot short\ leg$$

$$c = 2 \cdot 5\sqrt{3}\ ft$$

$$c = 10\sqrt{3}\ ft$$

HOMEWORK

Objective 1

Find the measure of each angle.

See Example 1.

1.

$x°$

$(x+20)°$ $(210-3x)°$

B   C

A

50
+10
+60

2.

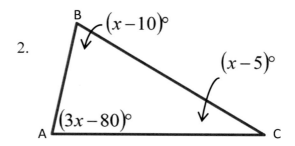

B $(x-10)°$

$(x-5)°$

$(3x-80)°$

A   C

3.

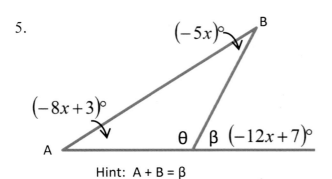

A $(x+15)°$

$(10x-20)°$ $(x+5)°$

C   B

4.

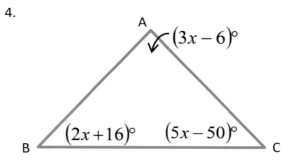

A $(3x-6)°$

$(2x+16)°$ $(5x-50)°$

B   C

5.

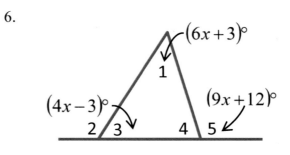

$(-5x)°$ B

$(-8x+3)°$

A θ β $(-12x+7)°$

Hint: A + B = β

6.

$(6x+3)°$

1

$(4x-3)°$ $(9x+12)°$

2 3   4 5

Objective 2

Solve. *See Examples 2, 3and 4.*

7.   An oak tree casts a shadow of 50 m long. At the same time a 2 m stick casts a shadow of 3 m. Find the height of the tree to the nearest tenth of a meter.

8.   The foot of a ladder is 2 feet from a fence that is 3 feet high. The ladder touches the fence and rests against a building that is 6 feet behind the fence. Determine the height of the building reached by the top of the ladder.

9.   A balloon is hovering over a group of people. The shadow of the balloon is lagging approximately 28 feet behind a point directly below the balloon. At the same time the shadow of a man that is 6 feet tall is 4 feet, what is the altitude (height) of the balloon?

10.   Find the height of the silo in the drawing. Round to the nearest meter.

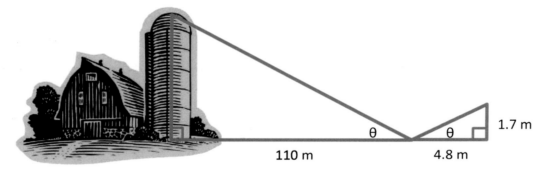

110 m θ θ 1.7 m

4.8 m

11.   In the figure $\triangle ABC \sim \triangle DEF$. Use properties of similar triangles to find the length of the missing side. Round to the nearest tenth if necessary.

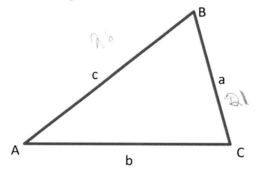
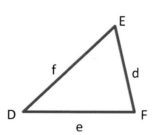

a. $a = 21m, c = 26m, d = 8m, f = ?$

b. $b = 19m, e = 8m, f = 7m, c = ?$

c. $c = 23in, d = 10in, f = 16in, a = ?$

d. $b = 12ft, c = 5.8ft, e = 1.6ft, f = ?$

12.   A building casts a shadow of 250 feet long. At the same time a 6 foot man casts a
       shadow of 11 feet. Find the height of the building to the nearest tenth of a foot.

Objective 3

Solve. *See Examples 5 and 6.*

13.   Find the measure of the missing
       side. Round to the nearest tenth of a
       inch.

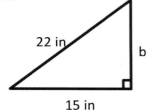

14.   Find the measure of the missing
       side. Round to the nearest tenth of a
       meter.

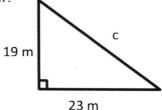

15.   A 25 foot guy wire is connected to the top of a light pole that is 18 feet tall. How far is
       the guy wire anchored from the base of the light pole? Round to the nearest tenth of a
       foot.

16.   A kite has a vertical height of 23 meters and is attached to a string that is 30 meters long.
       What is the horizontal distance of the kite from the anchor of the string? Round to the
       nearest tenth of a meter.

17.   Find, to the nearest tenth, the length of the diagonal of a rectangle that is 12 centimeters
       long and 8 centimeters wide.

Objective 4

Solve. *See Examples 7 and 8.*

18.   Find the missing sides. Give exact values.

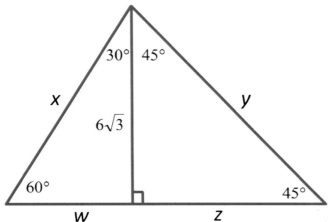

$w = \underline{\quad 6 \quad}$

$x = \underline{\quad 12 \quad}$

$y = \underline{\quad 12\sqrt{3} \quad}$

$z = \underline{\quad 6\sqrt{3} \quad}$

19.   A ladder on a fire truck extends 80 feet to a sixth floor window of a burning building. If the angle formed with the ladder and the wall of the building is 45°, how far is the fire truck from the building. Round to the nearest tenth of a foot.

20.   If the ramp of a building is 20 feet long and 10 feet high, what angle does the ramp make with the ground? How long, to the nearest tenth, is the base of the ramp?

21.   A guy wire makes a 60° angle with the ground. If the guy wire is 22 meters long and is attached to the top of a pole, how tall is the pole? Round to the nearest tenth of a meter.

22.   Find the measure of the diagonal of a square with a sides of 16 inches. Round to the nearest tenth of an inch.

ANSWERS

1. $A = 50°, B = 70°, C = 60°$ 2. $A = 85°, B = 45°, C = 50°$

3. $A = 30°, B = 20°, C = 130°$ 4. $A = 60°, B = 60°, C = 60°$

5. $A = 35°, B = 20°, \theta = 125°, \beta = 55°$

6. $1 = 75°, 2 = 135°, 3 = 45°, 4 = 60°, 5 = 120°$

7. 33.3 *m*   8. 12 *ft*   9. 42 *ft*

10. 39 *m*   11a. 9.9 *m*   11b. 16.6 *m*

11c. 14.4 *in*   11d. 0.8 *ft*   12. 136.4 *ft*

13. 16.1 *in*   14. 29.8 *m*   15. 17.3 *ft*

16. 19.3 *m*   17. 14.4 *cm*

18. $w = 6, x = 12, y = 6\sqrt{6}, z = 6\sqrt{3}$ 19. 56.6 *ft*

20. 30°, 17.3 *ft*   21. 19.1 *m*   22. 22.6 *in*

Section 4:   Right Triangle Trigonometry

Learning Outcomes:

- The student will correctly memorize and apply trigonometric formulas, definitions, identities, and properties.
- The student will correctly use mathematical symbols and mathematical structure to examine and solve real world applications.
- The student will solve trigonometric equations, right triangles, and oblique triangles.

Objectives: At the conclusion of this lesson you should be able to:

1. Determine the six trigonometric ratios for a given acute angle of a right triangle.
2. Use a calculator to find function values for any acute angle.
3. Use a calculator to find any acute angle given function values.
4. Solve right triangles given one side and one angle.
5. Solve right triangles given two sides.
6. Use cofunction identities to write equivalent expressions.

In the last section we looked at applications involving the 45°-45°-90° and 30°-60°-90° triangles using the relationships between their sides. We can carry this concept further to other triangles where each side is given a specific name depending on its location relative to a specified angle. Consider a right triangle with an acute angle θ, the three sides of the triangle are the adjacent side, opposite side, and the hypotenuse. The adjacent side is the side adjacent to the angle θ and the opposite side is the side opposite the angle θ. See **Figure 4.1.**

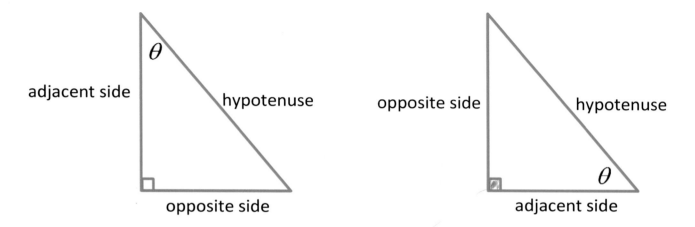

Figure 4.1

The six trigonometric functions are sine (sin), cosine (cos), tangent (tan), cosecant (csc), secant (sec), and cotangent (cot). We can define the six trigonometric functions of the acute angle θ using the lengths of the three sides as shown **Figure 4.2**.

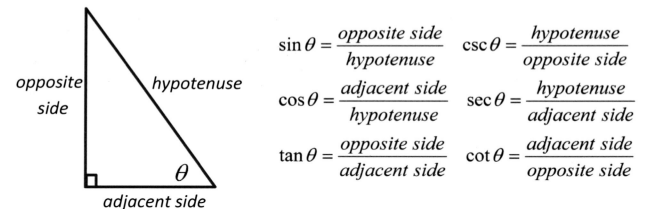

$$\sin \theta = \frac{opposite\ side}{hypotenuse} \qquad \csc \theta = \frac{hypotenuse}{opposite\ side}$$

$$\cos \theta = \frac{adjacent\ side}{hypotenuse} \qquad \sec \theta = \frac{hypotenuse}{adjacent\ side}$$

$$\tan \theta = \frac{opposite\ side}{adjacent\ side} \qquad \cot \theta = \frac{adjacent\ side}{opposite\ side}$$

Figure 4.2

Notice that sine and cosecant are reciprocals of each other, cosine and secant are reciprocals, as are tangent and cotangent.   The reciprocal identities hold for any angle θ .

*** Reciprocal Identities ***

$$\sin \theta = \frac{1}{\csc \theta} \qquad \cos \theta = \frac{1}{\sec \theta} \qquad \tan \theta = \frac{1}{\cot \theta}$$

$$\csc \theta = \frac{1}{\sin \theta} \qquad \sec \theta = \frac{1}{\cos \theta} \qquad \cot \theta = \frac{1}{\tan \theta}$$

► EXAMPLE 1

Use the triangle to find the exact values (no decimals) of the six trigonometric functions of θ.

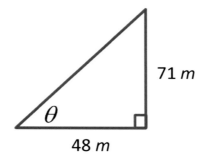

$\sin \theta = $ _____ $\csc \theta = $ _____

$\cos \theta = $ _____ $\sec \theta = $ _____

$\tan \theta = $ _____ $\cot \theta = $ _____

►►Solution

The adjacent side is 48 m and the opposite side is 71 m. We find the hypotenuse to be 86 m by using the Pythagorean Theorem. Applying the ratios we get the six trigonometric function values of θ. An exact value is a non-decimal answer left in fraction form, radical form, etc.

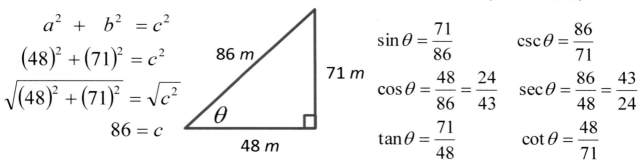

$$a^2 + b^2 = c^2$$
$$(48)^2 + (71)^2 = c^2$$
$$\sqrt{(48)^2 + (71)^2} = \sqrt{c^2}$$
$$86 = c$$

$$\sin \theta = \frac{71}{86}$$
$$\cos \theta = \frac{48}{86} = \frac{24}{43}$$
$$\tan \theta = \frac{71}{48}$$

$$\csc \theta = \frac{86}{71}$$
$$\sec \theta = \frac{86}{48} = \frac{43}{24}$$
$$\cot \theta = \frac{48}{71}$$

The following illustration gives us insight into the trigonometric values of complementary angles.

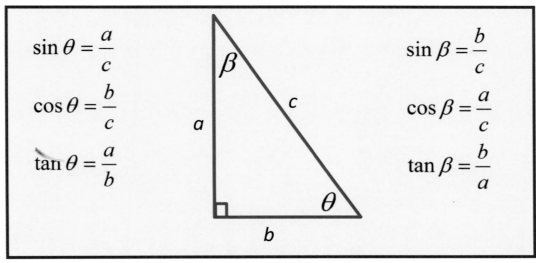

$$\sin \theta = \frac{a}{c}$$
$$\cos \theta = \frac{b}{c}$$
$$\tan \theta = \frac{a}{b}$$

$$\sin \beta = \frac{b}{c}$$
$$\cos \beta = \frac{a}{c}$$
$$\tan \beta = \frac{b}{a}$$

Recall the relationships of a 30°-60°-90° triangle. The short leg is half the hypotenuse and the long leg is $\sqrt{3}$ times the short leg. If we consider the 30° angle, the opposite side is the short leg and the adjacent side is the long leg as shown in **Figure 4.3**. If we find the sin, cos, and tan of the 30° angle we have what is shown in **Figure 4.4**.   Since all 30°-60°-90° triangles are similar and the ratios of corresponding sides is constant, the x cancels in each ratio reminding us the ratios are independent of the triangle's size.

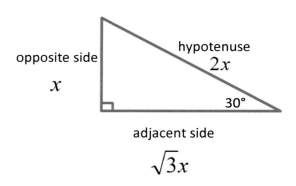

Figure 4.3

$$\sin 30° = \frac{opposite\ side}{hypotenuse} = \frac{x}{2x} = \frac{1}{2}$$

$$\cos 30° = \frac{adjacent\ side}{hypotenuse} = \frac{\sqrt{3}x}{2x} = \frac{\sqrt{3}}{2}$$

$$\tan 30° = \frac{opposite\ side}{adjacent\ side} = \frac{x}{\sqrt{3}x} = \frac{1}{\sqrt{3}} = \frac{\sqrt{3}}{3}$$

Figure 4.4

If we consider the 60° angle of a 30°-60°-90° triangle, the opposite side is the long leg and the adjacent side is the short leg as shown in **Figure 4.5**. If we find the sin, cos, and tan of the 60° angle we have what is shown in **Figure 4.6**.

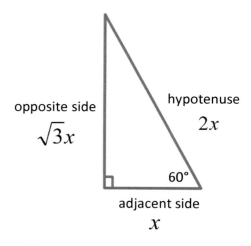

Figure 4.5

$$\sin 60° = \frac{opposite\ side}{hypotenuse} = \frac{\sqrt{3}x}{2x} = \frac{\sqrt{3}}{2}$$

$$\cos 60° = \frac{adjacent\ side}{hypotenuse} = \frac{x}{2x} = \frac{1}{2}$$

$$\tan 60° = \frac{opposite\ side}{adjacent\ side} = \frac{\sqrt{3}x}{x} = \sqrt{3}$$

Figure 4.6

$$\sin 30° = \sin \frac{\pi}{6} = \frac{1}{2}$$

$$\cos 30° = \cos \frac{\pi}{6} = \frac{\sqrt{3}}{2}$$

$$\tan 30° = \tan \frac{\pi}{6} = \frac{\sqrt{3}}{3}$$

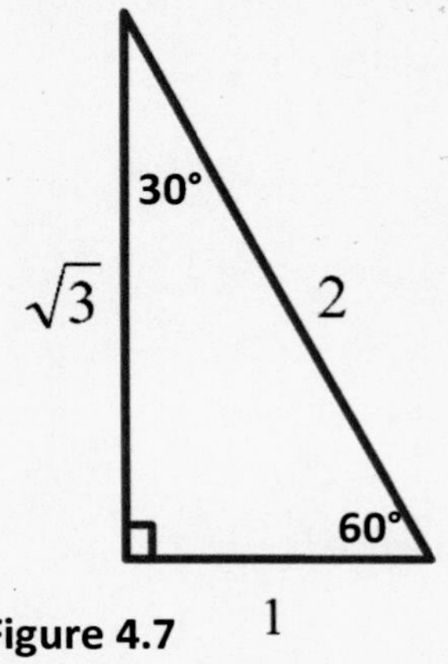

$$\sin 60° = \sin \frac{\pi}{3} = \frac{\sqrt{3}}{2}$$

$$\cos 60° = \cos \frac{\pi}{3} = \frac{1}{2}$$

$$\tan 60° = \tan \frac{\pi}{3} = \sqrt{3}$$

Figure 4.7

In **Figure 4.7** the 30° angle and the 60° angle are complements. Complementary angles in a right triangle have a unique relationship. Since the $\sin 60° = \frac{\sqrt{3}}{2}$ and the $\cos 30° = \frac{\sqrt{3}}{2}$ we can conclude that $\sin 60° = \cos 30°$.   We know that 90°-60°=30° so we can substitute and have $\sin 60° = \cos(90° - 60°)$. The sine of an angle is equal to the cosine of its complement and the cosine of an angle is equal to sine of its complement. The functions sine and cosine are called cofunctions. Tangent and cotangent are cofunctions as are secant and cosecant.

Cofunction Identities

$$\cos(90° - \theta) = \sin \theta \qquad \sec(90° - \theta) = \csc \theta \qquad \tan(90° - \theta) = \cot \theta$$

$$\cos\left(\frac{\pi}{2} - \theta\right) = \sin \theta \qquad \sec\left(\frac{\pi}{2} - \theta\right) = \csc \theta \qquad \tan\left(\frac{\pi}{2} - \theta\right) = \cot \theta$$

$$\sin(90° - \theta) = \cos \theta \qquad \csc(90° - \theta) = \sec \theta \qquad \cot(90° - \theta) = \tan \theta$$

$$\sin\left(\frac{\pi}{2} - \theta\right) = \cos \theta \qquad \csc\left(\frac{\pi}{2} - \theta\right) = \sec \theta \qquad \cot\left(\frac{\pi}{2} - \theta\right) = \tan \theta$$

▶EXAMPLE 2

1.)   Write each function in terms of its cofunction.

   a.) $\sin 52°$ b.) $\tan 67°$ c.) $\sec \dfrac{\pi}{5}$

2.)   Find the value of x that makes the given functions equal.

   a.) $\tan(3x)°, \cot(2x)°$ b.) $\sin(x+4)°, \cos(3x+2)°$

▶▶Solution

1.)   a.) $\sin 52° = \cos(90° - 52°)$ c.) $\sec \dfrac{\pi}{5} = \csc\left(\dfrac{\pi}{2} - \dfrac{\pi}{5}\right)$

$\sin 52° = \cos 38°$

$\sec \dfrac{\pi}{5} = \csc\left(\dfrac{5\pi}{10} - \dfrac{2\pi}{10}\right)$

   b.) $\tan 67° = \cos(90° - 67°)$

$\sin 67° = \cos 23°$ $\sec \dfrac{\pi}{5} = \csc \dfrac{3\pi}{10}$

2.)   a.) $3x + 2x = 90$ b.) $(x+4) + (3x+2) = 90$

$5x = 90$ $x + 4 + 3x + 2 = 90$

$\dfrac{5x}{5} = \dfrac{90}{5}$ $4x + 6 = 90$

$x = 18$ $4x + 6 - 6 = 90 - 6$

$4x = 84$

$\dfrac{4x}{4} = \dfrac{84}{4}$

$x = 21$

In a 45°-45°-90° triangle each of the two legs is equal to the hypotenuse divided by $\sqrt{2}$ and the hypotenuse is $\sqrt{2}$ times one of the legs. Since all 45°-45°-90° triangles are similar the ratios of the corresponding sides will be constant. The sin, cos, and tan of a 45° angle is shown in

Figure 4.8.

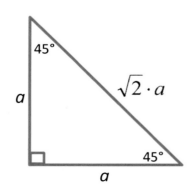

a

$\sqrt{2} \cdot a$

45°

45°

a

$\sin 45° = \dfrac{opposite\ side}{hypotenuse} = \dfrac{a}{\sqrt{2}a} = \dfrac{1}{\sqrt{2}} = \dfrac{\sqrt{2}}{2}$

$\cos 45° = \dfrac{adjacent\ side}{hypotenuse} = \dfrac{a}{\sqrt{2}a} = \dfrac{1}{\sqrt{2}} = \dfrac{\sqrt{2}}{2}$

$\tan 45° = \dfrac{opposite\ side}{adjacent\ side} = \dfrac{a}{a} = 1$

Figure 4.8

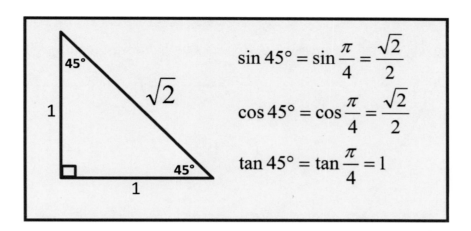

We are now ready to solve a right triangle which means to find the measures of all three sides and all three angles. In order for this to be accomplished we will use the properties of triangles, the Pythagorean Theorem, and the trigonometric ratios. Capital letters will represent the angles and each side is labeled using the related lowercase letter from the opposite angle.

Solving a Right Triangle

- The sum of the measures of the angles of a triangle is 180°. Use this property of a triangle to determine a third angle when two angles are known.

- Use the Pythagorean Theorem to find the measure of a third side when the measure of two sides is known.

- Use the trigonometric ratios to find missing sides or angles.

►EXAMPLE 3

Solve the triangle given. Round the measure of the sides to the nearest tenth.

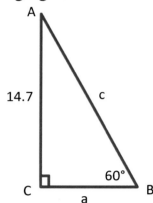

$A =$ _____ $a =$ _____

$B = 60°$ $b = 14.7$

$C =$ _____ $c =$ _____

►►Solution

- The box in the triangle for angle C means C is equal to 90°.
- Applying the property of a triangle that the three angles of a triangle add to equal 180° we get angle A is equal to 30°.
- We will use trigonometric ratios to find one of the missing sides. If we choose to find side c, we will use the given angle 60° and sine. We use sine because we know the opposite side and we are trying to find the hypotenuse.

$$\sin 60° = \frac{14.7}{c}$$

$$c \sin 60° = 14.7$$

$$c = \frac{14.7}{\sin 60°}$$

$$c = \frac{14.7}{\dfrac{\sqrt{3}}{2}}$$ ⟵ —— Recall that $\sin 60° = \dfrac{\sqrt{3}}{2}$, substitute $\dfrac{\sqrt{3}}{2}$ for $\sin 60°$.

$$c = 16.97$$

$$c = 17.0$$

- To find the remaining side a we will use the Pythagorean Theorem.

$$a^2 + (14.7)^2 = (17.0)^2$$

$$a^2 = (17.0)^2 - (14.7)^2$$

$$\sqrt{a^2} = \sqrt{(17.0)^2 - (14.7)^2}$$

$$a = 8.5$$

$A = 30°$	$a = 8.5$
$B = 60°$	$b = 14.7$
$C = 90°$	$c = 17.0$

Side a could also have been found using tangent.

► EXAMPLE 4

Solve the triangle given. Round the measure of the sides to the nearest tenth.

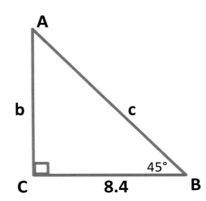

$A =$ _____ $a = 8.4$

$B = 45°$ $b =$ _____

$C =$ _____ $c =$ _____

►►Solution

- The box in the triangle for angle C means C is equal to 90°.
- Applying the property of a triangle that the three angles of a triangle add to equal 180° we get angle A is equal to 45°.
- We will use trigonometric ratios to find one of the missing sides. If we choose to find side c, we will use the given angle 45° and cosine. We use cosine because we know the adjacent side and we are trying to find the hypotenuse.

$$\cos 45° = \frac{8.4}{c}$$

$$c \cos 45° = 8.4$$

$$c = \frac{8.4}{\cos 45°}$$

$$c = \frac{8.4}{\frac{\sqrt{2}}{2}} \longleftarrow \text{Recall that } \cos 45° = \frac{\sqrt{2}}{2}, \text{ substitute } \frac{\sqrt{2}}{2} \text{ for } \cos 45°.$$

$$c = 11.87939$$

$$c = 11.9$$

- To find the remaining side b we will use the Pythagorean Theorem.

$$(8.4)^2 + b^2 = (11.9)^2$$

$$b^2 = (11.9)^2 - (8.4)^2$$

$$\sqrt{b^2} = \sqrt{(11.9)^2 - (8.4)^2}$$

$$b = 8.4$$

$A = 45°$	$a = 8.4$
$B = 45°$	$b = 8.4$
$C = 90°$	$c = 11.9$

In order to solve a right triangle that does not have standard angles (it is not a 30°-60°-90° triangle or 45°-45°-90° triangle) we need to practice using a calculator to find the six trigonometric ratios. The key strokes given in this book are based on a TI-84 graphing calculator. To find $\sin 68°$ make sure your calculator is in degree mode and press the following: [SIN] [6] [8] [)] [ENTER]. You will get 0.9271838546. To find $\csc 68°$ we recall that cosecant is the reciprocal of sine and we press the following keys: [1] [÷] [SIN] [6] [8] [)] [ENTER]. You will get 1.078534743.

▶ EXAMPLE 5

Use a calculator to find the value of each expression rounded to the nearest ten-thousandth.

a.) $\cos 79°$ b.) $\tan 58°$ c.) $\sin 43°18'$ d.) $\csc 59°$

e.) $\cot 55°$ f.) $\sec 36°$ g.) $\cos 29°42'$ h.) $\sec 63°15'$

▶▶ Solution

Make sure your calculator is in Degree Mode.

a.) $\cos 79° = 0.1908$ [COS] [7] [9] [)] [ENTER]

b.) $\tan 58° = 1.6003$ [TAN] [5] [8] [)] [ENTER]

c.) $\sin 43°18' = 0.6858$ [SIN] [4] [3] [+] [1] [8] [÷] [6] [0] [)] [ENTER]   or

[SIN] [4] [3] [2nd] [APPS] [1] [1] [8] [2nd] [APPS] [2] [)] [ENTER]

d.) $\csc 59° = 1.1666$ [1] [÷] [SIN] [5] [9] [)] [ENTER]

e.) $\cot 55° = 0.7002$ [1] [÷] [TAN] [5] [5] [)] [ENTER]

f.) $\sec 36° = 1.2361$ [1] [÷] [COS] [3] [6] [)] [ENTER]

g.) $\cos 29°42' = 0.8686$ [COS] [2] [9] [+] [4] [2] [÷] [6] [0] [)] [ENTER]   or

[COS] [2] [9] [2nd] [APPS] [1] [4] [2] [2nd] [APPS] [2] [)] [ENTER]

h.) $\sec 63°15' = 2.2217$ [1] [÷] [COS] [6] [3] [+] [1] [5] [÷] [6] [0] [)] [ENTER]   or

[1] [÷] [COS] [6] [3] [2nd] [APPS] [1] [1] [5] [2nd] [APPS] [2] [)] [ENTER]

▶ EXAMPLE 6

Solve the triangle given. Round the measure of the sides to the nearest tenth.

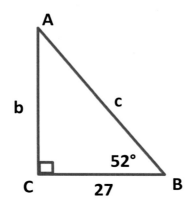

$A = \underline{\hspace{1.5cm}}$ $a = 27$

$B = 52°$ $b = \underline{\hspace{1.5cm}}$

$C = \underline{\hspace{1.5cm}}$ $c = \underline{\hspace{1.5cm}}$

▶▶ Solution

- The box in the triangle for angle C means C is equal to 90°.
- Applying the property of a triangle that the three angles of a triangle add to equal 180° we get angle A is equal to 38°. $180° - 52° - 90° = 38°$
- We will use trigonometric ratios to find one of the missing sides. If we choose to find side b, we will use the given angle 52° and tangent. We use tangent because we know the adjacent side and we are trying to find the opposite side.

$$\tan 52° = \frac{b}{27}$$ Make sure your calculator is in degree mode.

$$27 \tan 52° = b \longleftarrow \boxed{2}\boxed{7}\boxed{\text{TAN}}\boxed{5}\boxed{2}\boxed{)}\boxed{\text{ENTER}}$$

$$34.5584 = b$$

$$34.6 = b$$ Round to the nearest tenth.

- To find the remaining side b we will use the Pythagorean Theorem.

$$(27)^2 + (34.6)^2 = c^2$$

$$\sqrt{(27)^2 + (34.6)^2} = \sqrt{c^2}$$

$$43.88804 = c$$

$$43.9 = c$$

$A = 38°$	$a = 27$
$B = 52°$	$b = 34.6$
$C = 90°$	$c = 43.9$

►EXAMPLE 7

Solve the triangle given. Round the measure of the sides to the nearest tenth.

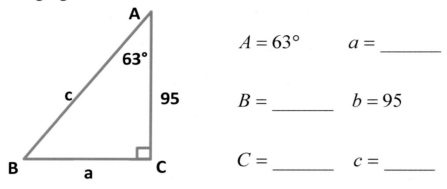

$$A = 63° \qquad a = \underline{\hspace{2cm}}$$

$$B = \underline{\hspace{2cm}} \qquad b = 95$$

$$C = \underline{\hspace{2cm}} \qquad c = \underline{\hspace{2cm}}$$

►►Solution

- The box in the triangle for angle C means C is equal to 90°.
- Applying the property of a triangle that the three angles of a triangle add to equal 180° we get angle B is equal to 27°. $180° - 63° - 90° = 27°$
- We will use trigonometric ratios to find one of the missing sides. If we choose to find side c, we will use the given angle 63° and cosine. We use cosine because we know the adjacent side and we are trying to find the hypotenuse.

$$\cos 63° = \frac{95}{c}$$

$$c \cos 63° = 95$$ Make sure your calculator is in degree mode.

$$c = \frac{95}{\cos 63°}$$ ← $\boxed{9}\,\boxed{5}\,\boxed{\div}\,\boxed{\text{COS}}\,\boxed{6}\,\boxed{3}\,\boxed{)}\,\boxed{\text{ENTER}}$

$$c = 209.2554801$$

$$c = 209.3$$ Round to the nearest tenth.

- To find the remaining side a we will use the Pythagorean Theorem.

$$a^2 + (95)^2 = (209.3)^2$$

$$a^2 = (209.3)^2 - (95)^2$$

$$\sqrt{a^2} = \sqrt{(209.3)^2 - (95)^2}$$

$$a = 186.4979625$$

$$a = 186.5$$

$$A = 63° \qquad a = 186.5$$

$$B = 27° \qquad b = 95$$

$$C = 90° \qquad c = 209.3$$

In Examples 6 and 7 an angle and a side were given. When two sides are given we will need to find an acute angle using trigonometric ratios because right triangles have one right angle and two acute angles. A solution to the equation $\sin\theta = \dfrac{\sqrt{3}}{2}$ is an angle whose sine is $\dfrac{\sqrt{3}}{2}$. Recall

$\sin 60° = \dfrac{\sqrt{3}}{2}$ therefore the acute angle solution for θ is $60°$. The function whose input is the sine

of an angle and whose output is the angle is the inverse sine function. The inverse sine notation

for $\sin 60° = \dfrac{\sqrt{3}}{2}$ is written $\sin^{-1}\dfrac{\sqrt{3}}{2} = 60°$ or $\arcsin\dfrac{\sqrt{3}}{2} = 60°$. Similarly, the inverse cosine and

inverse tangent functions pair numbers with angles. For example: the inverse cosine notation

for $\cos 60° = \dfrac{1}{2}$ is $\cos^{-1}\dfrac{1}{2} = 60°$ or $\arccos\dfrac{1}{2} = 60°$ and the inverse tangent notation for $\tan 60° = \sqrt{3}$

is $\tan^{-1}\sqrt{3} = 60°$ or $\arctan\sqrt{3} = 60°$. For angles that are not $30°$, $45°$, or $60°$ we use a calculator to

evaluate an inverse function. The calculator keys for inverse trigonometric functions are

$\boxed{\textbf{SIN}^{-1}}$, $\boxed{\textbf{COS}^{-1}}$ and $\boxed{\textbf{TAN}^{-1}}$. They are the second function keys for the sine, cosine, and

tangent keys. To evaluate $\cos\theta = \dfrac{4}{5}$ we make sure the calculator is in degree mode and press the

following keys: $\boxed{\textbf{2nd}}\ \boxed{\textbf{COS}}\ \boxed{4}\ \boxed{\div}\ \boxed{5}\ \boxed{)}\ \boxed{\textbf{ENTER}}$ and we get $36.86989765°$. Rounding to the
nearest tenth of a degree we get $36.9°$ and rounding to the nearest minute we get $36°52'$.   To
find $\sec\theta = 1.2536$ recall that secant is the reciprocal of cosine and press the following keys on
the calculator: $\boxed{\textbf{2nd}}\ \boxed{\textbf{COS}}\ \boxed{1}\ \boxed{\div}\ \boxed{1.2536}\ \boxed{)}\ \boxed{\textbf{ENTER}}$ and we get $37.08872497°$. Rounding to
the nearest tenth of a degree we get $37.1°$ and rounding to the nearest minute we get $37°5'$.

▶ **EXAMPLE 8**

Use a calculator to find the acute angle whose corresponding ratio is given. Round to the nearest
tenth of a degree.

a.) $\tan\theta = 2.1445$ b.) $\cos\theta = 0.8572$ c.) $\sin\theta = 0.9744$

d.) $\cot\theta = 0.8497$ e.) $\csc\theta = 2.4$ f.) $\sec\theta = 1.2369$

▶▶ **Solution**

Angle

Make sure your calculator is in degree mode.

a.) $\tan\theta = 2.1445$ $\boxed{\textbf{2nd}}\ \boxed{\textbf{TAN}}\ \boxed{2}\ \boxed{.}\ \boxed{1}\ \boxed{4}\ \boxed{4}\ \boxed{5}\ \boxed{)}\ \boxed{\textbf{ENTER}}$ $\theta = 65.0°$

b.) $\cos\theta = 0.8572$ $\boxed{\textbf{2nd}}\ \boxed{\textbf{COS}}\ \boxed{0}\ \boxed{.}\ \boxed{8}\ \boxed{5}\ \boxed{7}\ \boxed{2}\ \boxed{)}\ \boxed{\textbf{ENTER}}$ $\theta = 31.0°$

c.) $\sin\theta = 0.9744$ $\boxed{\textbf{2nd}}\ \boxed{\textbf{SIN}}\ \boxed{0}\ \boxed{.}\ \boxed{9}\ \boxed{7}\ \boxed{4}\ \boxed{4}\ \boxed{)}\ \boxed{\textbf{ENTER}}$ $\theta = 77.0°$

d.) $\cot\theta = 0.8497$ $\boxed{\textbf{2nd}}\ \boxed{\textbf{TAN}}\ \boxed{1}\ \boxed{\div}\ \boxed{0}\ \boxed{.}\ \boxed{8}\ \boxed{4}\ \boxed{9}\ \boxed{7}\ \boxed{)}\ \boxed{\textbf{ENTER}}$ $\theta = 49.6°$

e.) $\csc\theta = 2.4$ $\boxed{\textbf{2nd}}\ \boxed{\textbf{SIN}}\ \boxed{1}\ \boxed{\div}\ \boxed{2}\ \boxed{.}\ \boxed{4}\ \boxed{)}\ \boxed{\textbf{ENTER}}$ $\theta = 24.6°$

f.) $\sec\theta = 1.2369$ $\boxed{\textbf{2nd}}\ \boxed{\textbf{COS}}\ \boxed{1}\ \boxed{\div}\ \boxed{1}\ \boxed{.}\ \boxed{2}\ \boxed{3}\ \boxed{6}\ \boxed{9}\ \boxed{)}\ \boxed{\textbf{ENTER}}$ $\theta = 36.1°$

►EXAMPLE 9

Solve the triangle given. Round the measure of the sides to the nearest tenth and angles to the nearest minute.

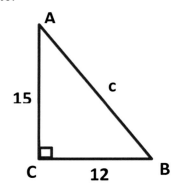

$A =$ _____ $a = 12$

$B =$ _____ $b = 15$

$C =$ _____ $c =$ _____

►►Solution

- The box in the triangle for angle C means C is equal to 90°.
- To find the remaining side c we will use the Pythagorean Theorem.

$$(12)^2 + (15)^2 = c^2$$

$$\sqrt{(12)^2 + (15)^2} = \sqrt{c^2}$$

$$19.20937 = c$$

$$19.2 = c$$

- We will use trigonometric ratios to find one of the angles. If we choose to find angle B, we will use tangent. We use tangent because we were given the adjacent side and the opposite side.

Make sure your calculator is in degree mode.

$$\tan B = \frac{15}{12} \quad \longleftarrow \quad \boxed{\text{2nd}}\ \boxed{\text{TAN}}\ \boxed{1}\ \boxed{5}\ \boxed{\div}\ \boxed{1}\ \boxed{2}\ \boxed{)}\ \boxed{\text{ENTER}}$$

$$B = 51.34019°$$

$$B = 51°20' \qquad \text{Round to the nearest minute.}$$

- Applying the property of a triangle that the three angles of a triangle add to equal 180° we get angle A is equal to 38°40'. $180° - 90° - 51°20' = 38°40'$

$A = 38°40'$	$a = 12$
$B = 51°20'$	$b = 15$
$C = 90°$	$c = 19.2$

► EXAMPLE 10

Solve the triangle given. Round the measure of the sides to the nearest tenth and angles to the nearest minute.

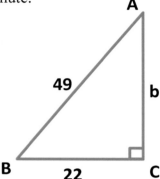

$A =$ _____ $a = 22$

$B =$ _____ $b =$ _____

$C =$ _____ $c = 49$

►►Solution

- The box in the triangle for angle C means C is equal to 90°.
- To find the remaining side b we will use the Pythagorean Theorem.

$$(22)^2 + b^2 = (49)^2$$
$$b^2 = (49)^2 - (22)^2$$
$$\sqrt{b^2} = \sqrt{(49)^2 - (22)^2}$$
$$b = 43.78356$$
$$b = 43.8$$

- We will use trigonometric ratios to find one of the angles. If we choose to find angle B, we will use cosine. We use cosine because we were given the adjacent side and the hypotenuse.

Make sure your calculator is in degree mode.

$$\cos B = \frac{22}{49} \quad \longleftarrow \quad \boxed{\text{2nd}}\ \boxed{\text{COS}}\ \boxed{2}\ \boxed{2}\ \boxed{\div}\ \boxed{4}\ \boxed{9}\ \boxed{)}\ \boxed{\text{ENTER}}$$

$$B = 63.321766°$$
$$B = 63°19' \qquad \text{Round to the nearest minute.}$$

- Applying the property of a triangle that the three angles of a triangle add to equal 180° we get angle A is equal to 26°41'. $180° - 90° - 63°19' = 26°41'$

$A = 26°41'$	$a = 22$
$B = 63°19'$	$b = 43.8$
$C = 90°$	$c = 49$

HOMEWORK

Objective 1

Find the exact values of the six trigonometric functions of the angle θ.

See Example 1.

1.
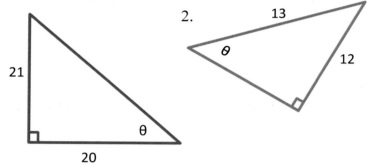
2.
3.

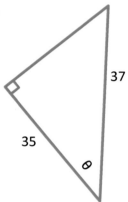

Objectives 4 and 5

Solve the right triangle shown. Round the measure of the sides to the nearest tenth and angles to the nearest minute.

See Examples 2, 3, 4, 6, 7, 9, 10.

4.　　$A = 45°$,　$b = 64\ cm$

5.　　$B = 60°$,　$a = 12\ m$

6.　　$A = 30°$,　$c = 16\ in$

7.　　$B = 32°15'$,　$a = 19\ m$

8.　　$B = 52°28'$,　$c = 22\ cm$

9.　　$a = 12\ m$,　$b = 17\ m$

10.　　$a = 12\ cm$,　$c = 23\ cm$

11.　　$b = 8\ m$,　$c = 19\ m$

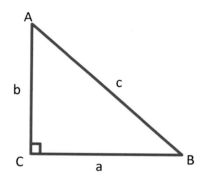

Objective 2

Use a calculator to evaluate each function. Round your answers to four decimal places.

See Example 5

12. $\sin 145°$ 13. $\tan 121°$ 14. $\cos 215°$ 15. $\cot 89°$

16. $\csc 129°$ 17. $\sec 66°$ 18. $\cos 97°22'$ 19. $\sin 63°48'$

Objective 3

Use a calculator to find the acute angle whose corresponding ratio is given. Round to the nearest tenth of a degree.

See Example 8

20. $\sin \theta = 0.9895$ 21. $\cos \theta = 0.7455$ 22. $\tan \theta = 2.3332$

23. $\cot \theta = 1.1667$ 24. $\csc \theta = 2.0183$ 25. $\sec \theta = 3.2188$

Objective 6

Write each function in terms of its cofunction.

See Example 2

26. $\sin 65°$ 27. $\tan 41°$ 28. $\cos 71°$ 29. $\cot 19°$

30. $\csc 39°$ 31. $\sec 26°$ 32. $\cos 14°32'$ 33. $\sin 33°18'$

34. $\sin \dfrac{\pi}{3}$ 35. $\cos \dfrac{2\pi}{5}$ 36. $\tan \dfrac{3\pi}{8}$ 37. $\sec \dfrac{2\pi}{7}$

Objective 6

Find the value of x that makes the given functions equal.

See Example 2

38. $\sin(2x)°, \cos(3x)°$ 39. $\sin(4x)°, \cos(5x)°$

40. $\tan(6x - 3)°, \cot(2x + 5)°$ 41. $\csc(6x - 1)°, \sec(5x + 3)°$

ANSWERS

1. $\sin\theta = \dfrac{21}{29}$ $\csc\theta = \dfrac{29}{21}$

    $\cos\theta = \dfrac{20}{29}$ $\sec\theta = \dfrac{29}{20}$

    $\tan\theta = \dfrac{21}{20}$ $\cot\theta = \dfrac{20}{21}$

2. $\sin\theta = \dfrac{12}{13}$ $\csc\theta = \dfrac{13}{12}$

    $\cos\theta = \dfrac{5}{13}$ $\sec\theta = \dfrac{13}{5}$

    $\tan\theta = \dfrac{12}{5}$ $\cot\theta = \dfrac{5}{12}$

3. $\sin\theta = \dfrac{12}{37}$ $\csc\theta = \dfrac{37}{12}$

    $\cos\theta = \dfrac{35}{37}$ $\sec\theta = \dfrac{37}{35}$

    $\tan\theta = \dfrac{12}{35}$ $\cot\theta = \dfrac{35}{12}$

4. $A = 45°$ $a = 64\,cm$

    $B = 45°$ $b = 64\,cm$

    $C = 90°$ $c = 90.5\,cm$

5. $A = 30°$ $a = 12\,m$

    $B = 60°$ $b = 20.8\,m$

    $C = 90°$ $c = 24\,m$

6. $A = 30°$ $a = 8\,in$

    $B = 60°$ $b = 13.9\,in$

    $C = 90°$ $c = 16\,in$

7. $A = 57°45'$ $a = 19\,m$

    $B = 32°15'$ $b = 12.0\,m$

    $C = 90°$ $c = 22.5\,m$

8. $A = 37°32'$ $a = 13.4\,cm$

    $B = 52°28'$ $b = 17.4\,cm$

    $C = 90°$ $c = 22\,cm$

9. $A = 35°13'$ $a = 12\,m$

    $B = 54°47'$ $b = 17\,m$

    $C = 90°$ $c = 20.8\,m$

10. $A = 31°27'$ $a = 12\,cm$

    $B = 58°33'$ $b = 19.6\,cm$

    $C = 90°$ $c = 23\,cm$

11. $A = 65°6'$ $a = 17.2\,m$

      $B = 24°54'$ $b = 8\,m$

      $C = 90°$ $c = 19\,cm$

12. 0.5736   13. -1.6643   14. -0.8192   15. 0.0175

16. 1.2868   17. 2.4586   18. -0.1282   19. 0.8973

20. 81.7°   21. 41.8°   22. 66.8°   23. 40.6°

24. 29.7°   25. 71.9°   26. $\cos 25°$ 27. $\cot 49°$

28. $\sin 19°$ 29. $\tan 71°$ 30. $\sec 51°$ 31. $\csc 64°$

32. $\sin 75°28'$ 33. $\cos 56°42'$ 34. $\cos \dfrac{\pi}{6}$ 35. $\sin \dfrac{\pi}{10}$

36. $\cot \dfrac{\pi}{8}$ 37. $\csc \dfrac{3\pi}{14}$ 38. $x = 18$ 39. $x = 10$

40. $x = 11$ 41 $x = 8$

Section 5:   Applications of Right Triangle Trigonometry

Learning Outcomes:

- The student will correctly memorize and apply trigonometric formulas, definitions, identities, and properties.
- The student will correctly use mathematical symbols and mathematical structure to examine and solve real world applications.
- The student will solve trigonometric equations, right triangles, and oblique triangles.

Objectives: At the conclusion of this lesson you should be able to:

1. Solve general applications involving right triangles.
2. Solve applications involving angles of elevations and depression.
3. Solve applications involving angles of rotation.
4. Solve applications involving bearing.

In the last section we developed the skills and techniques to solve right triangles. In this section we will use right triangles to model and solve applications. When solving applications follow the steps below.

Steps for Solving an Application

Step 1   Identify all given quantities. Make a sketch and label the quantities given.

Step 2   Label the quantity that you are trying to find with a variable.

Step 3   Use the information from the sketch to write an equation relating what is given to the variable.

Step 4   Solve the equation.

Step 5   Check your solution to make sure it makes sense.

►EXAMPLE 1

If an 18 ft ladder is placed against a building so that it reaches a window sill 7 ft off the ground, what is the acute angle θ between the ladder and the ground. Round to the nearest tenth of a degree.

►►Solution

Step 1:   We know the length of the ladder is 18ft and the height to the window sill is 7ft. We make a sketch and label what we know.

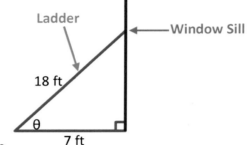

Step 2: We want to find the angle θ formed with the ground and the ladder.

Step 3: We are given the opposite side to θ and the hypotenuse which means we can use sine to find θ.

Step 4: $\sin\theta = \dfrac{7}{18}$

Make sure your calculator is in degree mode.

$\theta = 22.88538°$ ⟵——— 2nd SIN 7 ÷ 1 8) ENTER

$\theta = 22.9°$

Step 5: The angle $\theta = 22.9°$.

►EXAMPLE 2

A 60 ft ramp is inclined at an angle of 5° with the level ground. How high does the ramp rise above the ground? Round to the nearest tenth of a foot.

►►Solution

Step 1: We know the length of the ramp is 60 ft and the angle of incline is 5°. We make a sketch and label what we know.

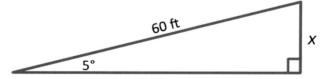

Step 2: We want to find the height of the ramp, labeled x in our sketch.

Step 3: We are given the angle and the hypotenuse and we want to find is the opposite side. We can use sine.

Step 4: $\sin 5° = \dfrac{x}{60}$ Step 5: The ramp is 5.2 ft above the ground.

$60\sin 5° = x$ Make sure your calculator is in degree mode.

$5.2293 = x$ ⟵——— 6 0 SIN 5) ENTER

$5.2\ ft = x$

Applications Involving Angles of Elevation/Depression

The angle of elevation of an object as seen by an observer is the acute angle between the horizontal and the line of sight from the observer's eye to the object.

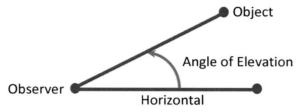

If the object is below the level of the observer, then the acute angle between the horizontal and the observer's line of sight is called the angle of depression.

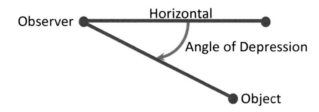

When two parallel lines are cut by a transversal, the angles inside the parallel lines and on opposite sides of the transversal are called alternate interior angles.   Since alternate interior angles are equal, the angle of depression and the angle of elevation are equal angles as shown in **Figure 5.1.**

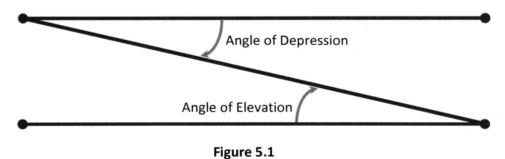

Figure 5.1

► EXAMPLE 3

A guy wire of length 120 meters runs from the ground to the top of a utility pole. If the angle of elevation formed with the ground and the guy wire to the top of the utility pole is 43°, how tall is the utility pole? Round to the nearest tenth of a meter.

►►Solution

Step 1: We know the length of the guy wire is 120 meters and the angle of elevation is 43°. We make a sketch and label what we know.

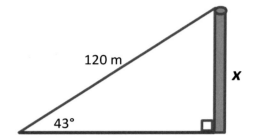

Step 2: We are trying to find the height of the utility pole. Let **x** represent the height of the utility pole.

Step 3: Since we know the hypotenuse and we are trying to find the opposite side, we will use sine: $\sin 43° = \dfrac{x}{120}$.

Step 4: $\sin 43° = \dfrac{x}{120}$

$$120 \cdot \sin 43° = 120 \cdot \dfrac{x}{120}$$

$$120 \sin 43° = x$$

$$81.839803 = x$$

$$81.8 = x$$

Step 5: The height of the utility pole is 81.8 m.

► EXAMPLE 4

A person standing on the top of a 300 meter building sees a parade some distance from the building. If the angle of depression from the person's eyes to the parade is 14.5°, how far from the base of the building is the parade? Round to the nearest whole meter.

►►Solution

Step 1: We know the height of the building is 300 meters and the angle of depression is 14.5°. We make a sketch and label what we know. Opposite sides of the rectangle are equal as shown.

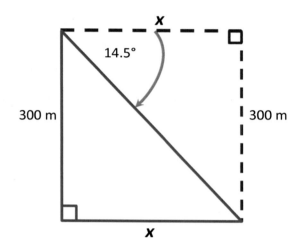

Step 2: We are trying to find the distance the parade is from the building. Let **x** represent the distance the parade is from the building.

Step 3: Since we know the opposite side and we are trying to find the adjacent side, we will use tangent: $\tan 14.5° = \dfrac{300}{x}$.

Step 4: $\tan 14.5° = \dfrac{300}{x}$

$x \cdot \tan 14.5° = x \cdot \dfrac{300}{x}$

$x \tan 14.5° = 300$

$\dfrac{x \tan 14.5°}{\tan 14.5°} = \dfrac{300}{\tan 14.5°}$

$x = 1160.0139$

$x = 1160$

Step 5: The distance the parade is from the building is 1160 m.

►EXAMPLE 5

A bird watching club has climbed to an observation post that is 25 feet off the ground and is observing a nest of birds located on the side of a nearby mountain. The angle of depression from the observation post to the base of the mountain is 12°, while the angle of elevation to the top of the mountain is 63°. The angle of elevation to the nest of birds is 57°. Use this information to find (a.) the distance between the observation post and the mountain, (b.) the height of the mountain, (c.) the height of the nest of birds. Round all measurements to the nearest tenth of a foot.

►►Solution

Step 1: We know the height of the observation post is 25 ft and the angle of depression to the base of the mountain is 12°. The angle of elevation to the top of the mountain is 63° and the angle of elevation to the nest is 57°. We make a sketch and label what we know.

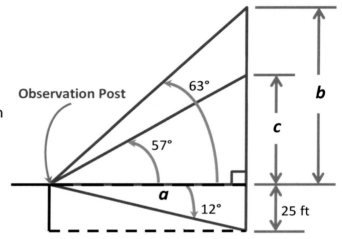

Step 2: We are trying to find the distances

Step 3: a.) We are trying to find the distance from the observation post to the mountain which is labeled **a** in the drawing. Since we know the opposite side and we are trying to find the adjacent side, we will use tangent: $\tan 12° = \dfrac{25}{a}$.

Step 4:   a.) $\tan 12° = \dfrac{25}{a}$

$a \cdot \tan 12° = a \cdot \dfrac{25}{a}$

$a \tan 12° = 25$

$\dfrac{a \tan 12°}{\tan 12°} = \dfrac{25}{\tan 12°}$

$a = 117.6$

Step 5: a.) The distance between the observation post and the mountain is 117.6 feet.

Step 3: b.) We are trying to find the height of the mountain which is **b** from the drawing plus 25 feet. In part a we found the distance between the observation post and the mountain which is the adjacent side. Since we are trying to find the opposite side, we will use tangent:

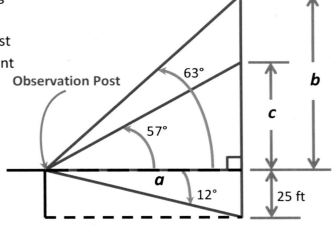

$$\tan 63° = \frac{b}{117.6}.$$

Step 4: b.) $\tan 63° = \dfrac{b}{117.6}$

$117.6 \cdot \tan 63° = b$

$117.6 \tan 63° = b$

$230.8 = b$

Step 5: b.) The height of the mountain is 230.8 feet plus 25 feet which is 255.8 feet.

Step 3: c.) We are trying to find the height of the nest which is **c** from the drawing plus 25 feet. In part a we found the distance between the observation post and the mountain which is the adjacent side. Since we are trying to find the opposite

side, we will use tangent: $\tan 57° = \dfrac{c}{117.6}.$

Step 4: c.) $\tan 57° = \dfrac{c}{117.6}$

$117.6 \cdot \tan 57° = c$

$117.6 \tan 57° = c$

$181.1 = c$

Step 5: c.) The height of the nest is 181.1 feet plus 25 feet which is 206.1 feet.

Angle of Rotation/Bearing

The angle of rotation is an angle from a fixed orientation to a fixed line of sight. Land surveyors use a special type of measure called bearings. In this case bearing indicates the acute angle the line of sight makes with a due north or due south line of reference. See **Figure 5.2.**

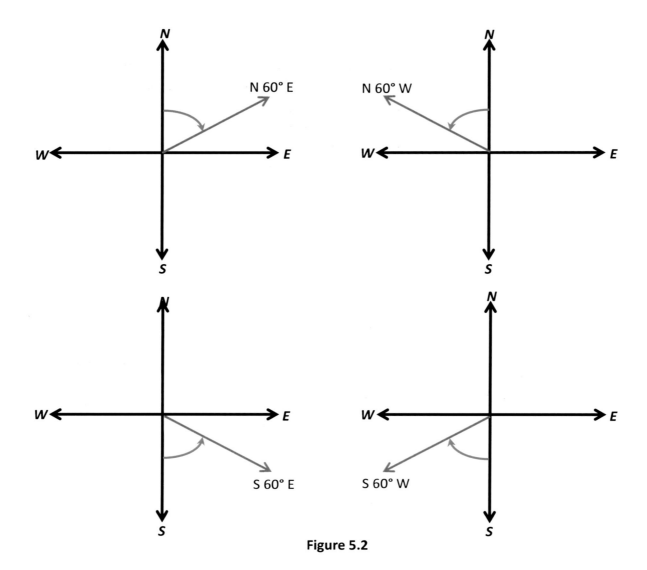

Figure 5.2

In air and sea navigation, bearing can also be given as a nonnegative angle less than 360° measured from the north in a clockwise direction. See **Figure 5.3.**

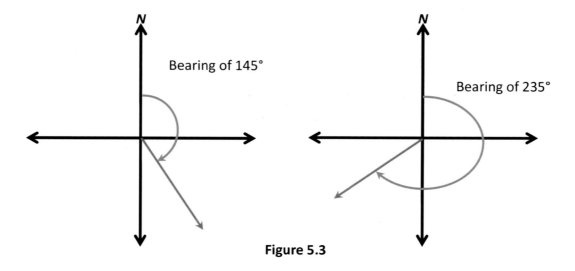

Figure 5.3

►EXAMPLE 6

A car is stuck in a line of cars 65 feet back from an intersection. The driver observes a fire truck with siren glaring approaching the intersection from his right. To pass the time the driver decides to estimate the speed of the fire truck. He spots a street lamp next to the road some distance away and figures the angle of rotation from the intersection to the street light is 80°. If it takes the fire truck 4 seconds to reach the street light from the intersection, how fast is the fire truck traveling in miles per hour? Round to the nearest whole number.

►►Solution

Step 1: We know the car is 65 feet back from the intersection and the angle of rotation is 80°.

Step 2: We are trying to find the speed of the fire truck. The rate something travels is equal to the distance divided by time. We will need to find the distance the fire truck travels, labeled **x** in the drawing.

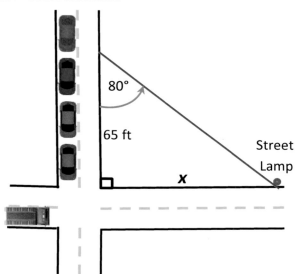

Step 3: Since we know the adjacent side and we are trying to find the opposite side, we will use tangent: $\tan 80° = \dfrac{x}{65}$.

Step 4: $\tan 80° = \dfrac{x}{65}$

$65 \cdot \tan 80° = 65 \cdot \dfrac{x}{65}$

$65 \tan 80° = x$

$r = \dfrac{d}{t}$

$rate = \dfrac{65 \tan 80° \, ft}{4 \sec} \cdot \dfrac{60 \text{ sec}}{1 \min} \cdot \dfrac{60 \min}{1 \, hour} \cdot \dfrac{1 \, mile}{5280 \, ft}$ *We need to convert to miles per hour.*

$rate = 62.8352247 \; mph$

$rate = 63 \; mph$

Step 5: The rate of the fire truck is 63 mph.

▶ EXAMPLE 7

The bearing from port A to port C is **S 35° E** and from port C to port B **N 55° E**. If a ship sails from A to C a distance of 80 kilometers, and then from C to B a distance of 56 kilometers, how far is it from port A to port B? Round to the nearest tenth of a kilometer.

▶▶ Solution

Step 1: We know the distance from A to C is 80 km and the distance from C to B is 56 km. We also know the angle from A to C and the angle from C to B as shown in the drawing.

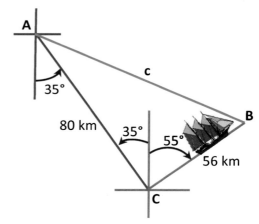

Step 2: We are trying to find the distance between ports A and B. Since this side is opposite angle C we will label it c.

Step 3: We are trying to find the distance between ports A and B which is labeled c in the drawing. Notice that at port C the angles sum to 90°. We can use the Pythagorean Theorem to find c.

Step 4:
$$a^2 + b^2 = c^2$$
$$(80)^2 + (56)^2 = c^2$$
$$6400 + 3136 = c^2$$
$$9536 = c^2$$
$$\sqrt{9536} = \sqrt{c^2}$$
$$97.65244 = c$$

Step 5: The distance between port A and port B is 97.7 km.

▶ **EXAMPLE 8**

A plane leaves an airport and flies on a bearing of 28° for 1.3 hours at 150 miles per hour. The plane then turns and flies on a bearing of 118° for 1.5 hours at the same speed. How far is the plane from its starting pointing? Round to the nearest whole number.

▶▶ **Solution**

Step 1: We know the bearings of the plane and we can sketch our drawing. Red represents the first part of the flight and green represents the second part of the flight.

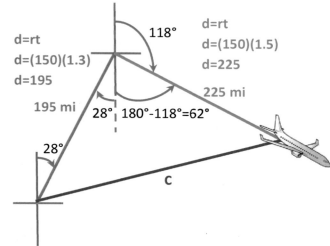

d=rt
d=(150)(1.3)
d=195

195 mi

d=rt
d=(150)(1.5)
d=225

225 mi

118°

28° 180°-118°=62°

28°

c

Step 2: We are trying to find the distance from the airport to where the plane is located. We will label this distance c.

Step 3: In order to find c we need to find the angle formed with the first part of the flight and the second part of the flight. Using alternate interior angles we have 28° and subtracting 118° from 180° gives 62°. Adding the angles together gives us 90°. Using the distance formula we are able to find the distance the plane travels for each part of the flight. We can use the Pythagorean Theorem to find c.

Step 4:
$$a^2 + b^2 = c^2$$
$$(195)^2 + (225)^2 = c^2$$
$$38025 + 50625 = c^2$$
$$88650 = c^2$$
$$\sqrt{88650} = \sqrt{c^2}$$
$$297.7415 = c$$

Step 5: The plane is 298 miles from its starting point.

▶ **EXAMPLE 9**

Riding your bicycle across flat land, you notice a mountain in the distance. The angle of elevation to the top of the mountain is 4°. After you ride your bike 15 miles closer to the mountain, the angle of elevation is 10°. Approximate the height of the mountain to the nearest tenth of a mile.

▶▶**Solution**

Step 1: At a distance of (15 +*x*) miles the angle of elevation to the top of the mountain is 4°. At a distance of *x* miles the angle of elevation to the top of the mountain is 10°.

Step 2: We are trying to find the height of the mountain labeled **h**.

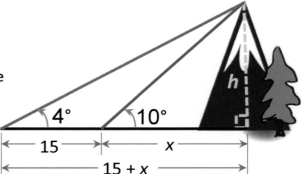

Step 3: To get the height of the mountain we can use tangent and the 4° angle of elevation:

$\tan 4° = \dfrac{h}{15+x}$. We have two unknowns, *x* and *h*. We need to get our variable *x* in terms of *h* since we are trying to find *h* (the height of the mountain). Using tangent and the 10° angle of elevation we get: $\tan 10° = \dfrac{h}{x}$. Solving for *x* gives us: $x = \dfrac{h}{\tan 10°}$. Substituting *x* into our first equation we have: $\tan 4° = \dfrac{h}{15 + \dfrac{h}{\tan 10°}}$.

Step 4:
$$\tan 4° = \frac{h}{15 + \dfrac{h}{\tan 10°}}$$

$$\left(15 + \frac{h}{\tan 10°}\right)\tan 4° = \left(15 + \frac{h}{\tan 10°}\right)\frac{h}{15 + \dfrac{h}{\tan 10°}}$$

$$15\tan 4° + \frac{\tan 4°}{\tan 10°}h = h$$

$$15\tan 4° = h - \frac{\tan 4°}{\tan 10°}h$$

$$15\tan 4° = h\left(1 - \frac{\tan 4°}{\tan 10°}\right)$$

$$\frac{15\tan 4°}{1 - \dfrac{\tan 4°}{\tan 10°}} = h$$

$$1.7m = h$$

Step 5: The height of the mountain 1.7 miles.

HOMEWORK

Objective 1

Solve. *See Examples 1 and 2.*

1.   A 24 ft ladder is leaning against a building. If the ladder makes an angle of 58° with the ground, how far up the building does the ladder reach? Round to the nearest tenth of a foot.

2.   A ramp for a wheelchair is to be built beside the main steps of a post office. The total vertical rise of the steps is 3 feet and the ramp will be inclined at angle of 12°. How long does the ramp need to be? Round to the nearest tenth of a foot.

Objective 2

Solve. *See Examples 3, 4, 5 and 9.*

3.   When the angle of elevation of the sun is 50°, a certain flagpole casts a shadow 30 feet long. How tall is the flagpole? Round to the nearest tenth of a foot.

4.   From a 100 ft tall lighthouse on the coast, an overturned sailboat is sighted. If the angle of depression is 9° from the lighthouse to the boat, how far is the boat from the lighthouse? Round to the nearest tenth of a foot.

5.   A 20 ft tall flagpole is mounted on the top of a building. A person standing level with the building observes the angle of elevation to the top of the flag pole is 65°. From the same spot the angle of elevation to the foot of the flagpole is 57°. How tall is the building? Round to the nearest tenth of a foot.

6.   An observer stops to look at a 500 year old tree and observes the angle of elevation to the top of the tree to be 23.5°. The observer walks 100 ft closer to the tree and observes the new angle of elevation to the top of the tree to be 46°. How tall is the tree? Round to the nearest tenth of a foot.

7.   From point A on the ground the angle of elevation to the top of a water tower is 19.9°. From a point B, 50 feet closer to the tower, the angle of elevation is 21.8°. What is the height of the tower? Round to the nearest tenth of a foot.

8.   A hiker is hiking down a mountain with a vertical height of 1500 feet. The distance from the top of the mountain to the base is 3000 feet. What is the angle of elevation from the base to the top of the mountain? Round to the nearest tenth of a degree.

9.   A guy wire of length 108 meters runs from the top of antenna to the ground. If the height of the antenna is 73 meters, what is the angle of elevation the guy wire makes with the ground? Round to the nearest tenth of a degree.

10.   A ladder 20 feet long leans against the side of a house. Find the height h from the top of ladder to ground if the angle of elevation of the ladder is $80°$. Round to the nearest tenth of a foot.

11.   A cellular telephone tower that is 150 feet tall is placed on top of a mountain that is 1220 feet above sea level. What is the angle of depression from the top of the tower to a cell phone user who is 5 horizontal miles away and 400 feet above sea level? Round to the nearest tenth of a degree.

12.   An antenna is on top of the center of a building. The angle of elevation from a point 100 feet from the center of the building to the top of the antenna is $24°23'$, and the angle of elevation to the bottom of the antenna is $20°12'$. Find the height of the antenna. Round to the nearest whole number.

13.   The angle of elevation from a point on the ground to the top of a pyramid is $35°30'$. The angle of elevation from a point 135 feet farther back to the top of the pyramid is $21°10'$. Find the height of the pyramid. Round to the nearest whole number.

14.   A bird watching club has climbed to an observation post that is 75 feet off the ground and is observing a nest of birds located on the side of a nearby mountain. The angle of depression from the observation post to the base of the mountain is $14°$, while the angle of elevation to the top of the mountain is $25°$. The angle of elevation to the nest of birds is $21°$. Use this information to find (a.) the distance between the observation post and the mountain, (b.) the height of the mountain, (c.) the height of the nest of birds. Round all measurements to the nearest tenth of a foot.

15.   From her apartment window on the sixth floor, Emma notices some window washers high above her on the building across the street. She estimates the buildings are 50 feet apart, the angle of elevation to the window washers is about $80°$, and the angle of depression to the base of the building is $50°$. (a.) How high above the ground is the window of Emma's apartment? (b.) How high above the ground are the window washers? Round to the nearest foot.

16.   The angle of elevation from the top of a small building to the top of a nearby taller building $49°$ and the angle of depression to the bottom of the building is $15°$. If the shorter building is 30 m high, how tall is the taller building. Round to the nearest tenth of a meter.

17.   From a window 30 ft above the street, the angle of elevation to the top of the building across the street is $53°$ and the angle of depression to the base of this building is $22°$. Find the height of the building across the street. Round to the nearest tenth of a foot.

Objectives 3 and 5

Solve. *See Examples 6, 7 and 8.*

18.   The bearing from point A to point B is *S 55°E*, and from point B to point C *N 35°E*. If a ship sails from A to B, a distance of 80 km, and then from B to C, a distance of 56 km, how far is it from A to C? Round to the nearest tenth of a kilometer.

19.   An observer attempts to an estimate the speed of a plane crop dusting. Standing 50 feet from the edge of the field he observes that the plane makes a pass across the field in 3.5 seconds. He estimates the angle of rotation from one end of the field to the other to be 85°. How fast is the plane flying in miles per hour (mph)? Round to the nearest tenth.

20.   Stuck in a line of cars at a railroad crossing 50 feet from the crossing, an observer spots a train in the distance. The observer notices a building beside the tracks and estimates the angle of rotation from the crossing to the building to be 75°. If the train takes 4 seconds to travel from the crossing to the building, how fast is the train traveling in miles per hour? Round to the nearest tenth.

21.   A ship travels 65 km on a bearing of 27° and then travels on a bearing of 117° for 140 km. Find the distance from the starting point to the ending point. Round to the nearest kilometer.

22.   Two lighthouses are located on a north-south line. From lighthouse A, the bearing of a ship 4800 m away is 129°43′. From lighthouse B, the bearing of the same ship is 39°43′. Find the distance between the two lighthouses. Round to the nearest meter.

23.   The bearing from A to C is N 64° W. The bearing from A to B is S 82° W. The bearing from B to C is N 26° E. A plane flying at 450 miles per hour takes 2 hours to go from A to B. Find the distance from B to C. Round to the nearest mile.

24.   The bearing from Atlanta, GA to Macon, GA is S 27° E and the bearing from Macon, GA to Augusta, GA is N 63° E. A car traveling at 62 miles per hour needs 1.25 hours to go from Atlanta to Macon and 1.75 hours to go from Macon to Augusta. Find the distance from Atlanta to Augusta. Round to the nearest mile.

ANSWERS

1.	20.4 ft	2.	14.4 ft	3.	35.8 ft	4.	631.4 ft
5.	50.9 ft	6.	75 ft	7.	190.6 ft	8.	30°
9.	42.5°	10.	19.7 ft	11.	2.1°	12.	9 ft
13.	114 ft	14.	300.8 ft, , 215.3 ft, 190.5 ft			15.	60 ft, 344 ft
16.	158.8 m	17.	128.6 ft	18.	97.7 km	19.	111.3 mph
20.	31.8 mph	21.	154 km	22.	7512 m	23.	503 mi
24.	133 mi						

Section 6: Trigonometric Functions of Any Angle

Learning Outcomes:

- The student will correctly memorize and apply trigonometric formulas, definitions, identities, and properties.
- The student will correctly use mathematical symbols and mathematical structure to examine and solve real world applications.
- The student will solve trigonometric equations, right triangles, and oblique triangles.

Objectives: At the conclusion of this lesson you should be able to:

1. Define the six trigonometric functions using the coordinates of a point on the terminal side of an angle.
2. Analyze the signs of the trigonometric functions of an angle based on the quadrant of the terminal side.
3. Evaluate the trigonometric functions of quadrantal angles.
4. Find reference angles of nonquadrantal angles.
5. Use reference angles to evaluate trigonometric functions.

Recall that an angle is in standard position when the vertex is at the origin and the initial side is on the x-axis as shown in **Figure 6.1**. Suppose we have an acute angle θ in standard position and we choose a point (x, y) on the terminal side as shown in **Figure 6.2**.

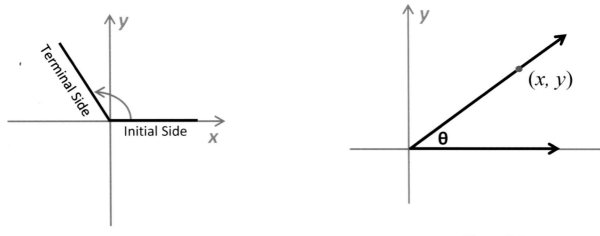

Figure 6.1   Figure 6.2

If we draw a perpendicular from the point on the terminal side to the initial side of angle θ we have the right triangle shown in **Figure 6.3.** The adjacent side to angle θ is *x* and the opposite side to angle θ is *y*.   Letting *r* represent the hypotenuse of the triangle, we know that by the Pythagorean Theorem $r = \sqrt{x^2 + y^2}$ as shown in **Figure 6.4**. The measure of *r* will always be positive.

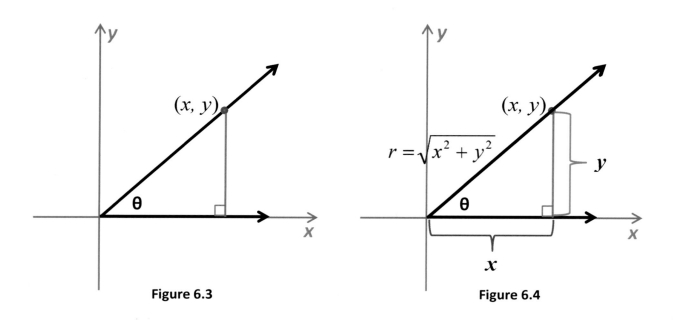

Figure 6.3   Figure 6.4

Applying the definitions of trigonometric functions from Section 4 to our drawing in **Figure 6.4** we have:

$$\sin \theta = \frac{opposite\ side}{hypotenuse} = \frac{y}{r} \qquad \csc \theta = \frac{hypotenuse}{opposite\ side} = \frac{r}{y}$$

$$\cos \theta = \frac{adjacent\ side}{hypotenuse} = \frac{x}{r} \qquad \sec \theta = \frac{hypotenuse}{adjacent\ side} = \frac{r}{x}$$

$$\tan \theta = \frac{opposite\ side}{adjacent\ side} = \frac{y}{x} \qquad \cot \theta = \frac{adjacent\ side}{opposite\ side} = \frac{x}{y}$$

The trigonometric functions of θ have been defined in terms of the coordinates of a point on the terminal side. This method suggests a way of defining the values of the six trigonometric functions for any angle in any quadrant.

Definition of the Trigonometric Functions of Any Angle

Let θ be any angle in standard position with (x, y) a point on the terminal side of θ and $r = \sqrt{x^2 + y^2}$, then

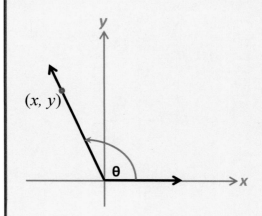

$$\sin \theta = \frac{y}{r} \qquad\qquad \csc \theta = \frac{r}{y}, (y \neq 0)$$

$$\cos \theta = \frac{x}{r} \qquad\qquad \sec \theta = \frac{r}{x}, (x \neq 0)$$

$$\tan \theta = \frac{y}{x}, (x \neq 0) \qquad \cot \theta = \frac{x}{y}, (y \neq 0)$$

Assuming that x and y are not simultaneously 0, r cannot be zero. This means that the sine and cosine functions are defined for any real value of θ. If $y = 0$, cosecant and cotangent are undefined and if $x = 0$, secant and tangent are undefined.

►EXAMPLE 1

Let $(-12,5)$ be a point on the terminal side of θ. Find the six trigonometric functions of θ.

►►Solution

$(-12,5)$ $x = -12$ and $y = 5$. We need to find r.

$r = \sqrt{x^2 + y^2}$.

$r = \sqrt{(-12)^2 + (5)^2}$

$r = \sqrt{144 + 25}$

$r = \sqrt{169}$

$r = 13$

$$\sin \theta = \frac{y}{r} = \frac{5}{13} \qquad\qquad \csc \theta = \frac{r}{y} = \frac{13}{5}$$

$$\cos \theta = \frac{x}{r} = -\frac{12}{13} \qquad\qquad \sec \theta = \frac{r}{x} = -\frac{13}{12}$$

$$\tan \theta = \frac{y}{x} = -\frac{5}{12} \qquad\qquad \cot \theta = \frac{x}{y} = -\frac{12}{5}$$

▶ **EXAMPLE 2**

Let $\left(\sqrt{3}, -1\right)$ be a point on the terminal side of θ. Find the six trigonometric functions of θ.

▶▶ **Solution**

$\left(\sqrt{3}, -1\right)$ $x = \sqrt{3}$ and $y = -1$. We need to find r.

$r = \sqrt{x^2 + y^2}$.

$r = \sqrt{\left(\sqrt{3}\right)^2 + \left(-1\right)^2}$

$r = \sqrt{3 + 1}$

$r = \sqrt{4}$

$r = 2$

$\sin\theta = \dfrac{y}{r} = -\dfrac{1}{2}$

$\cos\theta = \dfrac{x}{r} = \dfrac{\sqrt{3}}{2}$

$\tan\theta = \dfrac{y}{x} = -\dfrac{1}{\sqrt{3}}$

$\csc\theta = \dfrac{r}{y} = -\dfrac{2}{1}$

$\sec\theta = \dfrac{r}{x} = \dfrac{2}{\sqrt{3}}$

$\cot\theta = \dfrac{x}{y} = -\sqrt{3}$

Consider the angle θ and the line $y = \dfrac{4}{3}x$ where $x \in [0, \infty)$. If we pick $\left(3, 4\right)$ and $\left(6, 8\right)$, two distinct points on the terminal side of θ and drop the perpendiculars, we have two similar triangles as shown in **Figure 6.5**. Because corresponding sides of similar triangles are

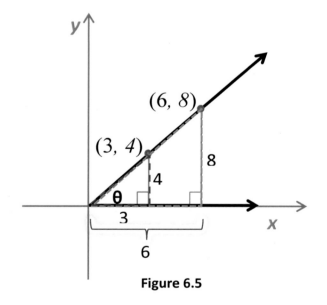

Figure 6.5

proportional, $\dfrac{4}{3} = \dfrac{8}{6}$, thus $\tan\theta = \dfrac{4}{3}$ and

$\tan\theta = \dfrac{8}{6} = \dfrac{4}{3}$ is the same no matter which point is picked on the line. This holds for the other 5 trigonometric functions.

If we are given a point (x, y) on the terminal side of θ, we can use the point (ax, ay) where a is multiple scalar of (x, y) to evaluate the six trigonometric functions.

►EXAMPLE 3

Let $\left(-\dfrac{3}{2},-\dfrac{3}{4}\right)$ be a point on the terminal side of θ. Find the six trigonometric functions of θ.

►►Solution

$\left(-\dfrac{3}{2},-\dfrac{3}{4}\right)$ multiply both x and y by 4 to eliminate the fractions $\left(4\left(-\dfrac{3}{2}\right),4\left(-\dfrac{3}{4}\right)\right)$. Since we

are multiplying both x and y by the same value, 4, the point $(-6,-3)$ is a point on the terminal side of θ.

$(-6,-3)$ $x=-6$ and $y=-3$. We need to find r.

$$r=\sqrt{x^2+y^2}$$

$$r=\sqrt{(-6)^2+(-3)^2}$$

$$r=\sqrt{36+9}$$

$$r=\sqrt{45}$$

$$r=\sqrt{9\cdot5}$$

$$r=3\sqrt{5}$$

$$\sin\theta=\frac{-3}{3\sqrt{5}}=-\frac{1}{\sqrt{5}}=-\frac{\sqrt{5}}{5} \qquad \csc\theta=\frac{3\sqrt{5}}{-3}=-\sqrt{5}$$

$$\cos\theta=\frac{-6}{3\sqrt{5}}=-\frac{2}{\sqrt{5}}=-\frac{2\sqrt{5}}{5} \qquad \sec\theta=-\frac{3\sqrt{5}}{-6}=-\frac{\sqrt{5}}{2}$$

$$\tan\theta=\frac{-3}{-6}=\frac{1}{2} \qquad\qquad\qquad \cot\theta=\frac{-6}{-3}=2$$

If we use the given point $\left(-\dfrac{3}{2},-\dfrac{3}{4}\right)$ $x=-\dfrac{3}{2}$ and $y=-\dfrac{3}{4}$. We need to find r.

$$r=\sqrt{x^2+y^2}\,.$$

$$r=\sqrt{\left(-\frac{3}{2}\right)^2+\left(-\frac{3}{4}\right)^2}$$

$$r=\sqrt{\frac{9}{4}+\frac{9}{16}}$$

$$r=\sqrt{\frac{36}{16}+\frac{9}{16}}$$

$$r=\sqrt{\frac{45}{16}}$$

$$r=\frac{\sqrt{9\cdot5}}{\sqrt{16}}$$

$$r=\frac{3\sqrt{5}}{4}$$

$$\sin\theta=\frac{-\dfrac{3}{4}}{\dfrac{3\sqrt{5}}{4}}=-\frac{1}{\sqrt{5}}=-\frac{\sqrt{5}}{5} \qquad \csc\theta=\frac{\dfrac{3\sqrt{5}}{4}}{-\dfrac{3}{4}}=-\sqrt{5}$$

$$\cos\theta=\frac{-\dfrac{3}{2}}{\dfrac{3\sqrt{5}}{4}}=-\frac{2}{\sqrt{5}}=-\frac{2\sqrt{5}}{5} \qquad \sec\theta=\frac{\dfrac{3\sqrt{5}}{4}}{-\dfrac{3}{2}}=-\frac{\sqrt{5}}{2}$$

$$\tan\theta=\frac{-\dfrac{3}{4}}{-\dfrac{3}{2}}=\frac{1}{2} \qquad\qquad \cot\theta=\frac{-\dfrac{3}{2}}{-\dfrac{3}{4}}=2$$

In the first quadrant both x and y are positive $(x > 0,\ y > 0)$, in the second quadrant x is negative $(x < 0)$ and y is positive $(y > 0)$, in the third quadrant both x and y are negative $(x < 0,\ y < 0)$, and in the fourth quadrant x is positive $(x > 0)$ and y is negative $(y < 0)$ as shown in **Figure 6.6**. The signs of the six trigonometric functions can be determined using the definitions of the trigonometric functions. Since x and y are both positive in Quadrant I, all six trigonometric functions are positive. In Quadrant II y is positive which means sine and its reciprocal cosecant are both positive since $\sin\theta = \dfrac{y}{r}$ and $\csc\theta = \dfrac{r}{y}$. In Quadrant III both x and y are negative which means tangent and its reciprocal cotangent are positive since $\tan\theta = \dfrac{y}{x}$ and $\cot\theta = \dfrac{x}{y}$. In Quadrant IV x is positive which means cosine and its reciprocal secant are both positive since $\cos\theta = \dfrac{x}{r}$ and $\sec\theta = \dfrac{r}{x}$.

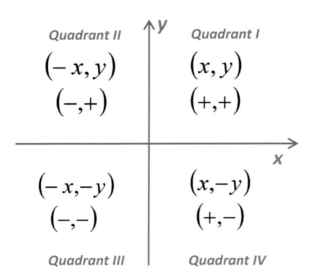

Figure 6.6

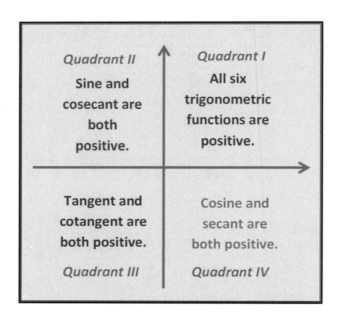

► EXAMPLE 4

State the quadrant in which θ lies if the $\sin\theta < 0$ and $\cos\theta > 0$.

►►Solution

Since y is negative in in both *Quadrant III* and *Quadrant IV,* $\sin\theta < 0$ (sine is negative) in these two quadrants. We know that $\cos\theta > 0$ (cosine is positive) in both *Quadrant I* and *Quadrant IV.* Therefore θ must be in *Quadrant IV.*

Quadrant II   *Quadrant I*

$\cos\theta > 0$

$\sin\theta < 0$ $\begin{array}{l}\sin\theta < 0 \\ \cos\theta > 0\end{array}$

Quadrant III   *Quadrant IV*

► EXAMPLE 5

State the quadrant in which θ lies if the $\sec\theta < 0$ and $\cot\theta > 0$.

►►Solution

Since x is negative in both *Quadrant II* and *Quadrant III,* $\sec\theta < 0$ (secant is negative) in these two quadrants. We know that $\cot\theta > 0$ (cotangent is positive) in *Quadrant I* because x and y are both positive. $\cot\theta > 0$ (cotangent is positive) in *Quadrant III* since x and y are both negative in *Quadrant III.* Therefore θ must be in *Quadrant III.*

Quadrant II   *Quadrant I*

$\sec\theta < 0$ $\cot\theta > 0$

$\begin{array}{l}\sec\theta < 0 \\ \cot\theta > 0\end{array}$

Quadrant III   *Quadrant IV*

► EXAMPLE 6

Given $\tan\theta = -\dfrac{15}{8}$ and θ is *Quadrant IV*, find the remaining 5 trigonometric functions.

►► Solution

Since we are in *Quadrant IV* we know that x is positive and y is negative. $\tan\theta = \dfrac{y}{x}$

so, for $\tan\theta = -\dfrac{15}{8}$ $x = 8,$ $y = -15$. We need to find r (recall that *r* is always

positive). We will use $r = \sqrt{x^2 + y^2}$.

$$r = \sqrt{(8)^2 + (-15)^2}$$
$$r = \sqrt{64 + 225}$$
$$r = \sqrt{289}$$
$$r = 17$$

$\sin\theta = -\dfrac{15}{17}$ $\qquad \csc\theta = -\dfrac{17}{15}$

$\cos\theta = \dfrac{8}{17}$ $\qquad \sec\theta = \dfrac{17}{8}$

$\tan\theta = -\dfrac{15}{8}$ $\qquad \cot\theta = -\dfrac{8}{15}$

► EXAMPLE 7

Given $\sec\theta = -\dfrac{13}{5}$ and the $\sin\theta > 0$, find the remaining 5 trigonometric functions.

►► Solution

1. We need to determine what quadrant θ is located in. Since *x* is negative in both *Quadrant II* and *Quadrant III*, $\sec\theta < 0$ (secant is negative) in these two quadrants. We know that $\sin\theta > 0$ (sine is positive) in *Quadrant I* and *Quadrant II*. Therefore θ must be in *Quadrant II* which means *y* will be positive.

2. $\sec\theta = \dfrac{r}{x}$ so, for $\sec\theta = -\dfrac{13}{5}$ $x = -5,$ $r = 13$ (recall that *r* is always positive).

    We need to find *y*. We can use $r = \sqrt{x^2 + y^2}$.

$$13 = \sqrt{(-5)^2 + y^2}$$
$$(13)^2 = \left(\sqrt{25 + y^2}\right)^2$$
$$169 = 25 + y^2$$
$$169 - 25 = y^2$$
$$144 = y^2$$
$$\sqrt{144} = \sqrt{y^2}$$
$$12 = y$$

$\sin\theta = \dfrac{12}{13}$ $\qquad \csc\theta = \dfrac{13}{12}$

$\cos\theta = -\dfrac{5}{13}$ $\qquad \sec\theta = -\dfrac{13}{5}$

$\tan\theta = -\dfrac{12}{5}$ $\qquad \cot\theta = -\dfrac{5}{12}$

Quadrantal Angles

A quadrantal angle is an angle in standard position with terminal side on the x-axis or the y-axis. Quadrantal angles are labeled and colored coded in **Figure 6.7**.

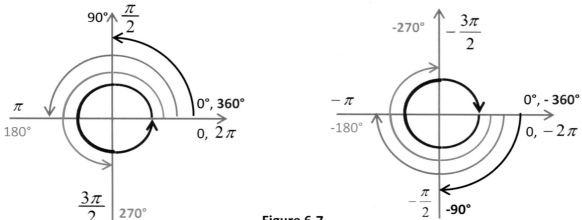

Figure 6.7

To find the sine, cosine and tangent for $90°$ or $\dfrac{\pi}{2}$ we can pick any point on the terminal side.

We can use the point $(0,3)$ where $x = 0$ and $y = 3$. Using $r = \sqrt{x^2 + y^2}$, $r = \sqrt{(0)^2 + (3)^2}$ gives us $r = 3$.

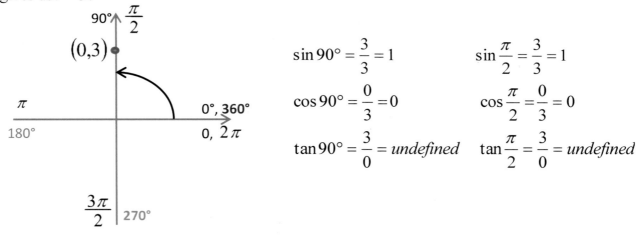

$$\sin 90° = \frac{3}{3} = 1 \qquad \sin \frac{\pi}{2} = \frac{3}{3} = 1$$

$$\cos 90° = \frac{0}{3} = 0 \qquad \cos \frac{\pi}{2} = \frac{0}{3} = 0$$

$$\tan 90° = \frac{3}{0} = undefined \qquad \tan \frac{\pi}{2} = \frac{3}{0} = undefined$$

Notice that if we pick any point on the terminal side for the quadrantal angle $90°$ or $\dfrac{\pi}{2}$ the

$\sin(90°) = 1$, $\sin\left(\dfrac{\pi}{2}\right) = 1$, $\cos(90°) = 0$, $\cos\left(\dfrac{\pi}{2}\right) = 0$, $\tan(90°) = undefined$ and

$\tan\left(\dfrac{\pi}{2}\right) = undefined$.

To find the sine, cosine and tangent for $180°$ or π we can pick any point on the terminal side. We can use the point $(-5,0)$ where $x = -5$ and $y = 0$. Using $r = \sqrt{x^2 + y^2}$,

$r = \sqrt{(-5)^2 + (0)^2}$ gives us $r = 5$.

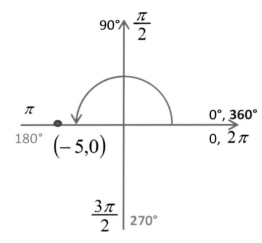

$$\sin 180° = \frac{0}{5} = 0 \qquad \sin \pi = \frac{0}{5} = 0$$

$$\cos 180° = \frac{-5}{5} = -1 \qquad \cos \pi = \frac{-5}{5} = -1$$

$$\tan 180° = \frac{0}{-5} = 0 \qquad \tan \pi = \frac{0}{-5} = 0$$

If we pick any point on the terminal side for the quadrantal angle $180°$ or π the $\sin(180°) = 0$, $\sin(\pi) = 0$, $\cos(180°) = -1$, $\cos(\pi) = -1$, $\tan(180°) = 0$ and $\tan(\pi) = 0$.

To find the sine, cosine and tangent for $270°$ or $\dfrac{3\pi}{2}$ we can pick any point on the terminal side. We can use the point $(0,-4)$ where $x = 0$ and $y = -4$. Using $r = \sqrt{x^2 + y^2}$, $r = \sqrt{(0)^2 + (-4)^2}$ gives us $r = 4$.

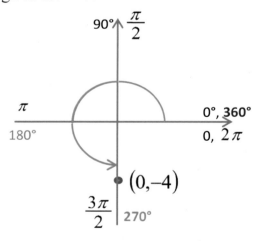

$$\sin 270° = \frac{-3}{3} = -1 \qquad \sin \frac{3\pi}{2} = \frac{-3}{3} = -1$$

$$\cos 270° = \frac{0}{3} = 0 \qquad \cos \frac{3\pi}{2} = \frac{0}{3} = 0$$

$$\tan 270° = \frac{-3}{0} = undefined \qquad \tan \frac{3\pi}{2} = \frac{-3}{0} = undefined$$

If we pick any point on the terminal side for the quadrantal angle 270° or $\dfrac{3\pi}{2}$ the

$\sin(270°) = -1$, $\sin\left(\dfrac{3\pi}{2}\right) = -1$, $\cos(270°) = 0$, $\cos\left(\dfrac{3\pi}{2}\right) = 0$, $\tan(270°) = undefined$ and

$\tan\left(\dfrac{3\pi}{2}\right) = undefined$.

To find the sine, cosine and tangent for 0°, 360°, 0, 2π we can pick any point on the terminal side. We can use the point $(2,0)$ where $x = 2$ and $y = 0$. Using $r = \sqrt{x^2 + y^2}$,

$r = \sqrt{(2)^2 + (0)^2}$ gives us $r = 2$.

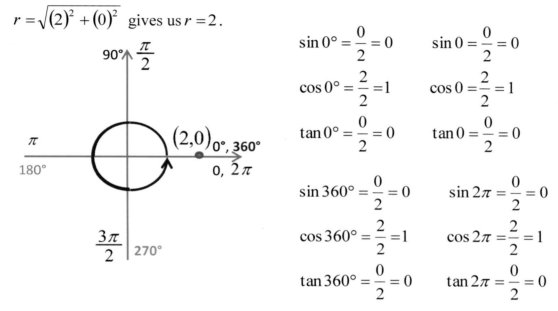

$$\sin 0° = \frac{0}{2} = 0 \qquad \sin 0 = \frac{0}{2} = 0$$

$$\cos 0° = \frac{2}{2} = 1 \qquad \cos 0 = \frac{2}{2} = 1$$

$$\tan 0° = \frac{0}{2} = 0 \qquad \tan 0 = \frac{0}{2} = 0$$

$$\sin 360° = \frac{0}{2} = 0 \qquad \sin 2\pi = \frac{0}{2} = 0$$

$$\cos 360° = \frac{2}{2} = 1 \qquad \cos 2\pi = \frac{2}{2} = 1$$

$$\tan 360° = \frac{0}{2} = 0 \qquad \tan 2\pi = \frac{0}{2} = 0$$

If we pick any point on the terminal side for the quadrantal angle 0°, 360°, 0, 2π the $\sin(0°) = 0$, $\sin(360°) = 0$, $\sin(0) = 0$, $\sin(2\pi) = 0$, $\cos(0°) = 1$, $\cos(360°) = 1$, $\cos(0) = 1$, $\cos(2\pi) = 1$, $\tan(0°) = 0$, $\tan(360°) = 0$, $\tan(0) = 0$ and $\tan(2\pi) = 0$.

Conditions for Undefined Function Values

- If the terminal side of the quadrantal angle lies along the *x*-axis, then the cotangent and the cosecant functions are undefined.
- If the terminal side of the quadrantal angle lies along the *y*-axis, then the tangent and the secant functions are undefined.

Summary of Function Values

$\sin 30° = \dfrac{1}{2}$ $\sin \dfrac{\pi}{6} = \dfrac{1}{2}$

$\cos 30° = \dfrac{\sqrt{3}}{2}$ $\cos \dfrac{\pi}{6} = \dfrac{\sqrt{3}}{2}$

$\tan 30° = \dfrac{\sqrt{3}}{3}$ $\tan \dfrac{\pi}{6} = \dfrac{\sqrt{3}}{3}$

$\sin 45° = \dfrac{\sqrt{2}}{2}$ $\sin \dfrac{\pi}{4} = \dfrac{\sqrt{2}}{2}$

$\cos 45° = \dfrac{\sqrt{2}}{2}$ $\cos \dfrac{\pi}{4} = \dfrac{\sqrt{2}}{2}$

$\tan 45° = 1$ $\tan \dfrac{\pi}{4} = 1$

$\sin 60° = \dfrac{\sqrt{3}}{2}$ $\sin \dfrac{\pi}{3} = \dfrac{\sqrt{3}}{2}$

$\cos 60° = \dfrac{1}{2}$ $\cos \dfrac{\pi}{3} = \dfrac{1}{2}$

$\tan 60° = \sqrt{3}$ $\tan \dfrac{\pi}{3} = \sqrt{3}$

$\sin 90° = 1$ $\sin \dfrac{\pi}{2} = 1$

$\cos 90° = 0$ $\cos \dfrac{\pi}{2} = 0$

$\tan 90° = und$ $\tan \dfrac{\pi}{2} = und$

$\sin 180° = 0$ $\sin \pi = 0$

$\cos 180° = -1$ $\cos \pi = -1$

$\tan 180° = 0$ $\tan \pi = 0$

$\sin 270° = -1$ $\sin \dfrac{3\pi}{2} = -1$

$\cos 270° = 0$ $\cos \dfrac{3\pi}{2} = 0$

$\tan 270° = und$ $\tan \dfrac{3\pi}{2} = und$

$\sin 0° = 0$ $\sin 360° = 0$ $\sin 0 = 0$ $\sin 2\pi = 0$

$\cos 0° = 1$ $\cos 360° = 1$ $\cos 0 = 1$ $\cos 2\pi = 1$

$\tan 0° = 0$ $\tan 360° = 0$ $\tan 0 = 0$ $\tan 2\pi = 0$

Reference Angles

We have defined the six trigonometric functions of angle θ in terms of x, y, and r where x and y are coordinates of a point on the terminal side of θ and r is the distance between the point and the origin. The values of the trigonometric functions of angles greater than $90°$, $\dfrac{\pi}{2}$ or less than $0°$, 0 can be determined from their values at corresponding acute angles called reference angles.

Definition of Reference Angles

If θ is a nonquadrantal angle in standard position, then the reference angle for θ is the positive acute angle θ_r (read "theta sub r") formed by the terminal side of θ and the horizontal x-axis.

Here are the reference angles for $\theta > 0$ in Quadrants I, II, III, and IV.

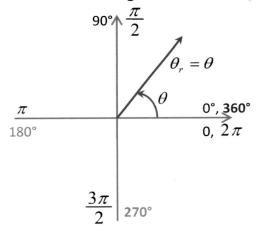

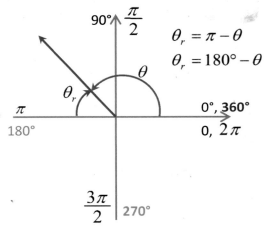

$\theta_r = \pi - \theta$

$\theta_r = 180° - \theta$

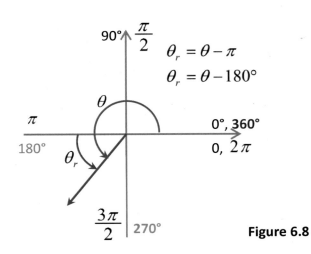

$\theta_r = \theta - \pi$

$\theta_r = \theta - 180°$

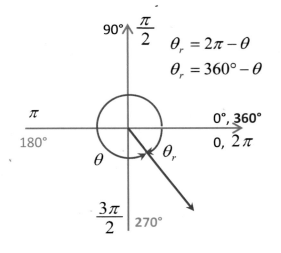

$\theta_r = 2\pi - \theta$

$\theta_r = 360° - \theta$

Figure 6.8

Reference Angles

Quadrant I $\theta_r = \theta$

Quadrant II $\theta_r = 180° - \theta$ or $\theta_r = \pi - \theta$

Quadrant III $\theta_r = \theta - 180°$ or $\theta_r = \theta - \pi$

Quadrant IV $\theta_r = 360° - \theta$ or $\theta_r = 2\pi - \theta$

►EXAMPLE 8

Find the reference angle θ_r.

a.) $\theta = 315°$　　　　b.) $\theta = 1.9$　　　　c.) $\theta = 480°$

►►Solution

a.)   315° is in Quadrant IV so we
will use $\theta_r = 360° - \theta$,

$\theta_r = 360° - 315° = 45°$,

$\theta_r = 45°$

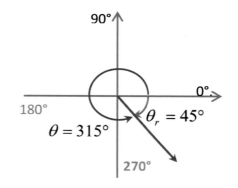

b.)   1.9 is in Quadrant II so we
will use $\theta_r = \pi - \theta$,

$\theta_r = \pi - 1.9 = 1.2$, $\theta_r = 1.2$

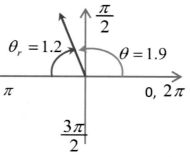

c.)   We need to find the smallest coterminal angle
to 480°, $480° - 360° = 120°$. 120° is in
Quadrant II so we will use $\theta_r = 180° - \theta$,

$\theta_r = 180° - 120° = 60°$, $\theta_r = 60°$

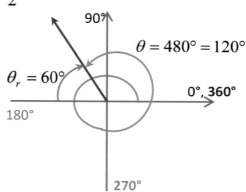

You can find the reference angle of a negative angle two ways. The first way is to find the smallest positive coterminal angle for the negative angle and apply the quadrant rules from **Figure 6.8.** The other method is to apply the reference angles for $\theta < 0$ in Quadrants I, II, III, and IV as shown in **Figure 6.9**.

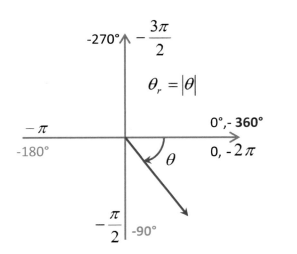

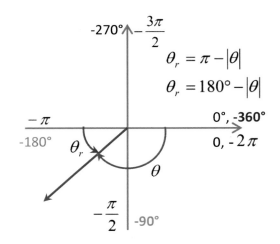

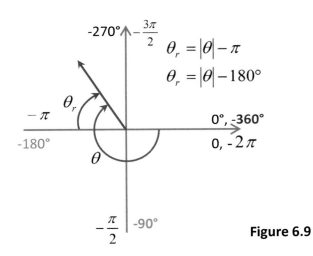

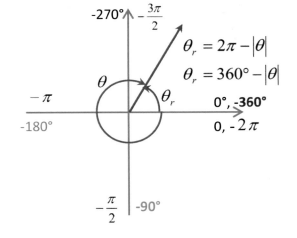

Figure 6.9

▶ EXAMPLE 9

Find the reference angle θ_r.

a.) $\theta = -150°$ b.) $\theta = -4.9$

▶▶ Solution

a.)   -150° is in Quadrant III. If we find the
        smallest positive coterminal angle,
        $-150° + 360° = 210°$. We will use
        $\theta_r = \theta - 180°$

        $\theta_r = 210° - 180° = 30°$

        $\theta_r = 30°$

        We can also use the rules from **Figure 6.9**.
        -150° is in Quadrant III so, we will use
        $\theta_r = |\theta| - 180°$

        $\theta_r = |-210°| - 180° = 210° - 180° = 30°$
        $\theta_r = 30°$

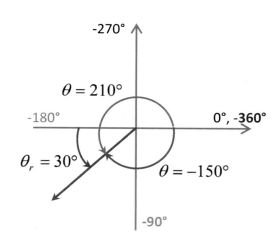

b.)   -4.9 is in Quadrant I. If we find the
        smallest positive coterminal angle,
        $-4.9 + 2\pi = 1.4$. Since we are in
        Quadrant I, $\theta_r = 1.4$.

        We can also use the rules from **Figure 6.9**.
        -4.9 is in Quadrant I so, we will use
        $\theta_r = 2\pi - |\theta|$
        $\theta_r = 2\pi - |-4.9| = 2\pi - 4.9 = 1.4$
        $\theta_r = 1.4$.

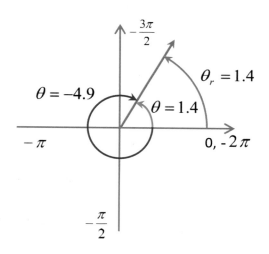

►EXAMPLE 10

Find the reference angle θ_r for $\theta = -\dfrac{7\pi}{6}$.

►►Solution

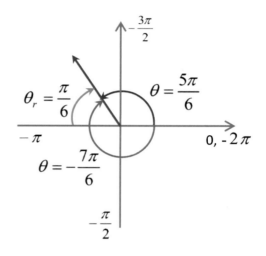

$-\dfrac{7\pi}{6}$ is in Quadrant II. If we find the

smallest positive coterminal angle,

$-\dfrac{7\pi}{6} + 2\pi = \dfrac{5\pi}{6}$. We will use $\theta_r = \pi - \theta$

$\theta_r = \pi - \dfrac{5\pi}{6} = \dfrac{\pi}{6}$

$\theta_r = \dfrac{\pi}{6}$

We can also use the rules from **Figure 6.9**.

$-\dfrac{7\pi}{6}$ is in Quadrant II so, we will use

$\theta_r = |\theta| - \pi$

$\theta_r = \left|-\dfrac{7\pi}{6}\right| - \pi = \dfrac{7\pi}{6} - \pi = \dfrac{\pi}{6}$

$\theta_r = \dfrac{\pi}{6}$

We use reference angles to evaluate trigonometric functions of any angle.

Evaluating Trigonometric Functions of Any Angle
1.   Determine the reference angle θ_r for the given angle θ.
2.   Determine the function value for the reference angle θ_r.
3.   Depending on the quadrant in which θ lies, affix the appropriate sign to the function value.

► EXAMPLE 11

Evaluate each trigonometric function. Give exact value.

a.) $\cos \dfrac{5\pi}{3}$ b.) $\sin 210°$ c.) $\tan \dfrac{11\pi}{6}$

►►Solution

a.)   1. $\dfrac{5\pi}{3}$ lies in Quadrant IV. The

reference angle is $\theta_r = 2\pi - \dfrac{5\pi}{3} = \dfrac{\pi}{3}$.

2. The $\cos \dfrac{\pi}{3} = \dfrac{1}{2}$.

3. Since cosine is positive in

Quadrant IV, the $\cos \dfrac{5\pi}{3} = \dfrac{1}{2}$.

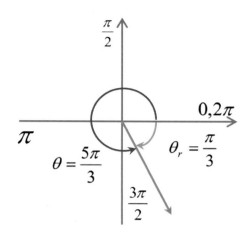

b.)   1. $210°$ lies in Quadrant III. The reference angle is
$\theta_r = 210° - 180° = 30°$.

2. The $\sin 30° = \dfrac{1}{2}$.

3. Since sine is negative in

Quadrant III, the $\sin 210° = \dfrac{1}{2}$.

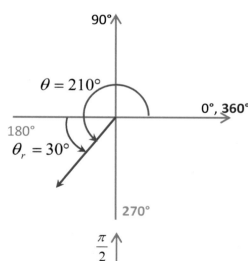

c.)   1. $\dfrac{11\pi}{6}$ lies in Quadrant IV. The

reference angle is

$\theta_r = 2\pi - \dfrac{11\pi}{6} = \dfrac{\pi}{6}$.

2. The $\tan \dfrac{\pi}{6} = \dfrac{\sqrt{3}}{3}$.

3. Since tangent is negative in

Quadrant IV, the $\tan \dfrac{11\pi}{6} = -\dfrac{\sqrt{3}}{3}$.

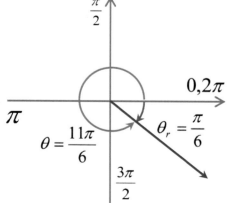

► EXAMPLE 12

Evaluate the sine, cosine, and tangent for $120°$ using reference angles. Give exact value.

►►Solution

1. $120°$ lies in Quadrant II. The reference angle is
    $\theta_r = 180° - 120° = 60°$.

2. The $\sin 60° = \dfrac{\sqrt{3}}{2}$, $\cos 60° = \dfrac{1}{2}$ and the
    $\tan 60° = \sqrt{3}$.

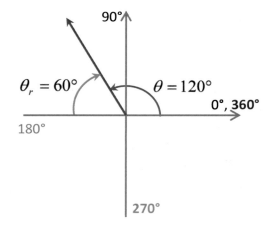

3. In Quadrant II, sine is positive, cosine is negative, and tangent is negative.
    $$\sin 120° = \frac{\sqrt{3}}{2}, \ \cos 120° = -\frac{1}{2}, \ \tan 120° = -\sqrt{3}$$

HOMEWORK

Objective 1

The point is on the terminal side of an angle in standard position. Determine the exact values of the six trigonometric functions of the angle.

See Examples 1, 2, and 3

1. $(-7,24)$ 2. $(8,-15)$ 3. $\left(-2\sqrt{3},2\right)$

4. $(-20,-21)$ 5. $\left(-\dfrac{5}{9},\dfrac{2}{3}\right)$ 6. $(-2.5,-3.25)$

7. $\left(-\dfrac{3}{4},-\dfrac{1}{2}\right)$ 8. $\left(6\sqrt{3},-6\right)$ 9. $(1.2,-0.5)$

Objective 2

Determine the quadrant for the terminal side of θ, using the information given.

See Example 4 and 5.

10. $\sin\theta < 0, \quad \cos\theta < 0$ 11. $\sin\theta > 0, \quad \cos\theta < 0$

12. $\tan\theta < 0, \quad \cos\theta > 0$ 13. $\sec\theta < 0, \quad \csc\theta < 0$

14. $\sec\theta < 0, \quad \cot\theta > 0$ 15. $\sec\theta > 0, \quad \tan\theta > 0$

Objectives 1 and 2

Use the given information to state the values of the six trigonometric functions.

See Example 6 and 7.

16. $\sin\theta = -\dfrac{4}{5}, \quad \theta$ lies in Quadrant IV

17. $\cos\theta = -\dfrac{20}{29}, \quad \theta$ lies in Quadrant III

18. $\tan\theta = \dfrac{5}{12}, \quad \sin\theta < 0$ 19. $\cos\theta = -\dfrac{5}{13}, \quad \tan\theta > 0$

20. $\csc\theta = \dfrac{37}{35}, \quad \cos\theta < 0$ 21. $\sec\theta = 3, \quad \csc\theta < 0$

Objective 4

Determine the reference angle for the given angle.

See Examples 9 and 10.

22. 240°   23. $\dfrac{7\pi}{6}$ 24. $-120°$ 25. 840°

26. $\dfrac{4\pi}{3}$ 27. 150°   28. $-\dfrac{5\pi}{3}$ 29. 200°

30. $\dfrac{7\pi}{12}$ 31. $-\dfrac{5\pi}{6}$ 32. $-140°$ 33. 640°

Objectives 3 and 5

Use reference angles to evaluate each trigonometric function. Give exact values.

See Example 11.

34. $\sin\dfrac{11\pi}{6}$ 35. $\cos\dfrac{5\pi}{6}$ 36. $\tan 60°$

37. $\tan\left(\dfrac{3\pi}{2}\right)$ 38 $\cos(-30°)$ 39. $\tan\dfrac{4\pi}{3}$

40. $\cos\dfrac{4\pi}{3}$ 41. $\sin\left(-\dfrac{\pi}{4}\right)$ 42. $\cos\left(-\dfrac{\pi}{3}\right)$

43. $\cos\dfrac{7\pi}{6}$ 44. $\tan\dfrac{\pi}{6}$ 45. $\sin\dfrac{5\pi}{6}$

46. $\tan\dfrac{\pi}{4}$ 47. $\tan(-\pi)$ 48. $\sin 120°$

Objective 5

Evaluate the sine, cosine, and tangent of the angle using reference angles. Give exact values.

See Example 12.

49. 240°   50. $\dfrac{3\pi}{2}$ 51. $-120°$ 52. 840°

53. $\dfrac{4\pi}{3}$ 54. 150°   55. $-\dfrac{5\pi}{3}$ 56. $-\dfrac{5\pi}{6}$

57. 180°   58. $\dfrac{2\pi}{3}$ 59. $\dfrac{7\pi}{6}$ 60. 300°

ANSWERS

1. $\sin\theta = \dfrac{24}{25}$ $\csc\theta = \dfrac{25}{24}$

    $\cos\theta = -\dfrac{7}{25}$ $\sec\theta = -\dfrac{25}{7}$

    $\tan\theta = -\dfrac{24}{7}$ $\cot\theta = -\dfrac{7}{24}$

2. $\sin\theta = -\dfrac{15}{17}$ $\csc\theta = -\dfrac{17}{15}$

    $\cos\theta = \dfrac{8}{17}$ $\sec\theta = \dfrac{17}{8}$

    $\tan\theta = -\dfrac{15}{8}$ $\cot\theta = -\dfrac{8}{15}$

3. $\sin\theta = \dfrac{1}{2}$ $\csc\theta = 2$

    $\cos\theta = -\dfrac{\sqrt{3}}{2}$ $\sec\theta = -\dfrac{2}{\sqrt{3}}$

    $\tan\theta = -\dfrac{1}{\sqrt{3}}$ $\cot\theta = -\sqrt{3}$

4. $\sin\theta = -\dfrac{21}{29}$ $\csc\theta = -\dfrac{29}{21}$

    $\cos\theta = -\dfrac{20}{29}$ $\sec\theta = -\dfrac{29}{20}$

    $\tan\theta = \dfrac{21}{20}$ $\cot\theta = \dfrac{20}{21}$

5. $\sin\theta = \dfrac{6}{\sqrt{61}}$ $\csc\theta = \dfrac{\sqrt{61}}{6}$

    $\cos\theta = -\dfrac{5}{\sqrt{61}}$ $\sec\theta = -\dfrac{\sqrt{61}}{5}$

    $\tan\theta = -\dfrac{6}{5}$ $\cot\theta = -\dfrac{5}{6}$

6. $\sin\theta = -\dfrac{13}{\sqrt{269}}$ $\csc\theta = -\dfrac{\sqrt{269}}{13}$

    $\cos\theta = -\dfrac{10}{\sqrt{269}}$ $\sec\theta = -\dfrac{\sqrt{269}}{10}$

    $\tan\theta = \dfrac{10}{13}$ $\cot\theta = \dfrac{13}{10}$

7. $\sin\theta = -\dfrac{2}{\sqrt{13}}$ $\csc\theta = -\dfrac{\sqrt{13}}{2}$

    $\cos\theta = -\dfrac{3}{\sqrt{13}}$ $\sec\theta = -\dfrac{\sqrt{13}}{3}$

    $\tan\theta = \dfrac{2}{3}$ $\cot\theta = \dfrac{3}{2}$

8. $\sin\theta = -\dfrac{1}{2}$ $\csc\theta = -2$

    $\cos\theta = \dfrac{\sqrt{3}}{2}$ $\sec\theta = \dfrac{2}{\sqrt{3}}$

    $\tan\theta = -\dfrac{1}{\sqrt{3}}$ $\cot\theta = -\sqrt{3}$

9. $\sin\theta = -\dfrac{5}{13}$ $\csc\theta = -\dfrac{13}{5}$

    $\cos\theta = \dfrac{12}{13}$ $\sec\theta = \dfrac{13}{12}$

    $\tan\theta = -\dfrac{5}{12}$ $\cot\theta = -\dfrac{12}{5}$

10. Quadrant III   11. Quadrant II   12. Quadrant IV

13. Quadrant III   14. Quadrant III   15. Quadrant I

16. $\sin\theta = -\dfrac{4}{5}$ $\csc\theta = -\dfrac{5}{4}$ 17. $\sin\theta = -\dfrac{21}{29}$ $\csc\theta = -\dfrac{29}{21}$

     $\cos\theta = \dfrac{3}{5}$ $\sec\theta = \dfrac{5}{3}$ $\cos\theta = -\dfrac{20}{29}$ $\sec\theta = -\dfrac{29}{20}$

     $\tan\theta = -\dfrac{4}{3}$ $\cot\theta = -\dfrac{3}{4}$ $\tan\theta = \dfrac{21}{20}$ $\cot\theta = \dfrac{20}{21}$

18. $\sin\theta = -\dfrac{5}{13}$ $\csc\theta = -\dfrac{13}{5}$ 19. $\sin\theta = -\dfrac{12}{13}$ $\csc\theta = -\dfrac{13}{12}$

     $\cos\theta = -\dfrac{12}{13}$ $\sec\theta = -\dfrac{13}{12}$ $\cos\theta = -\dfrac{5}{13}$ $\sec\theta = -\dfrac{13}{5}$

     $\tan\theta = \dfrac{5}{12}$ $\cot\theta = \dfrac{12}{5}$ $\tan\theta = \dfrac{12}{5}$ $\cot\theta = \dfrac{5}{12}$

20. $\sin\theta = \dfrac{35}{37}$ $\csc\theta = \dfrac{37}{35}$ 21. $\sin\theta = -\dfrac{2\sqrt{2}}{3}$ $\csc\theta = -\dfrac{3}{2\sqrt{2}}$

     $\cos\theta = -\dfrac{12}{37}$ $\sec\theta = -\dfrac{37}{12}$ $\cos\theta = \dfrac{1}{3}$ $\sec\theta = 3$

     $\tan\theta = -\dfrac{35}{12}$ $\cot\theta = -\dfrac{12}{35}$ $\tan\theta = -2\sqrt{2}$ $\cot\theta = -\dfrac{1}{2\sqrt{2}}$

22. $60°$ 23. $\dfrac{\pi}{6}$ 24. $60°$ 25. $60°$

26. $\dfrac{\pi}{3}$ 27. $30°$ 28. $\dfrac{\pi}{3}$ 29. $20°$

30. $\dfrac{\pi}{12}$ 31. $\dfrac{\pi}{6}$ 32. $40°$ 33. $80°$

34. $-\dfrac{1}{2}$ 35. $-\dfrac{\sqrt{3}}{2}$ 36. $\sqrt{3}$ 37. Undefined

38. $\dfrac{\sqrt{3}}{2}$ 39. $\sqrt{3}$ 40. $-\dfrac{1}{2}$ 41. $-\dfrac{\sqrt{2}}{2}$

42. $\dfrac{1}{2}$ 43. $-\dfrac{\sqrt{3}}{2}$ 44. $\dfrac{1}{\sqrt{3}}$ 45. $\dfrac{1}{2}$

46.   1   47.   0   48. $\dfrac{\sqrt{3}}{2}$

49. $\sin 240° = -\dfrac{\sqrt{3}}{2}$ $\csc 240° = -\dfrac{2}{\sqrt{3}}$

$\cos 240° = -\dfrac{1}{2}$ $\sec 240° = -2$

$\tan 240° = \sqrt{3}$ $\cot 240° = \dfrac{1}{\sqrt{3}}$

50. $\sin \dfrac{3\pi}{2} = -1$ $\csc \dfrac{3\pi}{2} = -1$

$\cos \dfrac{3\pi}{2} = 0$ $\sec \dfrac{3\pi}{2} = undefined$

$\tan \dfrac{3\pi}{2} = undefined$ $\cot \dfrac{3\pi}{2} = 0$

51. $\sin(-120°) = -\dfrac{\sqrt{3}}{2}$ $\csc(-120°) = -\dfrac{2}{\sqrt{3}}$

$\cos(-120°) = -\dfrac{1}{2}$ $\sec(-120°) = -2$

$\tan(-120°) = \sqrt{3}$ $\cot(-120°) = \dfrac{1}{\sqrt{3}}$

52. $\sin 120° = \dfrac{\sqrt{3}}{2}$ $\csc 120° = \dfrac{2}{\sqrt{3}}$ 53. $\sin \dfrac{4\pi}{3} = -\dfrac{\sqrt{3}}{2}$ $\csc \dfrac{4\pi}{3} = -\dfrac{2}{\sqrt{3}}$

$\cos 120° = -\dfrac{1}{2}$ $\sec 120° = -2$ $\cos \dfrac{4\pi}{3} = -\dfrac{1}{2}$ $\sec \dfrac{4\pi}{3} = -2$

$\tan 120° = -\sqrt{3}$ $\cot 120° = -\dfrac{1}{\sqrt{3}}$ $\tan \dfrac{4\pi}{3} = \sqrt{3}$ $\cot \dfrac{4\pi}{3} = \dfrac{1}{\sqrt{3}}$

54. $\sin 150° = \dfrac{1}{2}$ $\csc 150° = 2$ 55. $\sin\left(-\dfrac{5\pi}{3}\right) = \dfrac{\sqrt{3}}{2}$ $\csc\left(-\dfrac{5\pi}{3}\right) = \dfrac{2}{\sqrt{3}}$

$\cos 150° = -\dfrac{\sqrt{3}}{2}$ $\sec 150° = -\dfrac{2}{\sqrt{3}}$ $\cos\left(-\dfrac{5\pi}{3}\right) = \dfrac{1}{2}$ $\sec\left(-\dfrac{5\pi}{3}\right) = 2$

$\tan 150° = -\dfrac{1}{\sqrt{3}}$ $\cot 150° = -\sqrt{3}$ $\tan\left(-\dfrac{5\pi}{3}\right) = \sqrt{3}$ $\cot\left(-\dfrac{5\pi}{3}\right) = \dfrac{1}{\sqrt{3}}$

56. $\sin\left(-\dfrac{5\pi}{6}\right) = -\dfrac{1}{2}$ $\csc\left(-\dfrac{5\pi}{6}\right) = -2$

$\cos\left(-\dfrac{5\pi}{6}\right) = -\dfrac{\sqrt{3}}{2}$ $\sec\left(-\dfrac{5\pi}{6}\right) = -\dfrac{2}{\sqrt{3}}$

$\tan\left(-\dfrac{5\pi}{6}\right) = \dfrac{1}{\sqrt{3}}$ $\cot\left(-\dfrac{5\pi}{6}\right) = \sqrt{3}$

57. $\sin 180° = 0$ $\csc 180° = \mathit{undefined}$

$\cos 180° = -1$ $\sec 180° = -1$

$\tan 180° = 0$ $\cot 180° = \mathit{undefined}$

58. $\sin\dfrac{2\pi}{3} = \dfrac{\sqrt{3}}{2}$ $\csc\dfrac{2\pi}{3} = \dfrac{2}{\sqrt{3}}$

$\cos\dfrac{2\pi}{3} = -\dfrac{1}{2}$ $\sec\dfrac{2\pi}{3} = -2$

$\tan\dfrac{2\pi}{3} = -\sqrt{3}$ $\cot\dfrac{2\pi}{3} = -\dfrac{1}{\sqrt{3}}$

59. $\sin\dfrac{7\pi}{6} = -\dfrac{1}{2}$ $\csc\dfrac{7\pi}{6} = -2$

$\cos\dfrac{7\pi}{6} = -\dfrac{\sqrt{3}}{2}$ $\sec\dfrac{7\pi}{6} = -\dfrac{2}{\sqrt{3}}$

$\tan\dfrac{7\pi}{6} = \dfrac{1}{\sqrt{3}}$ $\cot\dfrac{7\pi}{6} = \sqrt{3}$

60. $\sin 300° = -\dfrac{\sqrt{3}}{2}$ $\csc 300° = -\dfrac{2}{\sqrt{3}}$

$\cos 300° = \dfrac{1}{2}$ $\sec 300° = 2$

$\tan 300° = -\sqrt{3}$ $\cot 300° = \dfrac{1}{\sqrt{3}}$

Section 7:   The Unit Circle

Learning Outcomes:

- The student will correctly memorize and apply trigonometric formulas, definitions, identities, and properties.
- The student will correctly use mathematical symbols and mathematical structure to examine and solve real world applications.

Objectives: At the conclusion of this lesson you should be able to:

1.   Verify points are on a unit circle.
2.   Determine the point associated with a real number *t*.
3.   Define the six trigonometric functions in terms of a point on the unit circle.
4.   Define trigonometric functions in terms of a real number *t*.
5.   Find the real number t corresponding to given values of the six trigonometric functions.

Recall that the standard form of the equation of a circle is $(x-h)^2 + (y-k)^2 = r^2$. When the center of the circle is at the origin we have what is called a central circle $x^2 + y^2 = r^2$ as shown in **Figure 7.1**. When the radius of a central circle is one, we have $x^2 + y^2 = 1$ which is called a unit circle as shown in **Figure 7.2**.

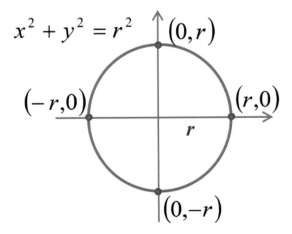

Figure 7.1

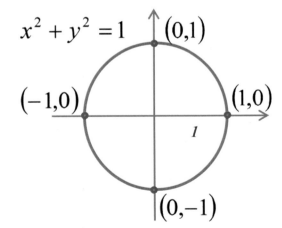

Figure 7.2

►EXAMPLE 1

Determine whether each of the following points is on the unit circle.

a.) $\left(\dfrac{1}{2}, -\dfrac{\sqrt{3}}{2}\right)$ b.) $\left(-\dfrac{\sqrt{2}}{2}, -\dfrac{\sqrt{2}}{2}\right)$ c.) $\left(\dfrac{1}{2}, -\dfrac{1}{2}\right)$

►►Solution

a.) $\left(\dfrac{1}{2}, -\dfrac{\sqrt{3}}{2}\right)$ $x = \dfrac{1}{2}, y = -\dfrac{\sqrt{3}}{2}$

Using the equation of the unit circle: $x^2 + y^2 = 1$, substitute for x and y and verify it equals 1.

$$\left(\dfrac{1}{2}\right)^2 + \left(-\dfrac{\sqrt{3}}{2}\right)^2 = 1$$

$$\dfrac{1}{4} + \dfrac{3}{4} = 1$$

$$\dfrac{4}{4} = 1 \quad \text{Yes!}$$

b.) $\left(-\dfrac{\sqrt{2}}{2}, -\dfrac{\sqrt{2}}{2}\right)$ $x = -\dfrac{\sqrt{2}}{2}, y = -\dfrac{\sqrt{2}}{2}$

Using the equation of the unit circle: $x^2 + y^2 = 1$, substitute for x and y and verify a sum of 1.

$$\left(-\dfrac{\sqrt{2}}{2}\right)^2 + \left(-\dfrac{\sqrt{2}}{2}\right)^2 = 1$$

$$\dfrac{2}{4} + \dfrac{2}{4} = 1$$

$$\dfrac{4}{4} = 1 \quad \text{Yes!}$$

c.) $\left(\dfrac{1}{2}, -\dfrac{1}{2}\right)$ $x = \dfrac{1}{2}, y = -\dfrac{1}{2}$

Using the equation of the unit circle: $x^2 + y^2 = 1$, substitute for x and y and verify a sum of 1.

$$\left(\dfrac{1}{2}\right)^2 + \left(-\dfrac{1}{2}\right)^2 = 1$$

$$\dfrac{1}{4} + \dfrac{1}{4} = 1$$

$$\dfrac{2}{4} = 1$$

$$\dfrac{1}{2} \neq 1 \quad \text{No!}$$

► EXAMPLE 2

The given point is on the unit circle. Complete the ordered pair (x, y) for the quadrant indicated. Give exact values.

a.) $\left(x, -\dfrac{\sqrt{2}}{2} \right)$, Quadrant III   b.) $\left(-\dfrac{1}{2}, y \right)$ Quadrant II

►► Solution

a.) $\left(x, -\dfrac{\sqrt{2}}{2} \right)$, Quadrant III

Using the equation of the unit circle:
$x^2 + y^2 = 1$, substitute for y and solve for x.

$$x^2 + \left(-\frac{\sqrt{2}}{2} \right)^2 = 1$$

$$x^2 + \frac{2}{4} = 1$$

$$x^2 + \frac{1}{2} = 1$$

$$x^2 = 1 - \frac{1}{2}$$

$$x^2 = \frac{1}{2}$$

$$\sqrt{x^2} = \pm\sqrt{\frac{1}{2}}$$

$$x = -\frac{1}{\sqrt{2}} \cdot \frac{\sqrt{2}}{\sqrt{2}} \quad \text{Since we are in Quadrant III, } x \text{ is negative.}$$

$$x = -\frac{\sqrt{2}}{2}$$

b.) $\left(-\dfrac{1}{2}, y \right)$ Quadrant II

Using the equation of the unit circle:
$x^2 + y^2 = 1$, substitute for x and solve for y.

$$\left(-\frac{1}{2} \right)^2 + y^2 = 1$$

$$\frac{1}{4} + y^2 = 1$$

$$y^2 = 1 - \frac{1}{4}$$

$$y^2 = \frac{3}{4}$$

$$\sqrt{y^2} = \pm\sqrt{\frac{3}{4}}$$

$$y = \frac{\sqrt{3}}{2} \quad \text{Since we are in Quadrant II, } y \text{ is positive.}$$

Consider the angle $30°, \dfrac{\pi}{6}$ and the point associated with this angle on the unit circle as shown in **Figure 7.3**.

To find the ordered pair we drop the perpendicular to the x-axis and we have a 30°-60°-90° right triangle as shown in **Figure 7.4**. In a 30°-60°-90 °triangle, the short leg is the side opposite the 30° angle. The short leg is half the measure of the hypotenuse. Since the radius is one, the hypotenuse is one. One-half of the hypotenuse is $\dfrac{1}{2}$. The opposite side is the y-value of the ordered pair.

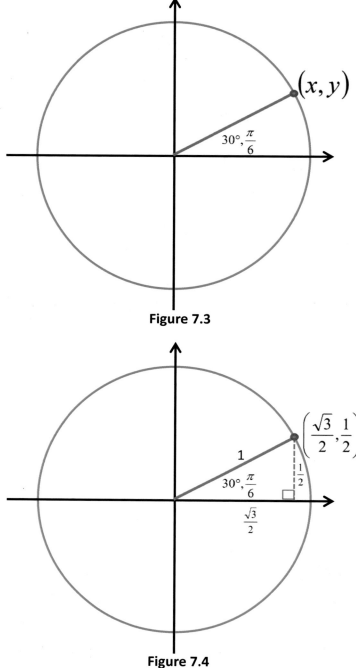

Figure 7.3

The other side of the triangle is the long leg which is $\sqrt{3}$ times the short leg. The long leg is $\dfrac{\sqrt{3}}{2}$ and this is the x-value of the ordered pair. The ordered pair associated with $30°, \dfrac{\pi}{6}$ is $\left(\dfrac{\sqrt{3}}{2}, \dfrac{1}{2} \right)$.

In the last section we learned that $\sin\theta = \dfrac{y}{r}$ and $\cos\theta = \dfrac{x}{r}$. If we apply that here the $\sin 30° = \dfrac{\frac{1}{2}}{1} = \dfrac{1}{2}$ and the

Figure 7.4

$\cos 30° = \dfrac{\frac{\sqrt{3}}{2}}{1} = \dfrac{\sqrt{3}}{2}$. We can conclude that on the unit circle the x-value of the ordered pair is cosine and the y-value of the ordered pair is sine. Since $\cos\theta = x$ and $\sin\theta = y$, the point where θ intersects the unit circle is $(\cos\theta, \sin\theta)$.

YOU TRY IT: Using this same process we used for $30°, \dfrac{\pi}{6}$, you find the ordered pairs for $45°, \dfrac{\pi}{4}$ and $60°, \dfrac{\pi}{3}$ and check your results with those shown in **Figure 7.5** and **Figure 7.6.**

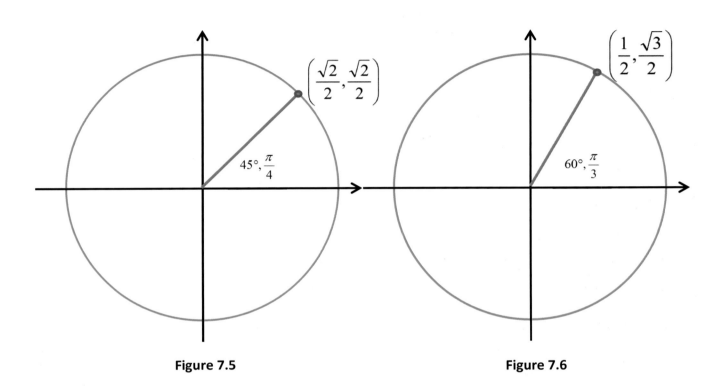

Figure 7.5   Figure 7.6

Additional points on the unit circle can be found using symmetry. We found the ordered pair associated with $30°, \dfrac{\pi}{6}$. Using symmetry with respect to the y-axis we get the ordered pair for $150°, \dfrac{5\pi}{6}$, $\left(-\dfrac{\sqrt{3}}{2}, \dfrac{1}{2}\right)$. Using symmetry with respect to the origin we get the ordered pair for $210°, \dfrac{7\pi}{6}$, $\left(-\dfrac{\sqrt{3}}{2}, -\dfrac{1}{2}\right)$. Using symmetry with respect to the x-axis we get the ordered pair for $330°, \dfrac{11\pi}{6}$, $\left(\dfrac{\sqrt{3}}{2}, -\dfrac{1}{2}\right)$. These are shown in **Figure 7.7.** We can do the same for $45°, \dfrac{\pi}{4}$ and $60°, \dfrac{\pi}{3}$ as shown in **Figure 7.8** and **Figure 7.9**.

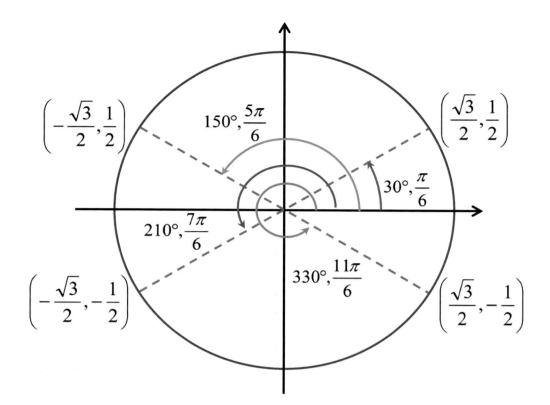

Figure 7.7

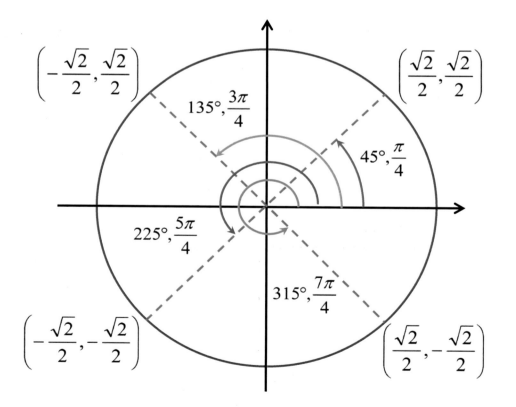

Figure 7.8

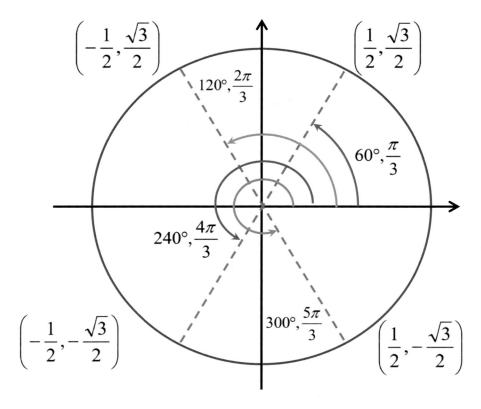

Figure 7.9

If we look at the quadrantal angles that we have talked about in Section 6, we have **Figure 7.10**.

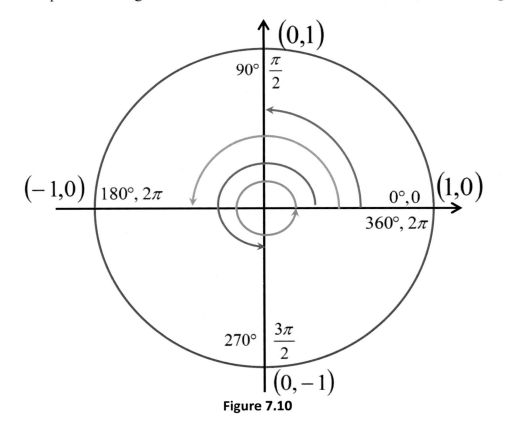

Figure 7.10

Putting all the angles and ordered pairs from **Figures 7.7, 7.8, 7.9, and 7.10** we get the unit circle as shown in **Figure 7.11**.

Unit Circle

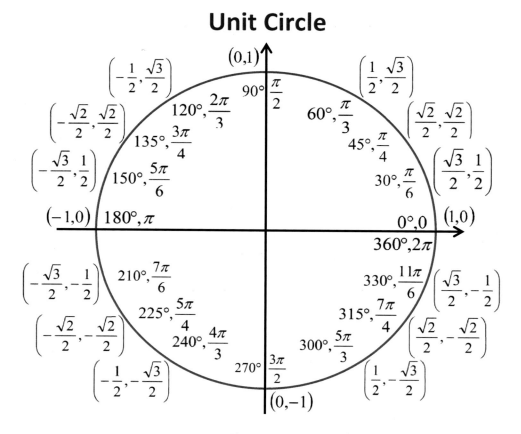

Figure 7.11

If the real number line is wrapped around the unit circle (positive numbers correspond to a counterclockwise wrapping and negative numbers correspond to a clockwise wrapping) each real number t corresponds to a point (x, y) on the unit circle, as shown in **Figure 7.12**. For example, the real numbers 0 and 2π corresponds to the point $(1,0)$. Each real number t also corresponds to a central angle θ (in standard position) whose radian measure is t. The real number t is the length of the arc intercepted by the angle θ, given in radians. Recall that $s = r\theta$, where θ is measured in radians, yields the measure of the arc length s. Since t is the measure of the arc length and the radius is equal to one on the unit circle we have $t = \theta$. Since $t = \theta$, we can define the trigonometric functions of the real number t to be the same as the trigonometric functions of angle θ.

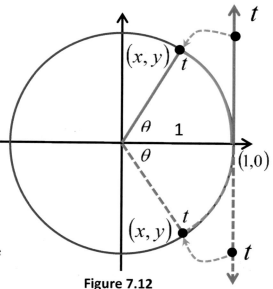

Figure 7.12

Definition of Trigonometric Functions of Real Numbers

Let t be a real number and let (x, y) be the point on the unit circle associated with t.

$$\cos t = x \qquad\qquad \sin t = y \qquad\qquad \tan t = \frac{y}{x}, \, x \neq 0$$

$$\sec t = \frac{1}{x}, \, x \neq 0 \qquad\qquad \csc t = \frac{1}{y}, \, y \neq 0 \qquad\qquad \cot t = \frac{x}{y}, \, y \neq 0$$

Given $\tan t = \frac{y}{x}$, we can substitute $\cos t$ for x and $\sin t$ for y and have $\tan t = \frac{\sin t}{\cos t}$ which is

called a ratio or quotient identity. We can replace sine and cosine with their reciprocals and have

$\tan t = \dfrac{\dfrac{1}{\csc t}}{\dfrac{1}{\sec t}}$ simplifying gives us $\tan t = \dfrac{\sec t}{\csc t}$. Cotangent is the reciprocal of tangent which

gives us $\cot t = \dfrac{\cos t}{\sin t}$ and $\cot t = \dfrac{\csc t}{\sec t}$.

Ratio/Quotient Identities

$$\tan t = \frac{\sin t}{\cos t} \qquad\qquad \tan t = \frac{\sec t}{\csc t}$$

$$\cot t = \frac{\cos t}{\sin t} \qquad\qquad \cot t = \frac{\csc t}{\sec t}$$

If we substitute $\cos t$ for x and $\sin t$ for y in the standard form of the equation of the unit circle $x^2 + y^2 = 1$ we have the first Pythagorean Identity: $\cos^2 t + \sin^2 t = 1$. We can manipulate this identity to arrive at the other Pythagorean Identities. As shown below.

$$\cos^2 t + \sin^2 t = 1 \qquad\qquad \cos^2 t + \sin^2 t = 1$$

$$\frac{1}{\sec^2 t} + \frac{1}{\csc^2 t} = 1 \qquad\qquad \frac{1}{\sec^2 t} + \frac{1}{\csc^2 t} = 1$$

$$\sec^2 t\left(\frac{1}{\sec^2 t} + \frac{1}{\csc^2 t}\right) = \sec^2 t(1) \qquad \csc^2 t\left(\frac{1}{\sec^2 t} + \frac{1}{\csc^2 t}\right) = \csc^2 t(1)$$

$$\frac{\sec^2 t}{\sec^2 t} + \frac{\sec^2 t}{\csc^2 t} = \sec^2 t \qquad\qquad \frac{\csc^2 t}{\sec^2 t} + \frac{\csc^2 t}{\csc^2 t} = \csc^2 t$$

$$1 + \tan^2 t = \sec^2 t \qquad\qquad \cot^2 t + 1 = \csc^2 t$$

Pythagorean Identities

$$\cos^2 t + \sin^2 t = 1 \qquad 1 + \tan^2 t = \sec^2 t \qquad \cot^2 t + 1 = \csc^2 t$$

▶ **EXAMPLE 3**

Find the associated point (x, y) on the unit circle that corresponds to the real number t.

a.) $t = \dfrac{\pi}{6}$ b.) $t = \dfrac{3\pi}{2}$ c.) $t = \dfrac{5\pi}{3}$ d.) $t = -\dfrac{3\pi}{4}$

▶▶ **Solution**

a.) $t = \dfrac{\pi}{6}$ corresponds to the point $\left(\dfrac{\sqrt{3}}{2}, \dfrac{1}{2}\right)$.

b.) $t = \dfrac{3\pi}{2}$ corresponds to the point $(0, -1)$.

c.) $t = \dfrac{5\pi}{3}$ corresponds to the point $\left(\dfrac{1}{2}, -\dfrac{\sqrt{3}}{2}\right)$. The reference angle for $\dfrac{5\pi}{3}$ is $\dfrac{\pi}{3}$. The

ordered pair that corresponds to $\dfrac{\pi}{3}$ is $\left(\dfrac{1}{2}, \dfrac{\sqrt{3}}{2}\right)$. Since $\dfrac{5\pi}{3}$ is in the fourth quadrant, y is

negative for $\dfrac{5\pi}{3}$.

d.) $t = -\dfrac{3\pi}{4}$ corresponds to the point $\left(-\dfrac{\sqrt{2}}{2}, -\dfrac{\sqrt{2}}{2}\right)$. The reference angle for $-\dfrac{3\pi}{4}$ is $\dfrac{\pi}{4}$.

The ordered pair that corresponds to $\dfrac{\pi}{4}$ is $\left(\dfrac{\sqrt{2}}{2}, \dfrac{\sqrt{2}}{2}\right)$. Since $-\dfrac{3\pi}{4}$ is in the third

quadrant, x and y are negative for $-\dfrac{3\pi}{4}$.

► EXAMPLE 4

Find the function value using reference angles and/or symmetry on the unit circle. Give exact values.

a.) $\cos \dfrac{5\pi}{6}$ b.) $\sin \dfrac{5\pi}{4}$ c.) $\sec \dfrac{5\pi}{3}$ d.) $\tan\left(-\dfrac{7\pi}{6}\right)$

►►**Solution**

a.) $\cos \dfrac{5\pi}{6} = -\dfrac{\sqrt{3}}{2}$. The angle $\dfrac{5\pi}{6}$ has y-axis symmetry to $\dfrac{\pi}{6}$ and the reference angle for

$\dfrac{5\pi}{6}$ is $\dfrac{\pi}{6}$. The ordered pair that corresponds to $\dfrac{\pi}{6}$ is $\left(\dfrac{\sqrt{3}}{2}, \dfrac{1}{2}\right)$. Since $\dfrac{5\pi}{6}$ is in the

second quadrant, x is negative for $\dfrac{5\pi}{6}$ and therefore cosine is negative.

b.) $\sin \dfrac{5\pi}{4} = -\dfrac{\sqrt{2}}{2}$. The angle $\dfrac{5\pi}{4}$ has origin symmetry to $\dfrac{\pi}{4}$ and the reference angle for $\dfrac{5\pi}{4}$

is $\dfrac{\pi}{4}$. The ordered pair that corresponds to $\dfrac{\pi}{4}$ is $\left(\dfrac{\sqrt{2}}{2}, \dfrac{\sqrt{2}}{2}\right)$. Since $\dfrac{5\pi}{4}$ is in the third

quadrant, y is negative for $\dfrac{5\pi}{4}$ and therefore sine is negative.

c.) $\sec \dfrac{5\pi}{3} = 2$. The angle $\dfrac{5\pi}{3}$ has x-axis symmetry to $\dfrac{\pi}{3}$ and the reference angle for $\dfrac{5\pi}{3}$ is

$\dfrac{\pi}{3}$. The ordered pair that corresponds to $\dfrac{\pi}{3}$ is $\left(\dfrac{1}{2}, \dfrac{\sqrt{3}}{2}\right)$. Secant is the reciprocal for

cosine or $\sec t = \dfrac{1}{x}$. Since $\dfrac{5\pi}{3}$ is in the fourth   quadrant, x is positive for $\dfrac{5\pi}{3}$ and

therefore cosine and its reciprocal secant are positive.

d.) $\tan\left(-\dfrac{7\pi}{6}\right) = -\dfrac{\sqrt{3}}{3}$. The angle $-\dfrac{7\pi}{6}$ has y-axis symmetry to $\dfrac{\pi}{6}$ and the reference angle

for $-\dfrac{7\pi}{6}$ is $\dfrac{\pi}{6}$. The ordered pair that corresponds to $\dfrac{\pi}{6}$ is $\left(\dfrac{\sqrt{3}}{2}, \dfrac{1}{2}\right)$. Since $-\dfrac{7\pi}{6}$ is in

the second quadrant, x is negative and y is positive, therefore tangent is negative since

$\tan t = \dfrac{y}{x}$.

▶EXAMPLE 5

Given (x, y) is a point on the unit circle corresponding to t, find the value of all six trigonometric functions of t. Give exact values. It may be necessary to rationalize the denominators.

a.) $\left(\dfrac{\sqrt{7}}{3}, -\dfrac{\sqrt{2}}{3}\right)$ b.) $\left(-\dfrac{1}{4}, \dfrac{\sqrt{15}}{4}\right)$

▶▶Solution

a.) $\left(\dfrac{\sqrt{7}}{3}, -\dfrac{\sqrt{2}}{3}\right)$

$$\sin t = -\frac{\sqrt{2}}{3}$$

$$\csc t = \frac{1}{-\dfrac{\sqrt{2}}{3}} = -\frac{3}{\sqrt{2}} \cdot \frac{\sqrt{2}}{\sqrt{2}} = -\frac{3\sqrt{2}}{2}$$

$$\cos t = \frac{\sqrt{7}}{3}$$

$$\sec t = \frac{1}{\dfrac{\sqrt{7}}{3}} = \frac{3}{\sqrt{7}} \cdot \frac{\sqrt{7}}{\sqrt{7}} = \frac{3\sqrt{7}}{7}$$

$$\tan t = \frac{-\dfrac{\sqrt{2}}{3}}{\dfrac{\sqrt{7}}{3}} = -\frac{\sqrt{2}}{\sqrt{7}} \cdot \frac{\sqrt{7}}{\sqrt{7}} = -\frac{\sqrt{14}}{7}$$

$$\cot t = \frac{\dfrac{\sqrt{7}}{3}}{-\dfrac{\sqrt{2}}{3}} = -\frac{\sqrt{7}}{\sqrt{2}} \cdot \frac{\sqrt{2}}{\sqrt{2}} = -\frac{\sqrt{14}}{2}$$

b.) $\left(-\dfrac{1}{4}, \dfrac{\sqrt{15}}{4}\right)$

$$\sin t = \frac{\sqrt{15}}{4}$$

$$\csc t = \frac{1}{\dfrac{\sqrt{15}}{4}} = \frac{4}{\sqrt{15}} \cdot \frac{\sqrt{15}}{\sqrt{15}} = \frac{4\sqrt{15}}{15}$$

$$\cos t = -\frac{1}{4}$$

$$\sec t = \frac{1}{-\dfrac{1}{4}} = -4$$

$$\tan t = \frac{\dfrac{\sqrt{15}}{4}}{-\dfrac{1}{4}} = -\sqrt{15}$$

$$\cot t = \frac{-\dfrac{1}{4}}{\dfrac{\sqrt{15}}{4}} = -\frac{1}{\sqrt{15}} \cdot \frac{\sqrt{15}}{\sqrt{15}} = -\frac{\sqrt{15}}{15}$$

▶ EXAMPLE 6

Without using a calculator, find the values of *t* in $[0,2\pi)$ that make each equation true.

a.) $\sin t = -\dfrac{\sqrt{3}}{2}$ b.) $\cos t = \dfrac{\sqrt{3}}{2}$ c.) $\tan t = -\dfrac{\sqrt{3}}{3}$ d.) $\csc t = -\sqrt{2}$

▶▶ Solution

a.) Recognize that $\dfrac{\sqrt{3}}{2}$ is the *y* value for $t = \dfrac{\pi}{3}$. Since sine is negative in Quadrant III we

have $t = \dfrac{4\pi}{3}$. Sine is also negative in Quadrant IV and we have $t = \dfrac{5\pi}{3}$.

b.) Recognize that $\dfrac{\sqrt{3}}{2}$ is the *x* value for $t = \dfrac{\pi}{6}$. Since cosine is positive in Quadrant I we

have $t = \dfrac{\pi}{6}$. Sine is also positive in Quadrant IV and we have $t = \dfrac{11\pi}{6}$.

c.) Recognize that $\dfrac{\sqrt{3}}{3}$ is the tangent value for $t = \dfrac{\pi}{6}$ since $\tan t = \dfrac{y}{x}$. Tangent is negative in

Quadrant II we have $t = \dfrac{5\pi}{6}$. Tangent is also negative in Quadrant IV and we have

$t = \dfrac{11\pi}{6}$.

d.) Recognize that $\sqrt{2}$ is the reciprocal value for *y* at $t = \dfrac{\pi}{4}$ since $\csc t = \dfrac{1}{y}$. Since sine is

negative in Quadrant III , cosecant is negative as well and we have $t = \dfrac{5\pi}{4}$. Sine and

cosecant are also negative in Quadrant IV and we have $t = \dfrac{7\pi}{4}$. WTF?

HOMEWORK

Objective 1

Determine if the given point is on the unit circle.

See Example 1.

1. $\left(\dfrac{\sqrt{3}}{2}, \dfrac{1}{2}\right)$

2. $\left(-\dfrac{\sqrt{6}}{3}, \dfrac{\sqrt{3}}{3}\right)$

3. $\left(\dfrac{\sqrt{7}}{8}, \dfrac{1}{8}\right)$

4. $\left(-\dfrac{\sqrt{11}}{6}, \dfrac{5}{6}\right)$

Objective 1

The given point is on the unit circle, complete the ordered pair (x, y) for the quadrant indicated.   Give exact values.

See Example 2.

5. $\left(\dfrac{12}{13}, y\right)$ Quadrant IV

6. $\left(x, -\dfrac{\sqrt{11}}{6}\right)$ Quadrant III

7. $\left(x, -\dfrac{8}{17}\right)$ Quadrant IV

8. $\left(-\dfrac{\sqrt{6}}{5}, y\right)$ Quadrant II

Objective 2

Find the associated point (x, y) on the unit circle that corresponds to the real number t.

See Example 3.

9. $t = \dfrac{11\pi}{6}$

10. $t = \dfrac{3\pi}{4}$

11. $t = -\dfrac{5\pi}{6}$

12. $t = -\dfrac{2\pi}{3}$

13. $t = \dfrac{7\pi}{6}$

14. $t = \dfrac{\pi}{2}$

15. $t = \dfrac{2\pi}{3}$

16. $t = -\dfrac{\pi}{4}$

Objective 3

Given (x, y) is a point on the unit circle corresponding to t, find the value of all six trigonometric functions of t. Give exact values.

See Example 5.

17. $\left(\dfrac{\sqrt{7}}{4}, -\dfrac{3}{4}\right)$

18. $\left(-\dfrac{1}{3}, \dfrac{2\sqrt{2}}{3}\right)$

19. $\left(-\dfrac{5}{13}, -\dfrac{12}{13}\right)$

20. $\left(\dfrac{15}{17}, -\dfrac{8}{17}\right)$

21. $\left(-\dfrac{1}{2}, -\dfrac{\sqrt{3}}{2}\right)$

22. $\left(\dfrac{2\sqrt{6}}{5}, \dfrac{1}{5}\right)$

Objective 4

Find the function value using reference angles and/or symmetry on the unit circle. Give exact values.

See Example 4.

23. $\csc\dfrac{11\pi}{6}$

24. $\sin\dfrac{5\pi}{3}$

25. $\cos\dfrac{5\pi}{4}$

26. $\sin\left(-\dfrac{5\pi}{6}\right)$

27. $\tan\dfrac{2\pi}{3}$

28. $\cot\dfrac{7\pi}{4}$

29. $\sec\dfrac{11\pi}{6}$

30. $\tan\left(-\dfrac{\pi}{2}\right)$

31. $\cos\left(-\dfrac{7\pi}{4}\right)$

32. $\csc\dfrac{4\pi}{3}$

33. $\cot\dfrac{7\pi}{6}$

34. $\sec\left(-\dfrac{2\pi}{3}\right)$

Objective 5

Without using a calculator, find the values of t in $[0, 2\pi)$ that make each equation true.

See Example 6.

35. $\tan t = \sqrt{3}$

36. $\sin t = -\dfrac{\sqrt{3}}{2}$

37. $\cot t = -\dfrac{\sqrt{3}}{3}$

38. $\sec t = -\sqrt{2}$

39. $\sin t = \dfrac{\sqrt{2}}{2}$

40. $\cos t = -\dfrac{\sqrt{3}}{2}$

41. $\csc t = -\dfrac{2\sqrt{3}}{3}$

42. $\cot t = undefined$

43. $\tan t = 0$

ANSWERS

1. Yes   2. Yes   3. No   4. Yes

5. $-\dfrac{5}{13}$ 6. $-\dfrac{5}{6}$ 7. $\dfrac{15}{17}$ 8. $\dfrac{\sqrt{19}}{5}$

9. $\left(\dfrac{\sqrt{3}}{2}, -\dfrac{1}{2}\right)$ 10. $\left(-\dfrac{\sqrt{2}}{2}, \dfrac{\sqrt{2}}{2}\right)$

11. $\left(-\dfrac{\sqrt{3}}{2}, -\dfrac{1}{2}\right)$ 12. $\left(-\dfrac{1}{2}, -\dfrac{\sqrt{3}}{2}\right)$

13. $\left(-\dfrac{\sqrt{3}}{2}, -\dfrac{1}{2}\right)$ 14. $(0,1)$

15. $\left(-\dfrac{1}{2}, \dfrac{\sqrt{3}}{2}\right)$ 16. $\left(\dfrac{\sqrt{2}}{2}, -\dfrac{\sqrt{2}}{2}\right)$

17. $\sin t = -\dfrac{3}{4}$ $\csc t = -\dfrac{4}{3}$ 18. $\sin t = \dfrac{2\sqrt{2}}{3}$ $\csc t = \dfrac{3}{2\sqrt{2}}$

    $\cos t = \dfrac{\sqrt{7}}{4}$ $\sec t = \dfrac{4}{\sqrt{7}}$ $\cos t = -\dfrac{1}{3}$ $\sec t = -3$

    $\tan t = -\dfrac{3}{\sqrt{7}}$ $\cot t = -\dfrac{\sqrt{7}}{3}$ $\tan t = -2\sqrt{2}$ $\cot t = -\dfrac{1}{2\sqrt{2}}$

19. $\sin t = -\dfrac{12}{13}$ $\csc t = -\dfrac{13}{12}$ 20. $\sin t = -\dfrac{8}{17}$ $\csc t = -\dfrac{17}{8}$

    $\cos t = -\dfrac{5}{13}$ $\sec t = -\dfrac{13}{5}$ $\cos t = \dfrac{15}{17}$ $\sec t = \dfrac{17}{15}$

    $\tan t = \dfrac{12}{5}$ $\cot t = \dfrac{5}{12}$ $\tan t = -\dfrac{8}{15}$ $\cot t = -\dfrac{15}{8}$

21. $\sin t = -\dfrac{\sqrt{3}}{2}$ $\csc t = -\dfrac{2}{\sqrt{3}}$ 22. $\sin t = \dfrac{1}{5}$ $\csc t = 5$

    $\cos t = -\dfrac{1}{2}$ $\sec t = -2$ $\cos t = \dfrac{2\sqrt{6}}{5}$ $\sec t = \dfrac{5}{2\sqrt{6}}$

    $\tan t = \sqrt{3}$ $\cot t = \dfrac{1}{\sqrt{3}}$ $\tan t = \dfrac{1\backslash}{2\sqrt{6}}$ $\cot t = 2\sqrt{6}$

23. -2

24. $-\dfrac{\sqrt{3}}{2}$

25. $-\dfrac{\sqrt{2}}{2}$

26. $-\dfrac{1}{2}$

27. $-\sqrt{3}$

28. -1

29. $\dfrac{2}{\sqrt{3}}$

30. *undefined*

31. $\dfrac{\sqrt{2}}{2}$

32. $-\dfrac{2}{\sqrt{3}}$

33. $\sqrt{3}$

34. -2

35. $\dfrac{\pi}{3}, \dfrac{4\pi}{3}$

36. $\dfrac{4\pi}{3}, \dfrac{5\pi}{3}$

37. $\dfrac{2\pi}{3}, \dfrac{5\pi}{3}$

38. $\dfrac{3\pi}{4}, \dfrac{5\pi}{4}$

39. $\dfrac{\pi}{4}, \dfrac{3\pi}{4}$

40. $\dfrac{5\pi}{6}, \dfrac{7\pi}{6}$

41. $\dfrac{4\pi}{3}, \dfrac{5\pi}{3}$

42. $0, \pi$

43. $0, \pi$

Section 8:   Graph of the Sine Function

Learning Outcomes:

- The student will correctly memorize and apply trigonometric formulas, definitions, identities, and properties.
- The student will illustrate and examine the graphs of: the six trigonometric functions and their inverses, parametric equations, and polar equations.

Objectives: At the conclusion of this lesson you should be able to:

1.   Determine the amplitude, period, phase shift, vertical shift and reflection of a sine function.
2.   Determine the fourths of a sine function.
3.   Use amplitude, period, phase shift, vertical shift, reflections and the fourths to graph the sine function over a specified period.

As with graphs of other functions, trigonometric graphs contribute toward the understanding of each function and its applications. Occurrences that repeat with a predictable pattern, such as phases of the moon, hours of daylight, heartbeats, and tides can be modeled by sine and cosine functions. These functions are called periodic functions.

Periodic Functions

A function f is said to be periodic if there is a positive number P such that

$$f(t + P) = f(t)$$

for all t in the domain. The smallest P for which this occurs is called the period of f.

The smallest value of P for sine and cosine is 2π because the circumference of the unit circle is 2π. Sine and cosine are periodic functions with period of 2π.

To graph the sine function $y = \sin x$ we will look at the behavior of the x-values over one period which is from 0 to 2π.

As x increases from 0 to $\frac{\pi}{2}$, the y-values of terminal points increase from 0 to 1 as shown in **Figure 8.1**.

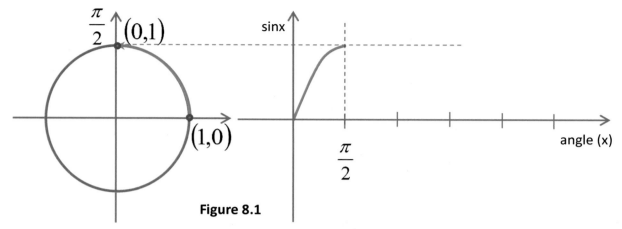

Figure 8.1

As x increases from $\frac{\pi}{2}$ to π, the y-values decrease from 1 to 0 as shown in **Figure 8.2**.

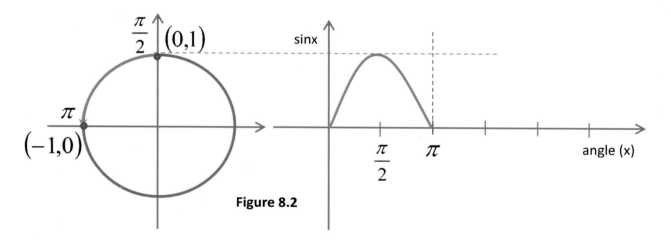

Figure 8.2

As x increases from π to $\frac{3\pi}{2}$, the y-values decrease from 0 to -1 as shown in **Figure 8.3**.

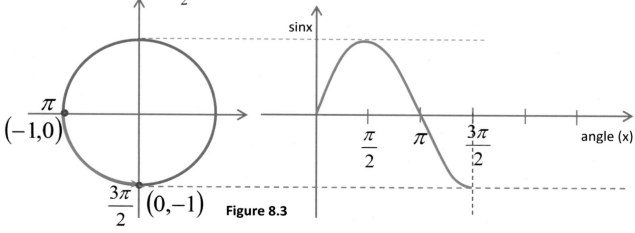

Figure 8.3

As x increases from $\dfrac{3\pi}{2}$ to 2π, the y-values increase from -1 to 0 as shown in **Figure 8.4**.

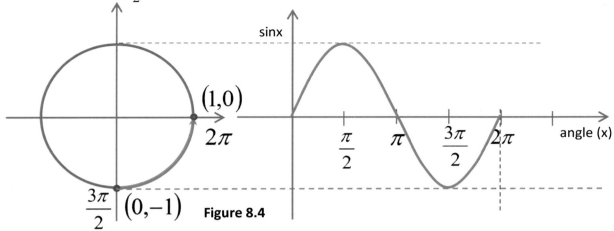

Figure 8.4

To get a more precise graph for sine we can construct a table:

x	$\dfrac{\pi}{6}$	$\dfrac{\pi}{4}$	$\dfrac{\pi}{3}$	$\dfrac{\pi}{2}$	$\dfrac{2\pi}{3}$	$\dfrac{3\pi}{4}$	$\dfrac{5\pi}{6}$	π	$\dfrac{7\pi}{6}$	$\dfrac{5\pi}{4}$	$\dfrac{4\pi}{3}$	$\dfrac{3\pi}{2}$	$\dfrac{5\pi}{3}$	$\dfrac{7\pi}{4}$	$\dfrac{11\pi}{6}$	2π
$\sin x$	$\dfrac{1}{2}$	$\dfrac{\sqrt{2}}{2}$	$\dfrac{\sqrt{3}}{2}$	1	$\dfrac{\sqrt{3}}{2}$	$\dfrac{\sqrt{2}}{2}$	$\dfrac{1}{2}$	0	$-\dfrac{1}{2}$	$-\dfrac{\sqrt{2}}{2}$	$-\dfrac{\sqrt{3}}{2}$	-1	$-\dfrac{\sqrt{3}}{2}$	$-\dfrac{\sqrt{2}}{2}$	$-\dfrac{1}{2}$	0

If we plot the points, we have the graph in **blue** below in **Figure 8.5**.

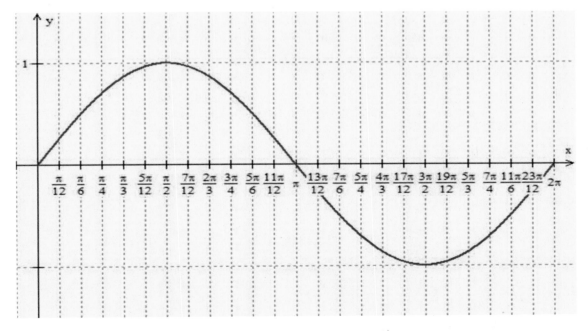

Figure 8.5

If we plot more points, as shown in **Figure 8.6**, notice the sine curve is continuous on the domain from $(-\infty, \infty)$. The sine curve has symmetry with respect to the origin which implies that $\sin(-x) = -\sin x$ and that sine is odd.

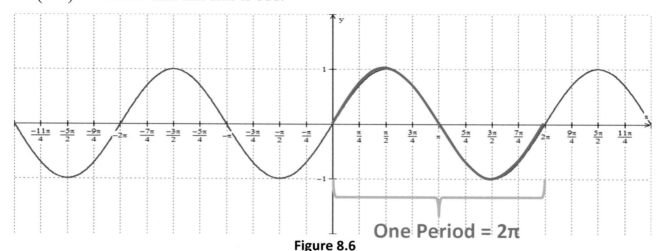

Figure 8.6

Now that we have graphed the basic sine curve, we will study the effects of the constants *a*, *b*, *c*, and *d* in the general sine equation: $y = a\sin(bx \pm c) \pm d$.

The Constant a

The *a* in the general sine equation represents the amplitude of the graph. The amplitude is the absolute value of half the difference between the maximum and minimum values. We use the absolute value since distance is always positive. If you have studied transformations, you may recognize amplitude as vertical stretching or shrinkage. The amplitude will be a vertical stretching of the basic sine curve if $|a| > 1$. The amplitude will be a vertical shrinking of the basic sine curve if $|a| < 1$. In **Figure 8.7** we have the graphs of $y = 2\sin x$ (red graph),

$y = \sin x$ (blue graph), and

$y = \frac{1}{2}\sin x$ (pink graph). The basic sine graph is the blue graph, $y = \sin x$. Notice the red graph, $y = 2\sin x$ has a vertical stretching of 2 and has an amplitude of 2. The pink graph, $y = \frac{1}{2}\sin x$ has a vertical shrinkage of $\frac{1}{2}$ and has an amplitude of $\frac{1}{2}$.

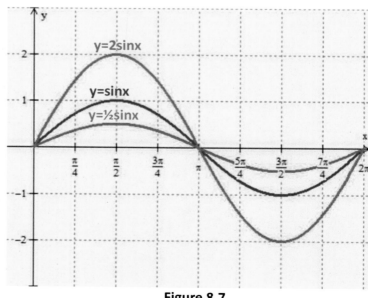

Figure 8.7

If a is negative, $a < 0$, the graph will have a reflection on the x-axis as shown in **Figure 8.8**. The basic sine graph is the **blue graph** $y = \sin x$ and the **green graph** is $y = -\sin x$. Notice that the **green graph** is a reflection on the x-axis of the **blue graph**.

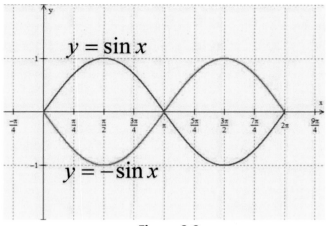

Figure 8.8

▶ EXAMPLE 1

Determine the amplitude of the following functions and determine if there is a reflection on the x-axis.

a.) $y = 3 \sin x$ b.) $y = -2 \sin x$ c.) $y = \frac{1}{3} \sin x$

▶▶ **Solution**

a.) $y = 3 \sin x$ $a = |3| = 3$ The amplitude is 3.   No reflection because $a > 0$.

b.) $y = -2 \sin x$ $a = |-2| = 2$ The amplitude is 2.   There is a reflection on the x-axis because $a < 0$.

c.) $y = \frac{1}{3} \sin x$ $a = \left|\frac{1}{3}\right| = \frac{1}{3}$ The amplitude is $\frac{1}{3}$.   No reflection because $a > 0$.

The Constant b

The period of $y = \sin x$ is 2π when $b = 1$. What happens to the period for $y = \sin 2x$? When $b = 1$ the period is $0 \leq x \leq 2\pi$ but here we have $b = 2$. Since $b = 2$ we will replace x with $2x$ and see how this effects the period. See **Figure 8.9**. One period of $y = \sin 2x$ is completed between 0 to π so, the period is π.

What happens to the period for $y = \sin \frac{1}{2} x$? See **Figure 8.10**. One period of $y = \sin \frac{1}{2} x$ is completed between 0 to 4π so, the period is 4π. Since the period is 2π when $b = 1$, we can find the period of other sine functions by using *Period* $= \frac{2\pi}{b}$, $p = \frac{2\pi}{b}$.

$$0 \leq 2x \leq 2\pi$$

$$\frac{0}{2} \leq \frac{2x}{2} \leq \frac{2\pi}{2}$$

$$0 \leq x \leq \pi$$

Figure 8.9

$$0 \leq \frac{1}{2} x \leq 2\pi$$

$$2(0) \leq 2\left(\frac{1}{2}x\right) \leq 2(2\pi)$$

$$0 \leq x \leq 4\pi$$

Figure 8.10

In **Figure 8.11** we have the graphs of $y = \sin x$ (**blue graph**), $y = \sin 2x$ (**green graph**) and $y = \sin \frac{1}{2}x$ (**red graph**). Notice how the sine graph is horizontally stretched or shrunk depending on the b value. If $b > 1$ we have horizontal shrinkage and if $b < 1$ we have horizontal stretching. The sine graph $y = \sin 2x$ (**green graph**) has a horizontal shrinkage of 2 and the sine graph $y = \sin \frac{1}{2}x$ (**red graph**) has a horizontal stretching of $\frac{1}{2}$.

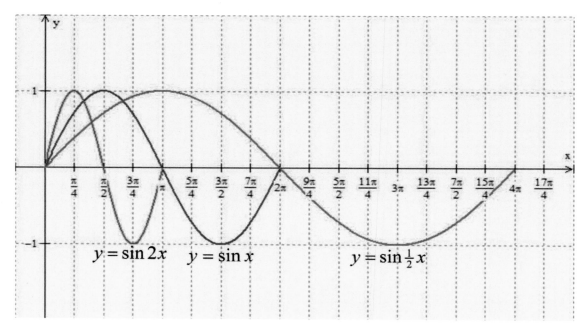

Figure 8.11

►EXAMPLE 2

Determine the period of the following sine functions.

a.) $y = \sin 6x$ b.) $y = \sin \frac{1}{4}x$ c.) $y = \sin 3x$

►►Solution

a.) $y = \sin 6x$ $p = \dfrac{2\pi}{6} = \dfrac{\pi}{3}$ The period is $\dfrac{\pi}{3}$.

b.) $y = \sin \frac{1}{4}x$ $p = \dfrac{2\pi}{\frac{1}{4}} = 2\pi \cdot 4 = 8\pi$ The period is 8π.

c.) $y = \sin 3x$ $p = \dfrac{2\pi}{3} = \dfrac{2\pi}{3}$ The period is $\dfrac{2\pi}{3}$.

The Constant c

To examine the effect of the constant c, consider the sine graphs, $y = \sin x$ (blue graph),
$y = \sin(x + \pi)$ (green graph), and $y = \sin(x - \pi)$ (red graph) shown in **Figure 8.12**. The
sine graph $y = \sin(x + \pi)$ (green graph) has a horizontal shift left of π and the sine graph
$y = \sin(x - \pi)$ (red graph) has a horizontal shift right of π.

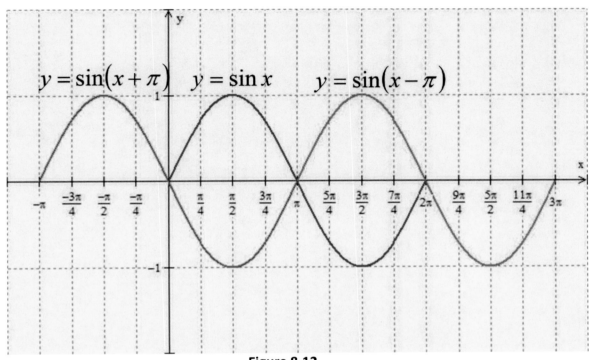

Figure 8.12

We call a horizontal shift a **phase shift** in trigonometry.   In **Figure 8.13** we have the graph of

$y = \sin x$ (blue graph), and the
graph of $y = \sin(2x - \pi)$ (red
graph) . Notice the red graph has
both a change in the period and a
horizontal shift.   If we look at the
beginning and the ending of
$y = \sin(2x - \pi)$ we have:

$0 \le 2x - \pi \le 2\pi$ which gives us
our new beginning and ending:

$\dfrac{\pi}{2} \le x \le \dfrac{3\pi}{2}$. Notice the phase

shift and the beginning are $\dfrac{\pi}{2}$.

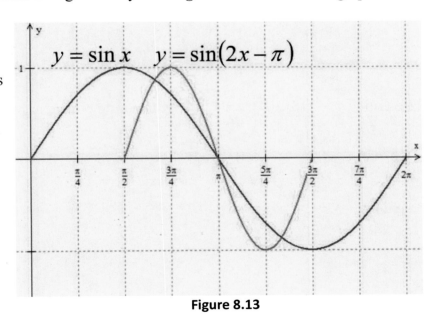

Figure 8.13

In the table below we can see the comparison between the beginning and the phase shift.

Sine Function	Period	Beginning	Ending	Phase Shift
$y = \sin x$	$0 \le x \le 2\pi$	0	2π	none
$y = \sin(x + \pi)$	$0 \le x + \pi \le 2\pi$ $-\pi \le x \le \pi$	$-\pi$	π	Left π
$y = \sin(x - \pi)$	$0 \le x - \pi \le 2\pi$ $\pi \le x \le 3\pi$	π	3π	Right π
$y = \sin\left(x - \dfrac{\pi}{3}\right)$	$0 \le x - \dfrac{\pi}{3} \le 2\pi$ $\dfrac{\pi}{3} \le x \le \dfrac{7\pi}{3}$	$\dfrac{\pi}{3}$	$\dfrac{7\pi}{3}$	Right $\dfrac{\pi}{3}$
$y = \sin(2x - \pi)$	$0 \le 2x - \pi \le 2\pi$ $\pi \le 2x \le 3\pi$ $\dfrac{\pi}{2} \le x \le \dfrac{3\pi}{2}$	$\dfrac{\pi}{2}$	$\dfrac{3\pi}{2}$	Right $\dfrac{\pi}{2}$
$y = \sin(4x + \pi)$	$0 \le 4x + \pi \le 2\pi$ $-\pi \le 4x \le \pi$ $-\dfrac{\pi}{4} \le x \le \dfrac{\pi}{4}$	$-\dfrac{\pi}{4}$	$\dfrac{\pi}{4}$	Left $\dfrac{\pi}{4}$
$y = \sin\left(3x - \dfrac{\pi}{4}\right)$	$0 \le 3x - \dfrac{\pi}{4} \le 2\pi$ $\dfrac{\pi}{4} \le 3x \le \dfrac{9\pi}{4}$ $\dfrac{\pi}{12} \le x \le \dfrac{3\pi}{4}$	$\dfrac{\pi}{12}$	$\dfrac{3\pi}{4}$	Right $\dfrac{\pi}{12}$

Rewriting $y = a\sin(bx \pm c) \pm d$ by factoring out the b, we have $y = a\sin\left[b\left(x \pm \dfrac{c}{b}\right)\right] \pm d$. $\dfrac{c}{b}$

is the phase shift. If $\dfrac{c}{b} > 0$, the phase shift is to the left $\dfrac{c}{b}$. If $\dfrac{c}{b} < 0$, the phase shift is to the

right $\dfrac{c}{b}$.

► EXAMPLE 3

Determine the phase shift of the following sine functions.

a.) $y = \sin(2x + \pi)$

b.) $y = \sin(3x - 2\pi)$

c.) $y = \sin\left(2x + \dfrac{\pi}{3}\right)$

d.) $y = \sin\left(3x - \dfrac{\pi}{6}\right)$

►► Solution

a.) $y = \sin(2x + \pi)$

$y = \sin\left[2\left(x + \dfrac{\pi}{2}\right)\right]$

The phase shift is $\dfrac{\pi}{2}$ left.

b.) $y = \sin(3x - 2\pi)$

$y = \sin\left[3\left(x - \dfrac{2\pi}{3}\right)\right]$

The phase shift is $\dfrac{2\pi}{3}$ right.

c.) $y = \sin\left(2x + \dfrac{\pi}{3}\right)$

$y = \sin\left[2\left(x + \dfrac{\pi}{6}\right)\right]$

The phase shift is $\dfrac{\pi}{6}$ left.

d.) $y = \sin\left(3x - \dfrac{\pi}{6}\right)$

$y = \sin\left[3\left(x - \dfrac{\pi}{18}\right)\right]$

The phase shift is $\dfrac{\pi}{18}$ right.

The Constant d

Our last constant, d tells us the vertical shift of the graph. If $d > 0$ we have a vertical shift up of d, if $d < 0$ we have vertical shift down of d. Consider the graphs in **Figure 8.14.**

The basic sine graph $y = \sin x$ is graphed in **blue**. The sine graph $y = \sin x + 3$ (**green graph**) has a vertical shift up of 3 since $3 > 0$ and the sine graph $y = \sin x - 3$ (**red graph**) has a vertical shift down of 3 since $-3 < 0$.

We have discussed all the constants in the general sine function:

$y = a\sin(bx \pm c) \pm d$. These constants have the effect of translating, reflecting, stretching, and shrinking the basic sine graph.

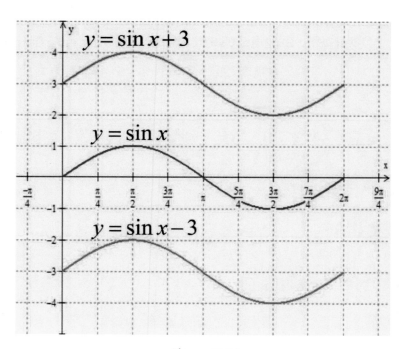

Figure 8.14

To help us sketch the graph of sine, we need to know the 5 key points over one period: the beginning, the maximum, the half period, the minimum, and the ending. We call these the fourths because they divide the graph into four intervals. See **Figure 8.15**.

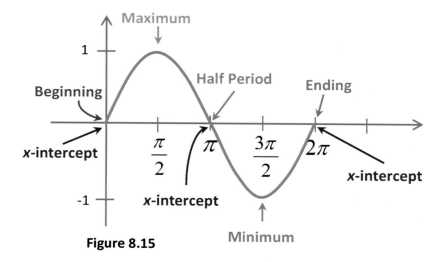

Figure 8.15

To find the fourths, divide the period by 4, $\dfrac{period}{4}$ and add the fourth to the beginning to get the maximum. Add to the maximum the fourth and get the half period. Adding the fourth to the half period gives the minimum and adding the fourth to the minimum gives us the ending.

► EXAMPLE 4

Determine the fourths for $y = \sin(2x + \pi)$.

►►Solution

Step 1:   Find the beginning and ending for $y = \sin(2x + \pi)$.

$$0 \le 2x + \pi \le 2\pi$$

$$-\pi \le 2x \le \pi$$

$$-\frac{\pi}{2} \le x \le \frac{\pi}{2}$$

The beginning is $-\dfrac{\pi}{2}$ and the ending is $\dfrac{\pi}{2}$.

Step 2:   Find the period: $p = \dfrac{2\pi}{2} = \pi$.

Step 3:   Find the fourths value: $\dfrac{\pi}{4}$.

Step 4:   Beginning: $-\dfrac{\pi}{2}$

Maximum: $-\dfrac{\pi}{2} + \dfrac{\pi}{4} = -\dfrac{2\pi}{4} + \dfrac{\pi}{4} = -\dfrac{\pi}{4}$

Half Period: $-\dfrac{\pi}{4} + \dfrac{\pi}{4} = 0$

Minimum: $0 + \dfrac{\pi}{4} = \dfrac{\pi}{4}$

Ending: $\dfrac{\pi}{4} + \dfrac{\pi}{4} = \dfrac{2\pi}{4} = \dfrac{\pi}{2}$

► EXAMPLE 5

Graph $y = 2\sin(3x)$ over one period. Label the fourths.

►►Solution

We will find the amplitude, period, fourths, beginning, ending, reflection, phase shift, and vertical shift.

Amplitude: $|2| = 2$ **Period:** $\dfrac{2\pi}{b} = \dfrac{2\pi}{3}$

Beginning and Ending: $0 \le 3x \le 2\pi$ Beginning: 0   Ending: $\dfrac{2\pi}{3}$

$$0 \le x \le \dfrac{2\pi}{3}$$

Phase Shift: None   **Vertical Shift:** None   **Reflection:** None

Fourths: $\dfrac{Period}{4} = \dfrac{\frac{2\pi}{3}}{4} = \dfrac{2\pi}{3} \cdot \dfrac{1}{4} = \dfrac{\pi}{6}$

Beginning: 0 $(0,0)$

Maximum: $0 + \dfrac{\pi}{6} = \dfrac{\pi}{6}$ $\left(\dfrac{\pi}{6}, 2\right)$

Half Period: $\dfrac{\pi}{6} + \dfrac{\pi}{6} = \dfrac{2\pi}{6} = \dfrac{\pi}{3}$ $\left(\dfrac{\pi}{3}, 0\right)$

Minimum: $\dfrac{\pi}{3} + \dfrac{\pi}{6} = \dfrac{2\pi}{6} + \dfrac{\pi}{6} = \dfrac{3\pi}{6} = \dfrac{\pi}{2}$ $\left(\dfrac{\pi}{2}, -2\right)$

Ending: $\dfrac{\pi}{2} + \dfrac{\pi}{6} = \dfrac{3\pi}{6} + \dfrac{\pi}{6} = \dfrac{4\pi}{6} = \dfrac{2\pi}{3}$ $\left(\dfrac{2\pi}{3}, 0\right)$

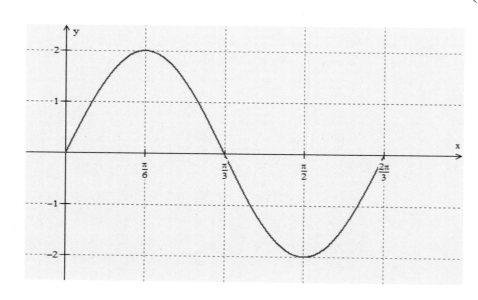

▶ EXAMPLE 6

Graph $y = -\frac{1}{2}\sin(2x)$ over one period. Label the fourths.

▶▶ Solution

We will find the amplitude, period, fourths, beginning, ending, reflection, phase shift, and vertical shift.

Amplitude: $\left|-\dfrac{1}{2}\right| = \dfrac{1}{2}$ **Period:** $\dfrac{2\pi}{b} = \dfrac{2\pi}{2} = \pi$

Beginning and Ending: $0 \le 2x \le 2\pi$ Beginning: 0   Ending: π

$0 \le x \le \pi$

Phase Shift: None   **Vertical Shift:** None   **Reflection:** x-axis

Fourths: $\dfrac{Period}{4} = \dfrac{\pi}{4} = \dfrac{\pi}{4}$

Beginning:	0	(0,0)
Maximum:	$0 + \dfrac{\pi}{4} = \dfrac{\pi}{4}$	$\left(\dfrac{\pi}{4}, -\dfrac{1}{2}\right)$
Half Period:	$\dfrac{\pi}{4} + \dfrac{\pi}{4} = \dfrac{2\pi}{4} = \dfrac{\pi}{2}$	$\left(\dfrac{\pi}{2}, 0\right)$
Minimum:	$\dfrac{\pi}{2} + \dfrac{\pi}{4} = \dfrac{2\pi}{4} + \dfrac{\pi}{4} = \dfrac{3\pi}{4}$	$\left(\dfrac{3\pi}{4}, \dfrac{1}{2}\right)$
Ending:	$\dfrac{3\pi}{4} + \dfrac{\pi}{4} = \dfrac{4\pi}{4} = \pi$	$(\pi, 0)$

Begin by dotting the graph of

$y = \frac{1}{2}\sin(2x)$ (red graph).

Reflect the red graph over the

x-axis to get the blue graph,

$y = -\frac{1}{2}\sin(2x)$.

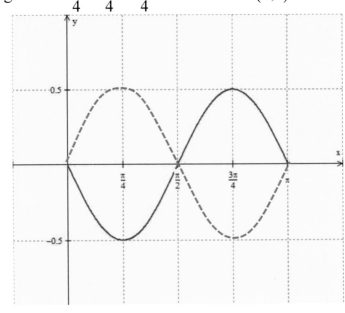

▶EXAMPLE 7

Graph $y = 3\sin\left(2x + \dfrac{\pi}{6}\right)$ over one period. **Label the fourths.**

▶▶Solution

We will find the amplitude, period, fourths, beginning, ending, reflection, phase shift, and vertical shift.

Amplitude: $\left|3\right| = 3$ **Period:** $\dfrac{2\pi}{b} = \dfrac{2\pi}{2} = \pi$

Beginning and Ending: $0 \le 2x + \dfrac{\pi}{6} \le 2\pi$ **Beginning:** $-\dfrac{\pi}{12}$ **Ending:** $\dfrac{11\pi}{12}$

$$-\dfrac{\pi}{12} \le x \le \dfrac{11\pi}{12}$$

Phase Shift: $\dfrac{\pi}{12}$ left   **Vertical Shift:** None   **Reflection:** None

Fourths: $\dfrac{Period}{4} = \dfrac{\pi}{4} = \dfrac{\pi}{4}$

Beginning: $-\dfrac{\pi}{12}$ $\left(-\dfrac{\pi}{12}, 0\right)$

Maximum: $-\dfrac{\pi}{12} + \dfrac{\pi}{4} = -\dfrac{\pi}{12} + \dfrac{3\pi}{12} = \dfrac{2\pi}{12} = \dfrac{\pi}{6}$ $\left(\dfrac{\pi}{6}, 3\right)$

Half Period: $\dfrac{\pi}{6} + \dfrac{\pi}{4} = \dfrac{2\pi}{12} + \dfrac{3\pi}{12} = \dfrac{5\pi}{12}$ $\left(\dfrac{5\pi}{12}, 0\right)$

Minimum: $\dfrac{5\pi}{12} + \dfrac{\pi}{4} = \dfrac{5\pi}{12} + \dfrac{3\pi}{12} = \dfrac{8\pi}{12} = \dfrac{2\pi}{3}$ $\left(\dfrac{2\pi}{3}, -3\right)$

Ending: $\dfrac{2\pi}{3} + \dfrac{\pi}{4} = \dfrac{8\pi}{12} + \dfrac{3\pi}{12} = \dfrac{11\pi}{12}$ $\left(\dfrac{11\pi}{12}, 0\right)$

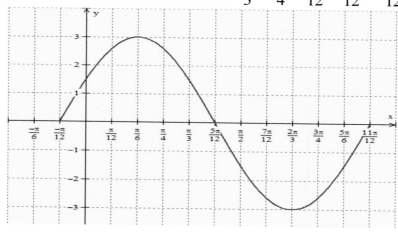

►EXAMPLE 8

Graph $y = 2\sin\left(\dfrac{1}{2}x - \pi\right) + 2$ over one period. **Label the fourths.**

►►Solution

We will find the amplitude, period, fourths, beginning, ending, reflection, phase shift, and vertical shift.

Amplitude: $|2| = 2$ **Period:** $\dfrac{2\pi}{b} = \dfrac{2\pi}{\dfrac{1}{2}} = 2\pi \cdot 2 = 4\pi$

Beginning and Ending: $0 \le \dfrac{1}{2}x - \pi \le 2\pi$ Beginning: 2π Ending: 6π

$$2\pi \le x \le 6\pi$$

Phase Shift: 2π right   **Vertical Shift:** Up 2   **Reflection:** None

Fourths: $\dfrac{Period}{4} = \dfrac{4\pi}{4} = \pi$

Beginning:	2π	$(2\pi,2)$
Maximum:	$2\pi + \pi = 3\pi$	$(3\pi,4)$
Half Period:	$3\pi + \pi = 4\pi$	$(4\pi,2)$
Minimum:	$4\pi + \pi = 5\pi$	$(5\pi,0)$
Ending:	$5\pi + \pi = 6\pi$	$(6\pi,2)$

Since we have a vertical shift up of 2, we will shift the x-axis up 2 and dot it in red. Graph

$y = 2\sin\left(\dfrac{1}{2}x - \pi\right) + 2$

using the red dotted line as the x-axis.

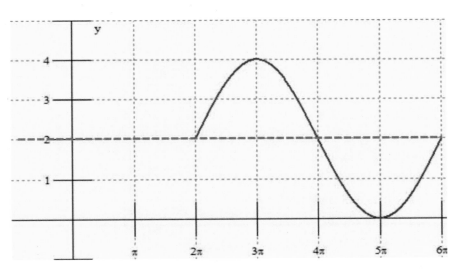

►EXAMPLE 9

Graph $y = 250\sin(4x - \pi) + 300$ **over one period. Label the fourths.**

►►**Solution**

We will find the amplitude, period, fourths, beginning, ending, reflection, phase shift, and vertical shift.

Amplitude: $|250| = 250$ **Period:** $\dfrac{2\pi}{b} = \dfrac{2\pi}{4} = \dfrac{\pi}{2}$

Beginning and Ending: $0 \le 4x - \pi \le 2\pi$ **Beginning:** $\dfrac{\pi}{4}$ **Ending:** $\dfrac{3\pi}{4}$

$$\dfrac{\pi}{4} \le x \le \dfrac{3\pi}{4}$$

Phase Shift: $\dfrac{\pi}{4}$ right   **Vertical Shift:** Up 300   **Reflection:** None

Fourths: $\dfrac{Period}{4} = \dfrac{\dfrac{\pi}{2}}{4} = \dfrac{\pi}{2} \cdot \dfrac{1}{4} = \dfrac{\pi}{8}$

Beginning: $\dfrac{\pi}{4}$ $\left(\dfrac{\pi}{4}, 300\right)$

Maximum: $\dfrac{\pi}{4} + \dfrac{\pi}{8} = \dfrac{2\pi}{8} + \dfrac{\pi}{8} = \dfrac{3\pi}{8}$ $\left(\dfrac{3\pi}{8}, 550\right)$

Half Period: $\dfrac{3\pi}{8} + \dfrac{\pi}{8} = \dfrac{4\pi}{8} = \dfrac{\pi}{2}$ $\left(\dfrac{\pi}{2}, 300\right)$

Minimum: $\dfrac{\pi}{2} + \dfrac{\pi}{8} = \dfrac{4\pi}{8} + \dfrac{\pi}{8} = \dfrac{5\pi}{8}$ $\left(\dfrac{5\pi}{8}, 50\right)$

Ending: $\dfrac{5\pi}{8} + \dfrac{\pi}{8} = \dfrac{6\pi}{8} = \dfrac{3\pi}{4}$ $\left(\dfrac{3\pi}{4}, 300\right)$

Since we have a vertical shift up of 300, we will shift the x-axis up 300 and dot it. The amplitude is 250, make equal marks from 300 on the y-axis, one above and one below the dotted line. Label your fourths and graph the sine curve.

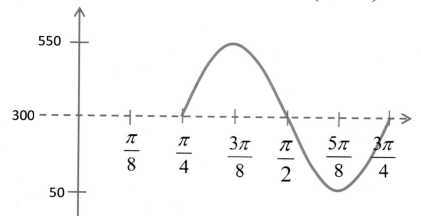

►EXAMPLE 10

Graph $y = 2\sin\left(2x - \dfrac{\pi}{4}\right) + 2$ **over two periods. Label the fourths.**

►►**Solution**

We will find the amplitude, period, fourths, beginning, ending, reflection, phase shift, and vertical shift for one period.

Amplitude: $|2| = 2$ **Period:** $\dfrac{2\pi}{b} = \dfrac{2\pi}{2} = \pi$

Beginning and Ending: $0 \le 2x - \dfrac{\pi}{4} \le 2\pi$ **Beginning:** $\dfrac{\pi}{8}$ **Ending:** $\dfrac{9\pi}{8}$

$$\dfrac{\pi}{8} \le x \le \dfrac{9\pi}{8}$$

Phase Shift: $\dfrac{\pi}{8}$ right   **Vertical Shift:** Up 2   **Reflection:** None

Fourths: $\dfrac{Period}{4} = \dfrac{\pi}{4}$

Beginning:	$\dfrac{\pi}{8}$	$\left(\dfrac{\pi}{8}, 2\right)$
Maximum:	$\dfrac{\pi}{8} + \dfrac{\pi}{4} = \dfrac{\pi}{8} + \dfrac{2\pi}{8} = \dfrac{3\pi}{8}$	$\left(\dfrac{3\pi}{8}, 4\right)$
Half Period:	$\dfrac{3\pi}{8} + \dfrac{\pi}{4} = \dfrac{3\pi}{8} + \dfrac{2\pi}{8} = \dfrac{5\pi}{8}$	$\left(\dfrac{5\pi}{8}, 2\right)$
Minimum:	$\dfrac{5\pi}{8} + \dfrac{\pi}{4} = \dfrac{5\pi}{8} + \dfrac{2\pi}{8} = \dfrac{7\pi}{8}$	$\left(\dfrac{7\pi}{8}, 0\right)$
Ending:	$\dfrac{7\pi}{8} + \dfrac{\pi}{4} = \dfrac{7\pi}{8} + \dfrac{2\pi}{8} = \dfrac{9\pi}{8}$	$\left(\dfrac{9\pi}{8}, 2\right)$

Use the information and graph the first period then continue the graph for one more period.

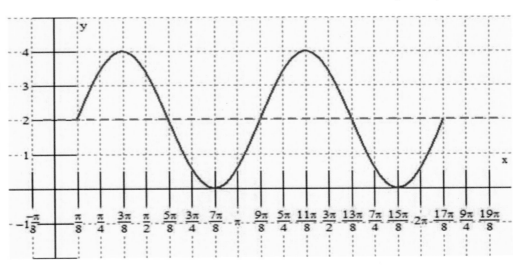

Section Wrap Up

$$y = \sin x$$

Domain: $(-\infty, \infty)$

Range: $[-1, 1]$

Odd Function
$$\sin(-x) = -\sin x$$

$$y = a\sin(bx \pm c) \pm d$$

$a < 0$ Reflection on the x-axis

$|a| < 1$ Vertical Shrinkage of a

$|a| > 1$ Vertical Stretching of a

$$Period = \frac{2\pi}{b}$$

$$Fourths = \frac{Period}{4}$$

$$Phase\ Shift = \frac{c}{b}$$

Beginning and Ending: $0 \le bx \pm c \le 2\pi$

Vertical Shift: $\pm d$

HOMEWORK

Objective 1

Determine the amplitude, period and phase shift of each sine function.

See Examples 1, 2and 3.

1. $y = -4\sin 5x$

2. $y = 3\sin\left(2x - \dfrac{\pi}{4}\right)$

3. $y = \frac{1}{4}\sin(x - 2\pi)$

4. $y = 10\sin\left(\dfrac{1}{3}x + \pi\right)$

5. $y = -25\sin\left(4x + \dfrac{\pi}{2}\right)$

6. $y = \frac{1}{2}\sin(3x - \pi)$

7. $y = 0.2\sin\left(\dfrac{1}{2}x + \dfrac{\pi}{8}\right)$

8. $y = 2\sin 8x$

Objective 2

Determine the fourths for each sine function. *See Example 4.*

9. $y = -\sin 2x$

10. $y = 2\sin\left(x + \dfrac{\pi}{6}\right)$

11. $y = \frac{1}{2}\sin(2x + \pi)$

12. $y = 10\sin\left(\dfrac{1}{2}x - 2\pi\right)$

Objective 3

Graph the following sine functions over one period. Label the fourths.

See Examples 5 - 9.

13. $y = -\sin 3x$

14. $y = -3\sin\left(2x - \dfrac{\pi}{2}\right)$

15. $y = \frac{1}{2}\sin(x + \pi)$

16. $y = \sin(x + \pi) + 1$

17. $y = 25\sin\left(2x + \dfrac{\pi}{2}\right) + 10$

18. $y = \frac{1}{2}\sin(3x - \pi)$

19. $y = 0.2\sin\left(\dfrac{1}{2}x + \dfrac{\pi}{8}\right)$

20. $y = 2\sin 4x$

21. $y = -\sin\left(2x + \dfrac{\pi}{6}\right)$

22. $y = -2\sin\left(x - \dfrac{\pi}{2}\right)$

23. $y = \tfrac{1}{2}\sin\left(x - \dfrac{\pi}{3}\right)$

24. $y = \sin(x + \pi) - 2$

25. $y = 100\sin\left(x - \dfrac{\pi}{4}\right) + 75$

26. $y = 2 + \tfrac{1}{2}\sin x$

27. $y = \sin\left(4x - \dfrac{\pi}{8}\right)$

28. $y = 3\sin 4x - 3$

Objective 3

Graph the following sine functions over two periods. Label the fourths.

See Example 10.

29. $y = 2\sin 2x$

30. $y = -\sin\left(x - \dfrac{\pi}{2}\right)$

31. $y = \tfrac{1}{2}\sin\left(x - \dfrac{\pi}{4}\right)$

32. $y = 2\sin(x + \pi) - 1$

33. $y = 3\sin\left(x + \dfrac{\pi}{2}\right) + 1$

34. $y = \tfrac{1}{2}\sin\left(2x - \dfrac{\pi}{4}\right)$

ANSWERS

1. amplitude: 4, period: $\frac{2\pi}{5}$, phase shift: None

2. amplitude: 3, period: π, phase shift: $\frac{\pi}{8}$ right

3. amplitude: $\frac{1}{4}$, period: 2π, phase shift: 2π right

4. amplitude: 10, period: 6π, phase shift: 3π left

5. amplitude: 25, period: $\frac{\pi}{2}$, phase shift: $\frac{\pi}{8}$ left

6. amplitude: $\frac{1}{2}$, period: $\frac{2\pi}{3}$, phase shift: $\frac{\pi}{3}$ right

7. amplitude: 0.2, period: 4π, phase shift: $\frac{\pi}{4}$ left

8. amplitude: 2, period: $\frac{\pi}{4}$, phase shift: None

9. $0, \frac{\pi}{4}, \frac{\pi}{2}, \frac{3\pi}{4}, \pi$

10. $-\frac{\pi}{6}, \frac{\pi}{3}, \frac{5\pi}{6}, \frac{4\pi}{3}, \frac{11\pi}{6}$

11. $-\frac{\pi}{2}, -\frac{\pi}{4}, 0, \frac{\pi}{4}, \frac{\pi}{2}$

12. $4\pi, 5\pi, 6\pi, 7\pi, 8\pi$

13.

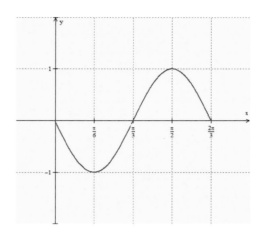

14.

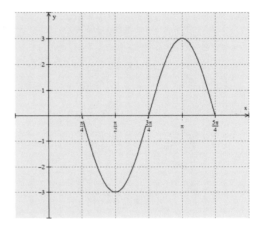

15.

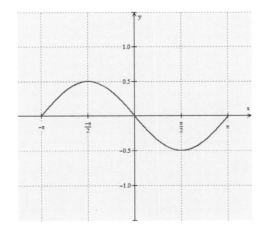

16.

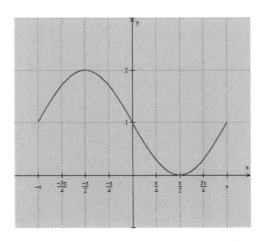

17.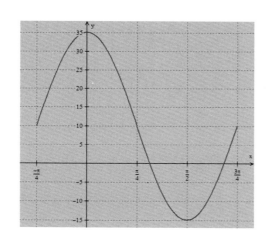

18.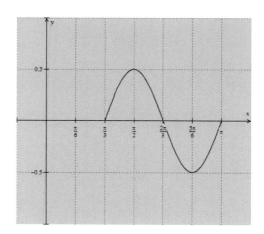

19.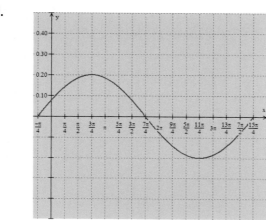

20.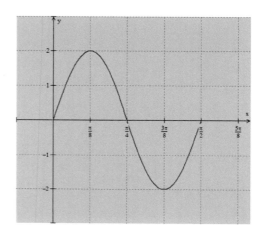

21.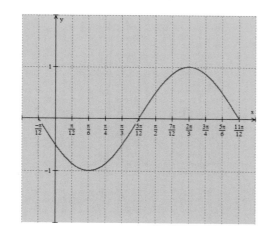

22.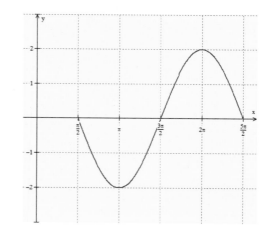

23.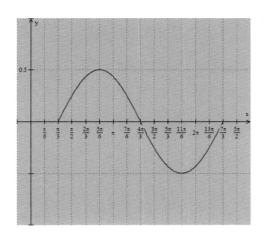

24.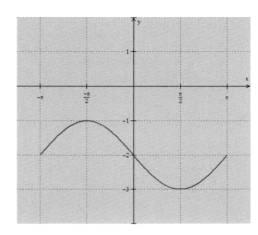

25.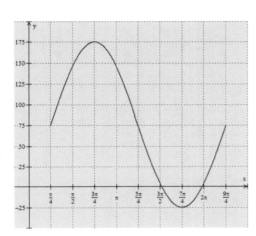

26.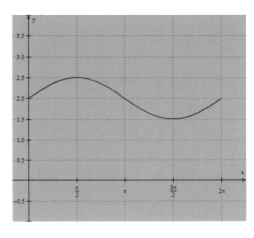

27.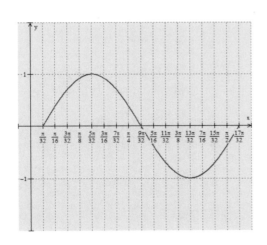

28.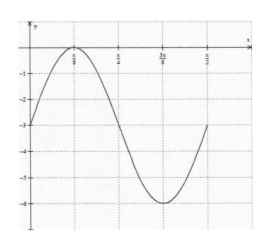

29.

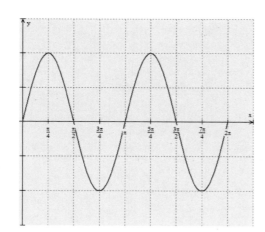

30.

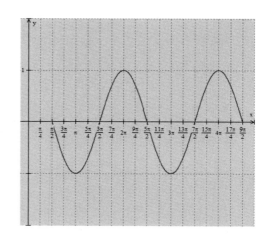

31.

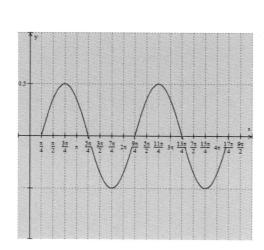

32.

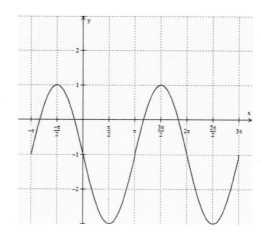

33.

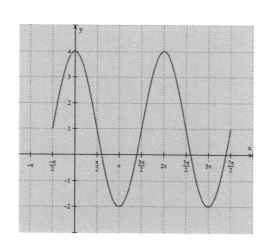

34.

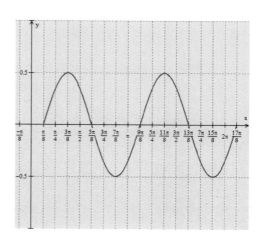

Section 9: Graph of the Cosine Function

Learning Outcomes:

- The student will correctly memorize and apply trigonometric formulas, definitions, identities, and properties.
- The student will illustrate and examine the graphs of: the six trigonometric functions and their inverses, parametric equations, and polar equations.

Objectives: At the conclusion of this lesson you should be able to:

1. Determine the amplitude, period, phase shift, vertical shift and reflection of a cosine function.
2. Determine the fourths of a cosine function.
3. Use amplitude, period, phase shift, vertical shift, reflections and the fourths to graph the cosine function over a specified period.

In the last section we talked about and graphed the sine function. In this section we are going to graph the cosine function. Like the sine function, cosine is a periodic function.

Periodic Functions

A function f is said to be periodic if there is a positive number P such that

$$f(t + P) = f(t)$$

for all t in the domain. The smallest P for which this occurs is called the period of f.

The smallest value of P for sine and cosine is 2π because the circumference of the unit circle is 2π. Sine and cosine are periodic functions with period of 2π.

To graph the cosine function $y = \cos x$ we will look at the behavior of the x-values over one period which is from 0 to 2π.

As x increases from 0 to $\frac{\pi}{2}$, the y-values of terminal points decrease from 1 to 0 as shown in **Figure 9.1**.

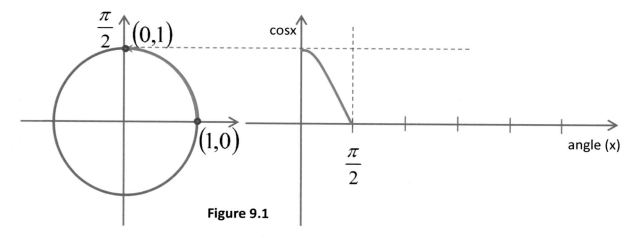

Figure 9.1

As x increases from $\frac{\pi}{2}$ to π, the y-values decrease from 0 to -1 as shown in **Figure 9.2**.

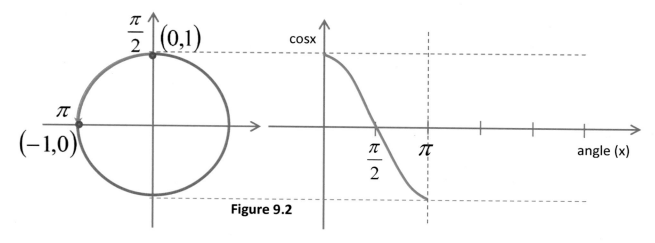

Figure 9.2

As x increases from π to $\frac{3\pi}{2}$, the y-values increase from -1 to 0 as shown in **Figure 9.3**.

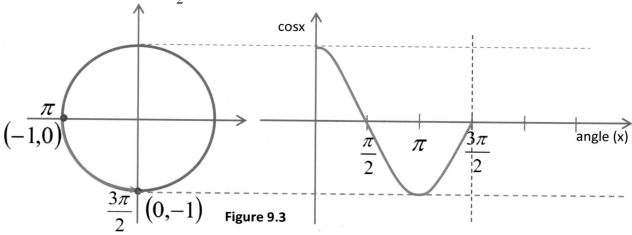

Figure 9.3

As x increases from $\dfrac{3\pi}{2}$ to 2π, the y-values increase from 0 to 1 as shown in **Figure 9.4**.

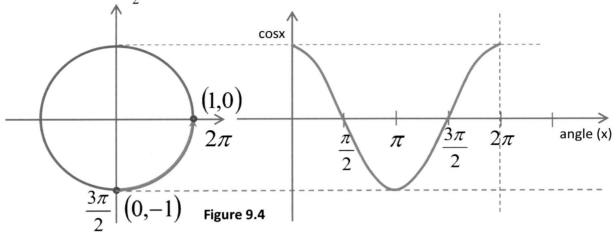

Figure 9.4

To get a more precise graph for cosine we can construct a table:

x	0	$\dfrac{\pi}{6}$	$\dfrac{\pi}{4}$	$\dfrac{\pi}{3}$	$\dfrac{\pi}{2}$	$\dfrac{2\pi}{3}$	$\dfrac{3\pi}{4}$	$\dfrac{5\pi}{6}$	π	$\dfrac{7\pi}{6}$	$\dfrac{5\pi}{4}$	$\dfrac{4\pi}{3}$	$\dfrac{3\pi}{2}$	$\dfrac{5\pi}{3}$	$\dfrac{7\pi}{4}$	$\dfrac{11\pi}{6}$	2π
$\cos x$	1	$\dfrac{\sqrt{3}}{2}$	$\dfrac{\sqrt{2}}{2}$	$\dfrac{1}{2}$	0	$-\dfrac{1}{2}$	$-\dfrac{\sqrt{2}}{2}$	$-\dfrac{\sqrt{3}}{2}$	-1	$-\dfrac{\sqrt{3}}{2}$	$-\dfrac{\sqrt{2}}{2}$	$-\dfrac{1}{2}$	0	$\dfrac{1}{2}$	$\dfrac{\sqrt{2}}{2}$	$\dfrac{\sqrt{3}}{2}$	1

If we plot the points, we have the graph in **blue** below in **Figure 9.5**.

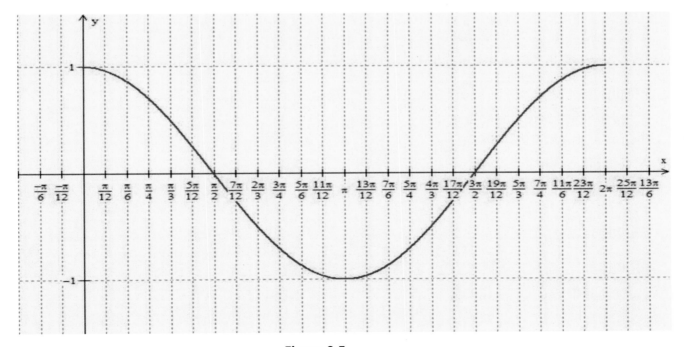

Figure 9.5

If we plot more points, as shown **Figure 9.6**, the cosine curve is continuous on the domain from $(-\infty, \infty)$. Notice that cosine has symmetry with respect to the y-axis which implies that $\cos(-x) = \cos x$ and that cosine is even.

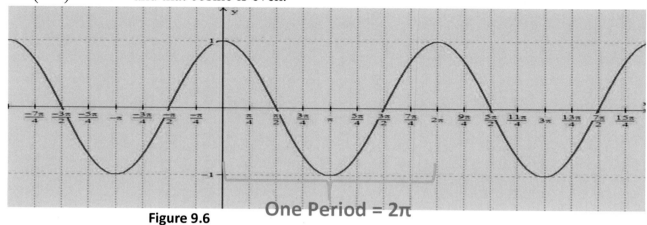

Figure 9.6

One Period = 2π

Now that we have graphed the basic cosine curve, we will study the effects of the constants *a, b,* *c,* and *d* in the general cosine equation: $y = a\cos(bx \pm c) \pm d$.

The Constant a

The *a* in the general cosine equation represents the amplitude of the graph. The amplitude for both sine and cosine is the absolute value of half the difference between the maximum and minimum values. The amplitude will be a vertical stretching of the basic cosine curve if $|a| > 1$.

The amplitude will be a vertical shrinking of the basic cosine curve if $|a| < 1$. In **Figure 9.7** we

have the graphs of $y = 2\cos x$ (red graph), $y = \cos x$ (blue graph), and $y = \frac{1}{2}\cos x$ (green graph). The basic sine graph is the blue graph, $y = \cos x$. Notice the red graph, $y = 2\cos x$ has a vertical stretching of 2 and has an amplitude of 2. The green graph, $y = \frac{1}{2}\cos x$ has a vertical shrinkage of $\frac{1}{2}$ and has an amplitude of $\frac{1}{2}$.

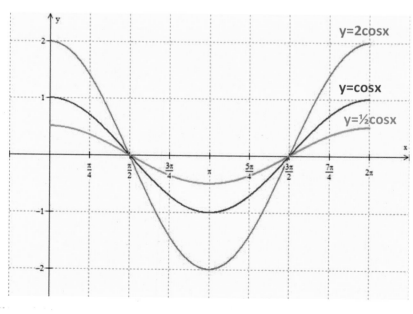

Figure 9.7

If a is negative, $a < 0$, the graph will have a reflection on the x-axis as shown in **Figure 9.8**. The basic sine graph is the **blue graph** $y = \cos x$ and the **green graph** is $y = -\cos x$. Notice that the **green graph** is a reflection on the x-axis of the **blue graph.**

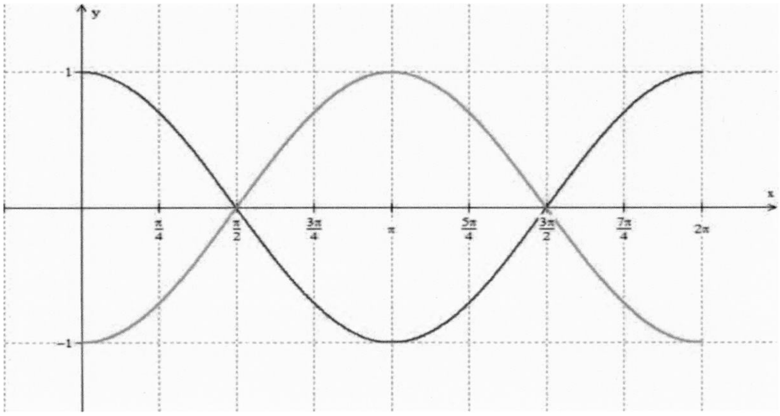

Figure 9.8

►EXAMPLE 1

Determine the amplitude of the following cosine functions and if there is a reflection on the *x*-axis.

a.) $y = 5\cos x$ b.) $y = -3\cos x$ c.) $y = \frac{1}{4}\cos x$

►►Solution

a.) $y = 5\cos x$ $a = |5| = 5$ The amplitude is 5. No reflection because $a > 0$.

b.) $y = -3\cos x$ $a = |-3| = 3$ The amplitude is 3. There is a reflection on the x-axis because $a < 0$.

c.) $y = \frac{1}{4}\cos x$ $a = \left|\frac{1}{4}\right| = \frac{1}{4}$ The amplitude is $\frac{1}{4}$. No reflection because $a > 0$.

The Constant b

The period of $y = \cos x$ is 2π when $b = 1$. What happens to the period for $y = \cos 3x$? When $b = 1$ the period is $0 \le x \le 2\pi$ but here we have $b = 3$. Since $b = 3$ we will replace x with $3x$ and see how this effects the period. See **Figure 9.9**. One period of $y = \cos 3x$ is completed between 0 and $\frac{2\pi}{3}$ which tells us the period is $\frac{2\pi}{3}$. What happens to the period for $y = \cos \frac{1}{3}x$? See **Figure 9.10**. One period of $y = \cos \frac{1}{3}x$ is completed between 0 and 6π which tells us the period is 6π. Since the period is 2π when $b = 1$, we can find the period of other cosine functions by using

$$Period = \frac{2\pi}{b},\ p = \frac{2\pi}{b}.$$

$$0 \le 3x \le 2\pi$$
$$\frac{0}{3} \le \frac{3x}{2} \le \frac{2\pi}{3}$$
$$0 \le x \le \frac{2\pi}{3}$$

Figure 9.9

$$0 \le \frac{1}{3}x \le 2\pi$$
$$3(0) \le 3\left(\frac{1}{3}x\right) \le 3(2\pi)$$
$$0 \le x \le 6\pi$$

Figure 9.10

In **Figure 9.11** we have the graphs of $y = \cos x$ (**blue graph**), $y = \cos 2x$ (**green graph**) and $y = \cos \frac{1}{2} x$ (**red graph**). Notice how the cosine graph is horizontally stretched or shrunk depending on the b value. If $b > 1$ we have horizontal shrinkage and if $b < 1$ we have horizontal stretching. The cosine graph $y = \cos 2x$ (**green graph**) has a horizontal shrinkage of 2 and the cosine graph $y = \cos \frac{1}{2} x$ (**red graph**) has a horizontal stretching of $\frac{1}{2}$.

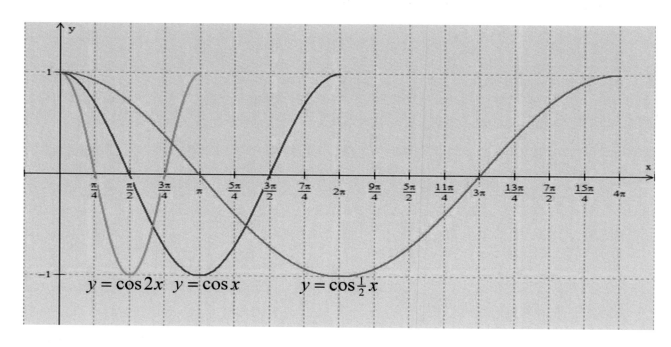

Figure 9.11

▶ EXAMPLE 2

Determine the period of the following sine functions.

a.) $y = \cos 6x$ b.) $y = \cos \frac{1}{4} x$ c.) $y = \cos 3x$

▶▶ Solution

a.) $y = \cos 6x$ $p = \dfrac{2\pi}{6} = \dfrac{\pi}{3}$ The period is $\dfrac{\pi}{3}$.

b.) $y = \cos \frac{1}{4} x$ $p = \dfrac{2\pi}{\frac{1}{4}} = 2\pi \cdot 4 = 8\pi$ The period is 8π.

c.) $y = \cos 3x$ $p = \dfrac{2\pi}{3} = \dfrac{2\pi}{3}$ The period is $\dfrac{2\pi}{3}$.

The Constant c

To examine the effect of the constant c, consider the sine graphs, $y = \cos x$ (**blue graph**), $y = \cos(x + \pi)$ (**green graph**), and $y = \cos(x - \pi)$ (**red graph**) shown in **Figure 9.12**. The cosine graph $y = \cos(x + \pi)$ (**green graph**) has a phase shift left of π and the cosine graph $y = \cos(x - \pi)$ (**red graph**) has a phase shift right of π.

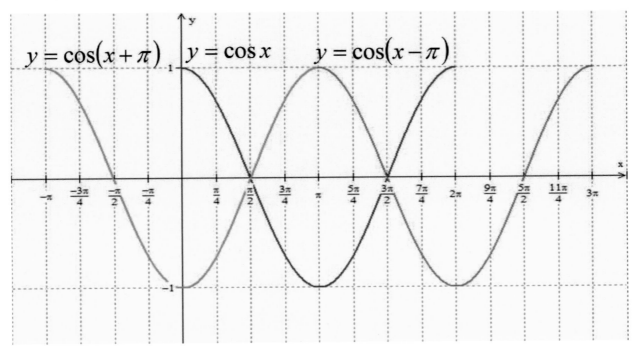

Figure 9.12

In **Figure 9.13** we have the graph of $y = \cos x$ (**blue graph**), and the graph of $y = \cos(2x - \pi)$ (**red graph**). Notice the **red graph** has both a change in the period and a horizontal shift. If we look at the beginning and the ending of $y = \cos(2x - \pi)$ we have: $0 \le 2x - \pi \le 2\pi$ which gives us our new beginning and ending: $\dfrac{\pi}{2} \le x \le \dfrac{3\pi}{2}$. Notice the phase shift and the beginning are $\dfrac{\pi}{2}$.

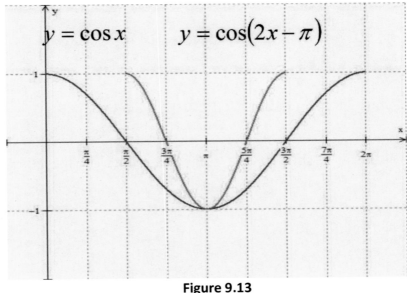

Figure 9.13

In the table below we can see the comparison between the beginning and the phase shift.

Sine Function	Period	Beginning	Ending	Phase Shift
$y = \cos x$	$0 \le x \le 2\pi$	0	2π	none
$y = \cos(x + 2\pi)$	$0 \le x + 2\pi \le 2\pi$ $-2\pi \le x \le 0$	-2π	0	Left 2π
$y = \cos(x - \pi)$	$0 \le x - \pi \le 2\pi$ $\pi \le x \le 3\pi$	π	3π	Right π
$y = \cos\left(x - \dfrac{\pi}{2}\right)$	$0 \le x - \dfrac{\pi}{2} \le 2\pi$ $\dfrac{\pi}{2} \le x \le \dfrac{5\pi}{2}$	$\dfrac{\pi}{2}$	$\dfrac{5\pi}{2}$	Right $\dfrac{\pi}{2}$
$y = \cos(2x - \pi)$	$0 \le 2x - \pi \le 2\pi$ $\pi \le 2x \le 3\pi$ $\dfrac{\pi}{2} \le x \le \dfrac{3\pi}{2}$	$\dfrac{\pi}{2}$	$\dfrac{3\pi}{2}$	Right $\dfrac{\pi}{2}$
$y = \cos(3x + \pi)$	$0 \le 3x + \pi \le 2\pi$ $-\pi \le 3x \le \pi$ $-\dfrac{\pi}{3} \le x \le \dfrac{\pi}{3}$	$-\dfrac{\pi}{3}$	$\dfrac{\pi}{3}$	Left $\dfrac{\pi}{3}$
$y = \cos\left(2x - \dfrac{\pi}{4}\right)$	$0 \le 2x - \dfrac{\pi}{4} \le 2\pi$ $\dfrac{\pi}{4} \le 2x \le \dfrac{9\pi}{4}$ $\dfrac{\pi}{8} \le x \le \dfrac{9\pi}{8}$	$\dfrac{\pi}{8}$	$\dfrac{9\pi}{8}$	Right $\dfrac{\pi}{8}$

Rewriting $y = a\cos(bx \pm c) \pm d$ by factoring out the b, we have $y = a\cos\left[b\left(x \pm \dfrac{c}{b}\right)\right] \pm d$.

$\dfrac{c}{b}$ is the phase shift. If $\dfrac{c}{b} > 0$, the phase shift is to the left $\dfrac{c}{b}$. If $\dfrac{c}{b} < 0$, the phase shift is to the right $\dfrac{c}{b}$.

► EXAMPLE 3

Determine the phase shift of the following sine functions.

a.) $\quad y = \cos(2x + \pi)$

b.) $\quad y = \cos(3x - 2\pi)$

c.) $\quad y = \cos\left(2x + \dfrac{\pi}{3}\right)$

d.) $\quad y = \cos\left(3x - \dfrac{\pi}{6}\right)$

►►Solution

a.) $\quad y = \cos(2x + \pi)$ The phase shift is $\dfrac{\pi}{2}$ left.

$$y = \cos\left[2\left(x + \dfrac{\pi}{2}\right)\right]$$

b.) $\quad y = \cos(3x - 2\pi)$ The phase shift is $\dfrac{2\pi}{3}$ right.

$$y = \cos\left[3\left(x - \dfrac{2\pi}{3}\right)\right]$$

c.) $\quad y = \cos\left(2x + \dfrac{\pi}{3}\right)$ The phase shift is $\dfrac{\pi}{6}$ left.

$$y = \cos\left[2\left(x + \dfrac{\pi}{6}\right)\right]$$

d.) $\quad y = \cos\left(3x - \dfrac{\pi}{6}\right)$ The phase shift is $\dfrac{\pi}{18}$ right.

$$y = \cos\left[3\left(x - \dfrac{\pi}{18}\right)\right]$$

The Constant d

Our last constant, d tells us the vertical shift of the graph. If $d > 0$ we have a vertical shift up of d, if $d < 0$ we have vertical shift down of d. Consider the graphs in **Figure 9.14.**

The basic cosine graph $y = \cos x$ is graphed in **blue**. The cosine graph $y = \cos x + 3$ (**green graph**) has a vertical shift up of 3 since $3 > 0$ and the cosine graph $y = \cos x - 3$ (**red graph**) has a vertical shift down of 3 since $-3 < 0$.

We have discussed all the constants in the general cosine function: $y = a\cos(bx \pm c) \pm d$. These constants have the effect of translating, reflecting, stretching, and shrinking the basic cosine graph.

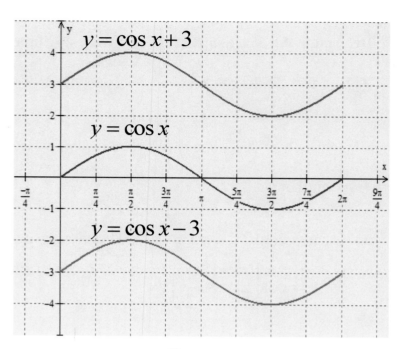

Figure 9.14

To help us sketch the graph of cosine, we need to know the 5 key points over one period: the beginning (maximum), inflection point (x-intercept), the minimum, inflection point (x-intercept), and the ending (maximum). We call these the fourths because they divide the graph into four intervals. See **Figure 9.15**.

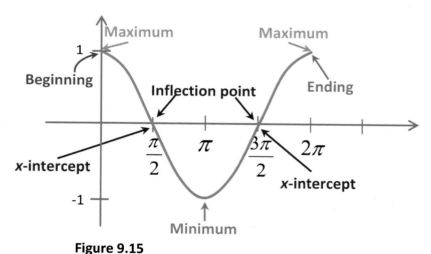

Figure 9.15

To find the fourths, divide the period by 4, $\dfrac{period}{4}$ and add the fourth to the beginning to get the first inflection point. Add to the first inflection point the fourth and get the minimum. Adding the fourth to the minimum gives the second inflection point and adding the fourth to the second inflection point gives us the ending.

► EXAMPLE 4

Determine the fourths for $y = \cos(2x + \pi)$.

►►Solution

Step 1:   Find the beginning and ending for $y = \cos(2x + \pi)$.

$$0 \le 2x + \pi \le 2\pi$$

$$-\pi \le 2x \le \pi$$

$$-\frac{\pi}{2} \le x \le \frac{\pi}{2}$$

The beginning is $-\dfrac{\pi}{2}$ and the ending is $\dfrac{\pi}{2}$.

Step 2:   Find the period: $p = \dfrac{2\pi}{2} = \pi$.

Step 3:   Find the fourths value: $\dfrac{period}{4} = \dfrac{\pi}{4}$.

Step 4:   Beginning: $-\dfrac{\pi}{2}$

1st Inflection Point: $-\dfrac{\pi}{2} + \dfrac{\pi}{4} = -\dfrac{2\pi}{4} + \dfrac{\pi}{4} = -\dfrac{\pi}{4}$

Minimum: $-\dfrac{\pi}{4} + \dfrac{\pi}{4} = 0$

2nd Inflection Point: $0 + \dfrac{\pi}{4} = \dfrac{\pi}{4}$

Ending: $\dfrac{\pi}{4} + \dfrac{\pi}{4} = \dfrac{2\pi}{4} = \dfrac{\pi}{2}$

►EXAMPLE 5

Graph $y = 2\cos(3x)$ over one period. Label the fourths.

►►Solution

We will find the amplitude, period, fourths, beginning, ending, reflection, phase shift, and vertical shift.

Amplitude: $|2| = 2$ **Period:** $\dfrac{2\pi}{b} = \dfrac{2\pi}{3}$

Beginning and Ending: $0 \le 3x \le 2\pi$ Beginning: 0   Ending: $\dfrac{2\pi}{3}$

$$0 \le x \le \dfrac{2\pi}{3}$$

Phase Shift: None   **Vertical Shift:** None   **Reflection:** None

Fourths:

$\dfrac{Period}{4} = \dfrac{\dfrac{2\pi}{3}}{4} = \dfrac{2\pi}{3} \cdot \dfrac{1}{4} = \dfrac{\pi}{6}$

Beginning: 0 $(0,2)$

1st Inflection Point: $0 + \dfrac{\pi}{6} = \dfrac{\pi}{6}$ $\left(\dfrac{\pi}{6}, 0\right)$

Minimum: $\dfrac{\pi}{6} + \dfrac{\pi}{6} = \dfrac{2\pi}{6} = \dfrac{\pi}{3}$ $\left(\dfrac{\pi}{3}, -2\right)$

2nd Inflection Point: $\dfrac{\pi}{3} + \dfrac{\pi}{6} = \dfrac{2\pi}{6} + \dfrac{\pi}{6} = \dfrac{3\pi}{6} = \dfrac{\pi}{2}$ $\left(\dfrac{\pi}{2}, 0\right)$

Ending: $\dfrac{\pi}{2} + \dfrac{\pi}{6} = \dfrac{3\pi}{6} + \dfrac{\pi}{6} = \dfrac{4\pi}{6} = \dfrac{2\pi}{3}$ $\left(\dfrac{2\pi}{3}, 2\right)$

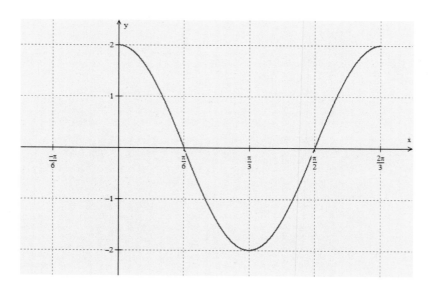

▶ EXAMPLE 6

Graph $y = -\frac{1}{2}\cos(2x)$ **over one period. Label the fourths.**

▶▶ **Solution**

We will find the amplitude, period, fourths, beginning, ending, reflection, phase shift, and vertical shift.

Amplitude: $\left|-\dfrac{1}{2}\right| = \dfrac{1}{2}$ **Period:** $\dfrac{2\pi}{b} = \dfrac{2\pi}{2} = \pi$

Beginning and Ending: $0 \le 2x \le 2\pi$ Beginning: 0   Ending: π

$$0 \le x \le \pi$$

Phase Shift: None   **Vertical Shift:** None   **Reflection:** x-axis

Fourths:

$\dfrac{Period}{4} = \dfrac{\pi}{4} = \dfrac{\pi}{4}$

Beginning:	0		$(0, -0.5)$
1st Inflection Point:	$0 + \dfrac{\pi}{4} = \dfrac{\pi}{4}$		$\left(\dfrac{\pi}{4}, 0\right)$
Minimum:	$\dfrac{\pi}{4} + \dfrac{\pi}{4} = \dfrac{2\pi}{4} = \dfrac{\pi}{2}$		$\left(\dfrac{\pi}{2}, 0.5\right)$
2nd Inflection Point:	$\dfrac{\pi}{2} + \dfrac{\pi}{4} = \dfrac{2\pi}{4} + \dfrac{\pi}{4} = \dfrac{3\pi}{4}$		$\left(\dfrac{3\pi}{4}, 0\right)$
Ending:	$\dfrac{3\pi}{4} + \dfrac{\pi}{4} = \dfrac{4\pi}{4} = \pi$		$(\pi, -0.5)$

Begin by plotting the graph of $y = -\frac{1}{2}\cos(2x)$ (**red graph**).

Reflect the **red graph** over the x-axis to get the **blue graph**, $y = -\frac{1}{2}\cos(2x)$.

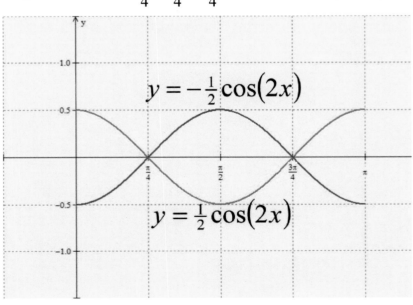

►EXAMPLE 7

Graph $y = 3\cos\left(2x + \dfrac{\pi}{6}\right)$ over one period. Label the fourths.

►►Solution

We will find the amplitude, period, fourths, beginning, ending, reflection, phase shift, and vertical shift.

Amplitude: $\left|3\right| = 3$ **Period:** $\dfrac{2\pi}{b} = \dfrac{2\pi}{2} = \pi$

Beginning and Ending: $0 \le 2x + \dfrac{\pi}{6} \le 2\pi$ Beginning: $-\dfrac{\pi}{12}$ Ending: $\dfrac{11\pi}{12}$

$$-\dfrac{\pi}{12} \le x \le \dfrac{11\pi}{12}$$

Phase Shift: $\dfrac{\pi}{12}$ left   **Vertical Shift:** None   **Reflection:** None

Fourths:

$y = \tfrac{1}{2}\cos(2x)$

Beginning:	$-\dfrac{\pi}{12}$	$\left(-\dfrac{\pi}{12}, 3\right)$
1ˢᵗ Inflection Point:	$-\dfrac{\pi}{12} + \dfrac{\pi}{4} = -\dfrac{\pi}{12} + \dfrac{3\pi}{12} = \dfrac{2\pi}{12} = \dfrac{\pi}{6}$	$\left(\dfrac{\pi}{6}, 0\right)$
Minimum:	$\dfrac{\pi}{6} + \dfrac{\pi}{4} = \dfrac{2\pi}{12} + \dfrac{3\pi}{12} = \dfrac{5\pi}{12}$	$\left(\dfrac{5\pi}{12}, -3\right)$
2ⁿᵈ Inflection Point:	$\dfrac{5\pi}{12} + \dfrac{\pi}{4} = \dfrac{5\pi}{12} + \dfrac{3\pi}{12} = \dfrac{8\pi}{12} = \dfrac{2\pi}{3}$	$\left(\dfrac{2\pi}{3}, 0\right)$
Ending:	$\dfrac{2\pi}{3} + \dfrac{\pi}{4} = \dfrac{8\pi}{12} + \dfrac{3\pi}{12} = \dfrac{11\pi}{12}$	$\left(\dfrac{11\pi}{12}, 3\right)$

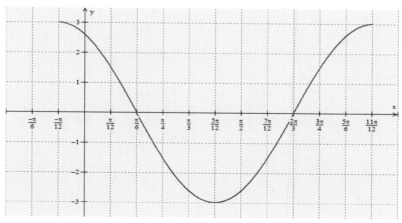

▶ **EXAMPLE 8**

Graph $y = 2\cos\left(\dfrac{1}{2}x - \pi\right) + 2$ **over one period. Label the fourths.**

▶▶ **Solution**

We will find the amplitude, period, fourths, beginning, ending, reflection, phase shift, and vertical shift.

Amplitude: $|2| = 2$ **Period:** $\dfrac{2\pi}{b} = \dfrac{2\pi}{\dfrac{1}{2}} = 2\pi \cdot 2 = 4\pi$

Beginning and Ending:
$$0 \le \frac{1}{2}x - \pi \le 2\pi$$
$$2\pi \le x \le 6\pi$$

Beginning: 2π Ending: 6π

Phase Shift: 2π right   **Vertical Shift:** Up 2   **Reflection:** None

Fourths:

$$\frac{Period}{4} = \frac{4\pi}{4} = \pi$$

Beginning:	2π	$(2\pi, 4)$
1st Inflection Point:	$2\pi + \pi = 3\pi$	$(3\pi, 2)$
Minimum:	$3\pi + \pi = 4\pi$	$(4\pi, 0)$
2nd Inflection Point:	$4\pi + \pi = 5\pi$	$(5\pi, 2)$
Ending:	$5\pi + \pi = 6\pi$	$(6\pi, 4)$

Since we have a vertical shift up of 2, we will shift the *x*-axis up 2 and dot it in red. Graph

$$y = 2\cos\left(\tfrac{1}{2}x - \pi\right) + 2$$

using the red dotted line as the *x*-axis.

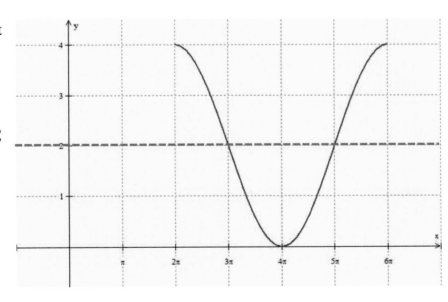

►EXAMPLE 9

Graph $y = 150\cos(4x - \pi) + 250$ over one period. Label the fourths.

►►Solution

We will find the amplitude, period, fourths, beginning, ending, reflection, phase shift, and vertical shift.

Amplitude: $|150| = 150$ **Period:** $\dfrac{2\pi}{b} = \dfrac{2\pi}{4} = \dfrac{\pi}{2}$

Beginning and Ending: $0 \le 4x - \pi \le 2\pi$ **Beginning:** $\dfrac{\pi}{4}$ **Ending:** $\dfrac{3\pi}{4}$

$$\dfrac{\pi}{4} \le x \le \dfrac{3\pi}{4}$$

Phase Shift: $\dfrac{\pi}{4}$ right   **Vertical Shift:** Up 250   **Reflection:** None

Fourths:

$$\dfrac{Period}{4} = \dfrac{\dfrac{\pi}{2}}{4} = \dfrac{\pi}{2} \cdot \dfrac{1}{4} = \dfrac{\pi}{8}$$

Beginning: $\dfrac{\pi}{4}$ $\left(\dfrac{\pi}{4}, 400\right)$

1st Inflection Point: $\dfrac{\pi}{4} + \dfrac{\pi}{8} = \dfrac{2\pi}{8} + \dfrac{\pi}{8} = \dfrac{3\pi}{8}$ $\left(\dfrac{3\pi}{8}, 250\right)$

Minimum: $\dfrac{3\pi}{8} + \dfrac{\pi}{8} = \dfrac{4\pi}{8} = \dfrac{\pi}{2}$ $\left(\dfrac{\pi}{2}, 100\right)$

2nd Inflection Point: $\dfrac{\pi}{2} + \dfrac{\pi}{8} = \dfrac{4\pi}{8} + \dfrac{\pi}{8} = \dfrac{5\pi}{8}$ $\left(\dfrac{5\pi}{8}, 250\right)$

Ending: $\dfrac{5\pi}{8} + \dfrac{\pi}{8} = \dfrac{6\pi}{8} = \dfrac{3\pi}{4}$ $\left(\dfrac{3\pi}{4}, 400\right)$

Since we have a vertical shift up of 250, we will shift the x-axis up 250 and dot it. The amplitude is 150, make equal marks from 250 on the y-axis, one above and one below the dotted line. Label your fourths and graph the cosine curve.

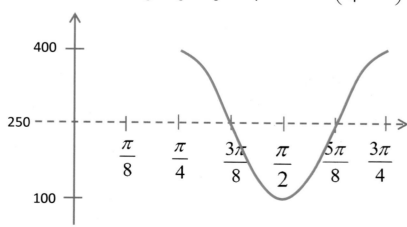

► EXAMPLE 10

Graph $y = 2\cos\left(2x - \dfrac{\pi}{4}\right) + 2$ over two periods. Label the fourths.

►►Solution

We will find the amplitude, period, fourths, beginning, ending, reflection, phase shift, and vertical shift for one period.

Amplitude: $|2| = 2$ **Period:** $\dfrac{2\pi}{b} = \dfrac{2\pi}{2} = \pi$

Beginning and Ending: $0 \leq 2x - \dfrac{\pi}{4} \leq 2\pi$ Beginning: $\dfrac{\pi}{8}$ Ending: $\dfrac{9\pi}{8}$

$$\dfrac{\pi}{8} \leq x \leq \dfrac{9\pi}{8}$$

Phase Shift: $\dfrac{\pi}{8}$ right   **Vertical Shift:** Up 2   **Reflection:** None

Fourths: Beginning: $\dfrac{\pi}{8}$ $\left(\dfrac{\pi}{8}, 4\right)$

$\dfrac{Period}{4} = \dfrac{\pi}{4}$ 1ˢᵗ Inflection Point: $\dfrac{\pi}{8} + \dfrac{\pi}{4} = \dfrac{\pi}{8} + \dfrac{2\pi}{8} = \dfrac{3\pi}{8}$ $\left(\dfrac{3\pi}{8}, 2\right)$

Minimum: $\dfrac{3\pi}{8} + \dfrac{\pi}{4} = \dfrac{3\pi}{8} + \dfrac{2\pi}{8} = \dfrac{5\pi}{8}$ $\left(\dfrac{5\pi}{8}, 0\right)$

2ⁿᵈ Inflection Point: $\dfrac{5\pi}{8} + \dfrac{\pi}{4} = \dfrac{5\pi}{8} + \dfrac{2\pi}{8} = \dfrac{7\pi}{8}$ $\left(\dfrac{7\pi}{8}, 2\right)$

Ending: $\dfrac{7\pi}{8} + \dfrac{\pi}{4} = \dfrac{7\pi}{8} + \dfrac{2\pi}{8} = \dfrac{9\pi}{8}$ $\left(\dfrac{9\pi}{8}, 4\right)$

Use the information and graph the first period then continue the graph for one more period.

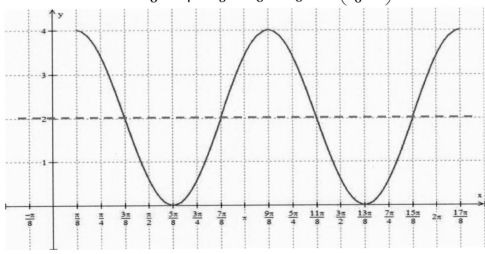

Section Wrap Up

$$y = \cos x$$

Domain: $(-\infty, \infty)$

Range: $[-1, 1]$

Even Function
$$\cos(-x) = \cos x$$

$$y = a\cos(bx \pm c) \pm d$$

$a < 0$ Reflection on the x-axis

$|a| < 1$ Vertical Shrinkage of a

$|a| > 1$ Vertical Stretching of a

$$Period = \frac{2\pi}{b}$$

$$Fourths = \frac{Period}{4}$$

$$Phase\ Shift = \frac{c}{b}$$

Beginning and Ending: $0 \le bx \pm c \le 2\pi$

Vertical Shift: $\pm d$

HOMEWORK

Objective 1

Determine the amplitude, period and phase shift of each sine function.

See Examples 1, 2and 3.

1. $y = -4\cos 5x$

2. $y = 3\cos\left(2x - \dfrac{\pi}{4}\right)$

3. $y = \tfrac{1}{4}\cos(x - 2\pi)$

4. $y = 10\cos\left(\dfrac{1}{3}x + \pi\right)$

5. $y = -25\cos\left(4x + \dfrac{\pi}{2}\right)$

6. $y = \tfrac{1}{2}\cos(3x - \pi)$

7. $y = 0.2\cos\left(\dfrac{1}{2}x + \dfrac{\pi}{8}\right)$

8. $y = 2\cos 8x$

Objective 2

Determine the fourths for each sine function. *See Example 4.*

9. $y = -\cos 2x$

10. $y = 2\cos\left(x + \dfrac{\pi}{6}\right)$

11. $y = \tfrac{1}{2}\cos(2x + \pi)$

12. $y = 10\cos\left(\dfrac{1}{2}x - 2\pi\right)$

Objective 3

Graph the following sine functions over one period. Label the fourths. *See Examples 5 - 9.*

13. $y = -\cos 3x$

14. $y = -3\cos\left(2x - \dfrac{\pi}{2}\right)$

15. $y = \tfrac{1}{2}\cos(x + \pi)$

16. $y = \cos(x + \pi) + 1$

17. $y = 25\cos\left(2x + \dfrac{\pi}{2}\right) + 10$

18. $y = \tfrac{1}{2}\cos(3x - \pi)$

19. $y = 0.2\cos\left(\dfrac{1}{2}x + \dfrac{\pi}{8}\right)$

20. $y = 2\cos 4x$

21. $y = -\cos\left(2x + \dfrac{\pi}{6}\right)$

22. $y = -2\cos\left(x - \dfrac{\pi}{2}\right)$

23. $y = \tfrac{1}{2}\cos\left(x - \dfrac{\pi}{3}\right)$

24. $y = \cos(x + \pi) - 2$

25. $y = 100\cos\left(x - \dfrac{\pi}{4}\right) + 75$

26. $y = \tfrac{1}{2}\cos x + 2$

27. $y = \cos\left(4x - \dfrac{\pi}{8}\right)$

28. $y = 3\cos 4x - 3$

Objective 3

Graph the following sine functions over two periods. Label the fourths. *See Example 10.*

29. $y = 2\cos 2x$

30. $y = -\cos\left(x - \dfrac{\pi}{2}\right)$

31. $y = \tfrac{1}{2}\cos\left(x - \dfrac{\pi}{4}\right)$

32. $y = 2\cos(x + \pi) - 1$

33. $y = 3\cos\left(x + \dfrac{\pi}{2}\right) + 1$

34. $y = \tfrac{1}{2}\cos\left(2x - \dfrac{\pi}{4}\right)$

ANSWERS

1. amplitude: 4, period: $\dfrac{2\pi}{5}$, phase shift: None

2. amplitude: 3, period: π, phase shift: $\dfrac{\pi}{8}$ right

3. amplitude: $\dfrac{1}{4}$, period: 2π, phase shift: 2π right

4. amplitude: 10, period: 6π, phase shift: 3π left

5. amplitude: 25, period: $\dfrac{\pi}{2}$, phase shift: $\dfrac{\pi}{8}$ left

6. amplitude: $\dfrac{1}{2}$, period: $\dfrac{2\pi}{3}$, phase shift: $\dfrac{\pi}{3}$ right

7. amplitude: 0.2, period: 4π, phase shift: $\dfrac{\pi}{4}$ left

8. amplitude: 2, period: $\dfrac{\pi}{4}$, phase shift: None

9. $0, \dfrac{\pi}{4}, \dfrac{\pi}{2}, \dfrac{3\pi}{4}, \pi$

10. $-\dfrac{\pi}{6}, \dfrac{\pi}{3}, \dfrac{5\pi}{6}, \dfrac{4\pi}{3}, \dfrac{11\pi}{6}$

11. $-\dfrac{\pi}{2}, -\dfrac{\pi}{4}, 0, \dfrac{\pi}{4}, \dfrac{\pi}{2}$

12. $4\pi, 5\pi, 6\pi, 7\pi, 8\pi$

13.

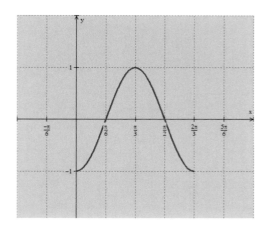

14.

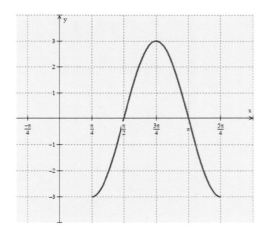

15.

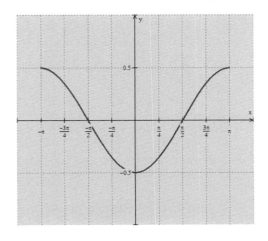

16.

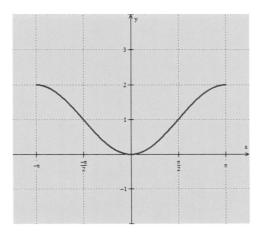

17.

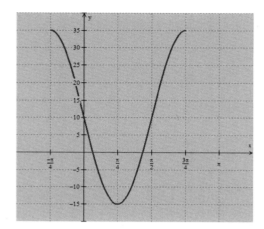

18.

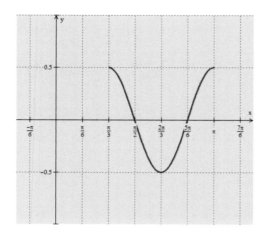

19.

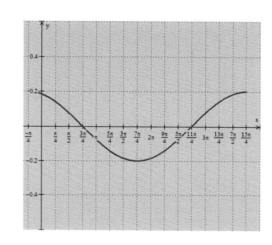

20.

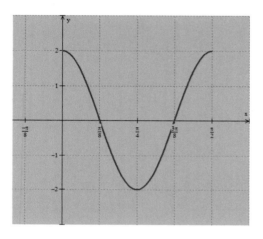

21.

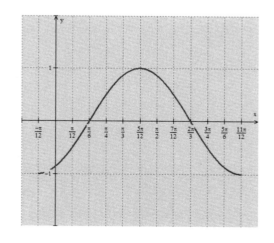

22.

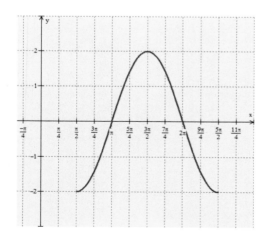

23.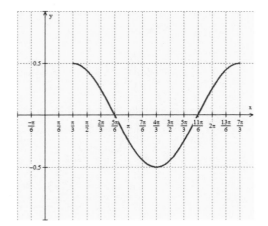

24.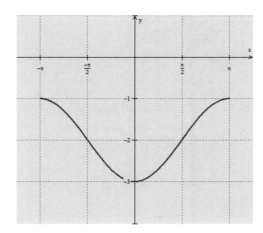

25.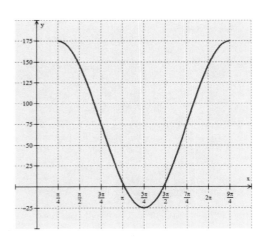

26.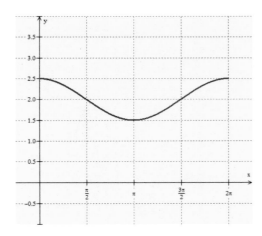

27.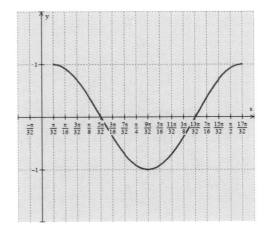

28.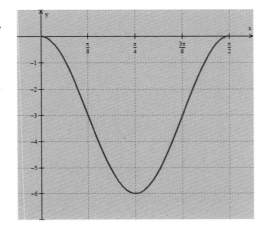

29.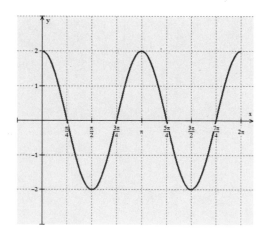

30.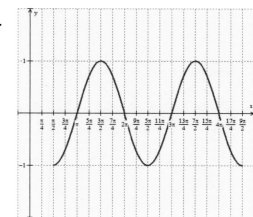

31.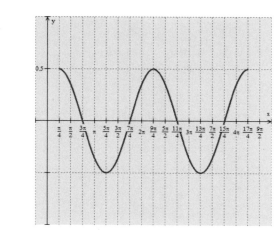

32.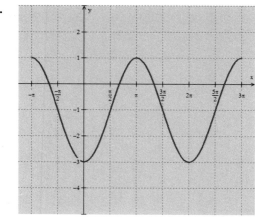

33.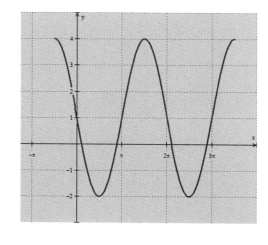

34.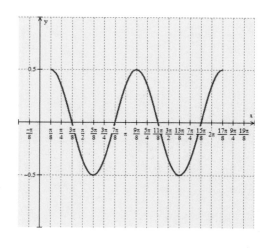

Section 10: Graph of the Secant and Cosecant Functions

Learning Outcomes:

- The student will correctly memorize and apply trigonometric formulas, definitions, identities, and properties.
- The student will illustrate and examine the graphs of: the six trigonometric functions and their inverses, parametric equations, and polar equations.

Objectives: At the conclusion of this lesson you should be able to:

1. Graph the cosecant function.
2. Graph the secant function.

The graphs of secant and cosecant can be obtained from the graphs of the sine and cosine functions using the reciprocal identities $\csc x = \dfrac{1}{\sin x}$ and $\sec x = \dfrac{1}{\cos x}$. At given values of x, the y-values of cosecant are the reciprocal y-values of sine and the y-values of secant are the reciprocal y-values of cosine. When $\sin x = 0$, the reciprocal is undefined (does not exist), the cosecant graph will have vertical asymptotes. The same applies when $\cos x = 0$, the reciprocal is undefined (does not exist), the secant graph will have vertical asymptotes.

To sketch the graph of a cosecant or secant function, begin by graphing its reciprocal function. In **Figure 10.1** we have graphed the sine function over one period. **In Figure 10.2** we have graphed the sine function in blue and the cosecant function in red. Notice the asymptotes occur where sine touches or crosses the x-axis.

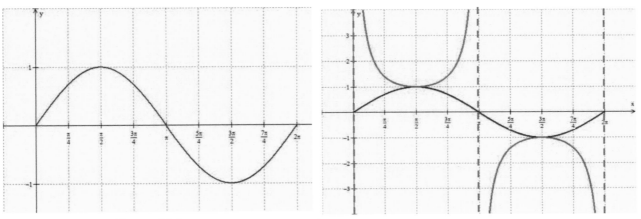

Figure 10.1   **Figure 10.2**

To sketch the graph of a secant function, begin by graphing its reciprocal function cosine. In **Figure 10.3** we have graphed the cosine function over one period. **In Figure 10.4** we have graphed the cosine function in blue and the secant function in red. Notice that the vertical asymptotes occur where cosine crosses or touches the *x*-axis.

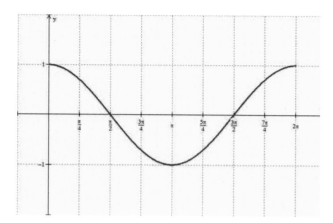

Figure 10.3

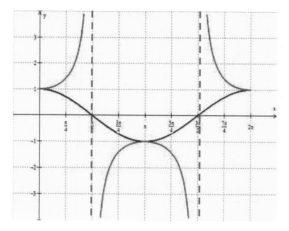

Figure 10.4

If we look at a table comparison for sine and its reciprocal cosecant and cosine and its reciprocal secant, we have the following table. In the table, U=undefined.

x	0	$\dfrac{\pi}{6}$	$\dfrac{\pi}{4}$	$\dfrac{\pi}{3}$	$\dfrac{\pi}{2}$	$\dfrac{2\pi}{3}$	$\dfrac{3\pi}{4}$	$\dfrac{5\pi}{6}$	π	$\dfrac{7\pi}{6}$	$\dfrac{5\pi}{4}$	$\dfrac{4\pi}{3}$	$\dfrac{3\pi}{2}$	$\dfrac{5\pi}{3}$	$\dfrac{7\pi}{4}$	$\dfrac{11\pi}{6}$	2π
$\sin x$	0	$\dfrac{1}{2}$	$\dfrac{\sqrt{2}}{2}$	$\dfrac{\sqrt{3}}{2}$	1	$\dfrac{\sqrt{3}}{2}$	$\dfrac{\sqrt{2}}{2}$	$\dfrac{1}{2}$	0	$-\dfrac{1}{2}$	$-\dfrac{\sqrt{2}}{2}$	$-\dfrac{\sqrt{3}}{2}$	-1	$-\dfrac{\sqrt{3}}{2}$	$-\dfrac{\sqrt{2}}{2}$	$-\dfrac{1}{2}$	0
$\csc x$	U	2	$\sqrt{2}$	$\dfrac{2\sqrt{3}}{3}$	1	$\dfrac{2\sqrt{3}}{3}$	$\sqrt{2}$	2	U	-2	$-\sqrt{2}$	$-\dfrac{2\sqrt{3}}{3}$	-1	$-\dfrac{2\sqrt{3}}{3}$	$-\sqrt{2}$	-2	U
$\cos x$	1	$\dfrac{\sqrt{3}}{2}$	$\dfrac{\sqrt{2}}{2}$	$\dfrac{1}{2}$	0	$-\dfrac{1}{2}$	$-\dfrac{\sqrt{2}}{2}$	$-\dfrac{\sqrt{3}}{2}$	-1	$-\dfrac{\sqrt{3}}{2}$	$-\dfrac{\sqrt{2}}{2}$	$-\dfrac{1}{2}$	0	$\dfrac{1}{2}$	$\dfrac{\sqrt{2}}{2}$	$\dfrac{\sqrt{3}}{2}$	1
$\sec x$	1	$\dfrac{2\sqrt{3}}{3}$	$\sqrt{2}$	2	U	-2	$-\sqrt{2}$	$-\dfrac{2\sqrt{3}}{3}$	-1	$-\dfrac{2\sqrt{3}}{3}$	$-\sqrt{2}$	-2	U	2	$\sqrt{2}$	$\dfrac{2\sqrt{3}}{3}$	1

When we compare the graphs and table of the cosecant and secant functions with those of their reciprocals sine and cosine, notice the local maximums and local minimums are interchanged. For example, a local maximum on the sine curve becomes a local minimum on the cosecant curve and a local minimum on the sine curve becomes a local maximum on the cosecant curve.

► EXAMPLE 1

Sketch the graph of $y = 2\sec\left(3x + \dfrac{\pi}{2}\right)$ **over one period.**

►►Solution

Begin by graphing the cosine function: $y = 2\cos\left(3x + \dfrac{\pi}{2}\right)$.

Amplitude: $|2| = 2$ 　　　　　　　**Period:** $\dfrac{2\pi}{b} = \dfrac{2\pi}{3}$

Beginning and Ending: $0 \le 3x + \dfrac{\pi}{2} \le 2\pi$ 　　　Beginning: $-\dfrac{\pi}{6}$ 　　Ending: $\dfrac{\pi}{2}$

$$-\dfrac{\pi}{6} \le x \le \dfrac{\pi}{2}$$

Phase Shift: $\dfrac{\pi}{6}$ left 　　　　**Vertical Shift:** None 　　　　**Reflection:** None

Fourths:

$\dfrac{Period}{4} = \dfrac{\dfrac{2\pi}{3}}{4} = \dfrac{\pi}{6}$

Beginning: 　　　　　　$-\dfrac{\pi}{6}$ 　　　　　　$\left(-\dfrac{\pi}{6}, 2\right)$

1st Inflection Point: 　$-\dfrac{\pi}{6} + \dfrac{\pi}{6} = 0$ 　　　$(0, 0)$

Minimum: 　　　　　　$0 + \dfrac{\pi}{6} = \dfrac{\pi}{6}$ 　　　　$\left(\dfrac{\pi}{6}, -2\right)$

2nd Inflection Point: 　$\dfrac{\pi}{6} + \dfrac{\pi}{6} = \dfrac{2\pi}{6} = \dfrac{\pi}{3}$ 　　$\left(\dfrac{\pi}{3}, 0\right)$

Ending: 　　　　　　$\dfrac{\pi}{3} + \dfrac{\pi}{6} = \dfrac{2\pi}{3} + \dfrac{\pi}{6} = \dfrac{3\pi}{6} = \dfrac{\pi}{2}$ 　$\left(\dfrac{\pi}{2}, 2\right)$

The red graph is the graph of

$y = 2\cos\left(3x + \dfrac{\pi}{2}\right)$. We

obtain the blue graph by
graphing the asymptotes
where cosine is equal to zero
and sketching the graph in the
opposite direction of the red
graph. The blue graph is the

graph of $y = 2\sec\left(3x + \dfrac{\pi}{2}\right)$.

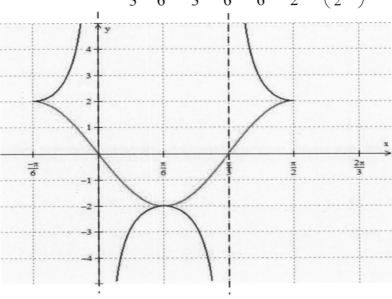

►EXAMPLE 2

Graph $y = 2\csc\left(2x - \dfrac{\pi}{3}\right)$ **over one period. Label the fourths.**

►►**Solution**

Begin by graphing the cosine function: $y = 2\sin\left(2x - \dfrac{\pi}{3}\right)$.

Amplitude: $|2| = 2$ **Period:** $\dfrac{2\pi}{b} = \dfrac{2\pi}{2} = \pi$

Beginning and Ending: $0 \le 2x - \dfrac{\pi}{3} \le 2\pi$ **Beginning:** $\dfrac{\pi}{6}$ **Ending:** $\dfrac{7\pi}{6}$

$$\dfrac{\pi}{6} \le x \le \dfrac{7\pi}{6}$$

Phase Shift: $\dfrac{\pi}{6}$ right   **Vertical Shift:** None   **Reflection:** None

Fourths:

$\dfrac{Period}{4} = \dfrac{\pi}{4} = \dfrac{\pi}{4}$

Beginning: $\dfrac{\pi}{6}$ $\left(\dfrac{\pi}{6}, 0\right)$

Maximum: $\dfrac{\pi}{6} + \dfrac{\pi}{4} = \dfrac{2\pi}{12} + \dfrac{3\pi}{12} = \dfrac{5\pi}{12}$ $\left(\dfrac{5\pi}{12}, 2\right)$

Half Period: $\dfrac{5\pi}{12} + \dfrac{\pi}{4} = \dfrac{5\pi}{12} + \dfrac{3\pi}{12} = \dfrac{8\pi}{12} = \dfrac{2\pi}{3}$ $\left(\dfrac{2\pi}{3}, 0\right)$

Minimum: $\dfrac{2\pi}{3} + \dfrac{\pi}{4} = \dfrac{8\pi}{12} + \dfrac{3\pi}{12} = \dfrac{11\pi}{12}$ $\left(\dfrac{11\pi}{12}, -2\right)$

Ending: $\dfrac{11\pi}{12} + \dfrac{\pi}{4} = \dfrac{11\pi}{12} + \dfrac{3\pi}{12} = \dfrac{14\pi}{12} = \dfrac{7\pi}{6}$ $\left(\dfrac{7\pi}{6}, 0\right)$

The red graph is the graph of

$y = 2\sin\left(2x - \dfrac{\pi}{3}\right)$. We obtain

the blue graph by graphing
the asymptotes where sine is
equal to zero and sketching
the graph in the opposite
direction of the red graph.
The blue graph is the graph

of $y = 2\csc\left(2x - \dfrac{\pi}{3}\right)$.

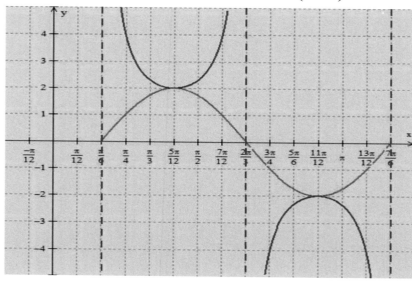

▶ EXAMPLE 3

Sketch the graph of $y = -2\sec\left(x - \dfrac{\pi}{4}\right)$ **over one period.**

▶▶ **Solution**

Begin by graphing the cosine function: $y = -2\cos\left(x - \dfrac{\pi}{4}\right)$.

Amplitude: $\left|-2\right| = 2$ **Period:** $\dfrac{2\pi}{b} = \dfrac{2\pi}{1} = 2\pi$

Beginning and Ending: $0 \le x - \dfrac{\pi}{4} \le 2\pi$ Beginning: $\dfrac{\pi}{4}$ Ending: $\dfrac{9\pi}{4}$

$$\dfrac{\pi}{4} \le x \le \dfrac{9\pi}{4}$$

Phase Shift: $\dfrac{\pi}{4}$ right **Vertical Shift:** None **Reflection:** x-axis

Fourths: Beginning: $\dfrac{\pi}{4}$ $\left(\dfrac{\pi}{4}, -2\right)$

$\dfrac{Period}{4} = \dfrac{2\pi}{4} = \dfrac{\pi}{2}$ 1st Inflection Point: $\dfrac{\pi}{4} + \dfrac{\pi}{2} = \dfrac{\pi}{4} + \dfrac{2\pi}{4} = \dfrac{3\pi}{4}$ $\left(\dfrac{3\pi}{4}, 0\right)$

Maximum: $\dfrac{3\pi}{4} + \dfrac{\pi}{2} = \dfrac{3\pi}{4} + \dfrac{2\pi}{4} = \dfrac{5\pi}{4}$ $\left(\dfrac{5\pi}{4}, 2\right)$

2nd Inflection Point: $\dfrac{5\pi}{4} + \dfrac{\pi}{2} = \dfrac{5\pi}{4} + \dfrac{2\pi}{4} = \dfrac{7\pi}{4}$ $\left(\dfrac{7\pi}{4}, 0\right)$

Ending: $\dfrac{7\pi}{4} + \dfrac{\pi}{2} = \dfrac{7\pi}{4} + \dfrac{2\pi}{4} = \dfrac{9\pi}{4}$ $\left(\dfrac{9\pi}{4}, -2\right)$

The red graph is the graph of

$y = -2\cos\left(x - \dfrac{\pi}{4}\right)$. We obtain

the blue graph by graphing the asymptotes where cosine is equal to zero and sketching the graph in the opposite direction of the red graph. The blue graph is the graph of

$y = -2\sec\left(x - \dfrac{\pi}{4}\right)$.

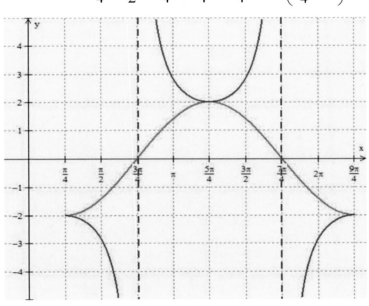

► EXAMPLE 4

Graph $y = -\dfrac{1}{2}\csc\left(x+\dfrac{\pi}{4}\right)$ **over one period. Label the fourths.**

►►**Solution**

Begin by graphing the cosine function: $y = -\dfrac{1}{2}\sin\left(x+\dfrac{\pi}{4}\right)$.

Amplitude: $\left|-\dfrac{1}{2}\right| = \dfrac{1}{2}$

Period: $\dfrac{2\pi}{b} = \dfrac{2\pi}{1} = 2\pi$

Beginning and Ending: $0 \le x + \dfrac{\pi}{4} \le 2\pi$

$-\dfrac{\pi}{4} \le x \le \dfrac{7\pi}{4}$

Beginning: $-\dfrac{\pi}{4}$ **Ending:** $\dfrac{7\pi}{4}$

Phase Shift: $\dfrac{\pi}{4}$ left   **Vertical Shift:** None   **Reflection:** x-axis

Fourths:

$\dfrac{Period}{4} = \dfrac{2\pi}{4} = \dfrac{\pi}{2}$

Beginning: $-\dfrac{\pi}{4}$ $\left(-\dfrac{\pi}{4},0\right)$

Minimum: $-\dfrac{\pi}{4} + \dfrac{\pi}{2} = -\dfrac{\pi}{4} + \dfrac{2\pi}{4} = \dfrac{\pi}{4}$ $\left(\dfrac{\pi}{4},-0.5\right)$

Half Period: $\dfrac{\pi}{4} + \dfrac{\pi}{2} = \dfrac{\pi}{4} + \dfrac{2\pi}{4} = \dfrac{3\pi}{4}$ $\left(\dfrac{3\pi}{4},0\right)$

Maximum: $\dfrac{3\pi}{4} + \dfrac{\pi}{2} = \dfrac{3\pi}{4} + \dfrac{2\pi}{4} = \dfrac{5\pi}{4}$ $\left(\dfrac{5\pi}{4},0.5\right)$

Ending: $\dfrac{5\pi}{4} + \dfrac{\pi}{2} = \dfrac{5\pi}{4} + \dfrac{2\pi}{4} = \dfrac{7\pi}{4}$ $\left(\dfrac{7\pi}{4},0\right)$

The red graph is the graph of

$y = -\dfrac{1}{2}\sin\left(x+\dfrac{\pi}{4}\right)$. We obtain

the blue graph by graphing the
asymptotes where sine is equal
to zero and sketching the graph
in the opposite direction of the
red graph. The blue graph is the

graph of $y = -\dfrac{1}{2}\csc\left(x+\dfrac{\pi}{4}\right)$.

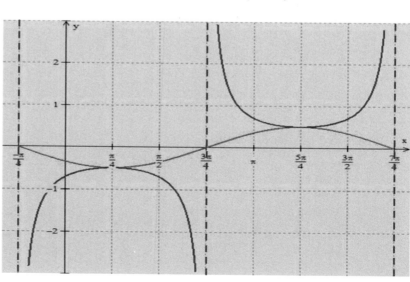

► EXAMPLE 5

Graph $y = 2\csc\left(x - \dfrac{\pi}{6}\right) + 4$ **over one period. Label the fourths.**

►►**Solution**

Begin by graphing the cosine function: $y = 2\sin\left(x - \dfrac{\pi}{6}\right) + 4$.

Amplitude: $|2| = 2$ **Period:** $\dfrac{2\pi}{b} = \dfrac{2\pi}{1} = 2\pi$

Beginning and Ending: $0 \le x - \dfrac{\pi}{6} \le 2\pi$ **Beginning:** $\dfrac{\pi}{6}$ **Ending:** $\dfrac{13\pi}{6}$

$$\dfrac{\pi}{6} \le x \le \dfrac{13\pi}{6}$$

Phase Shift: $\dfrac{\pi}{6}$ right **Vertical Shift:** up 4 **Reflection:** None

Fourths:

$\dfrac{Period}{4} = \dfrac{2\pi}{4} = \dfrac{\pi}{2}$

Beginning: $\dfrac{\pi}{6}$ $\left(\dfrac{\pi}{6}, 4\right)$

Maximum: $\dfrac{\pi}{6} + \dfrac{\pi}{2} = \dfrac{\pi}{6} + \dfrac{3\pi}{6} = \dfrac{4\pi}{6} = \dfrac{2\pi}{3}$ $\left(\dfrac{2\pi}{3}, 6\right)$

Half Period: $\dfrac{2\pi}{3} + \dfrac{\pi}{2} = \dfrac{4\pi}{6} + \dfrac{3\pi}{6} = \dfrac{7\pi}{6}$ $\left(\dfrac{7\pi}{6}, 4\right)$

Minimum: $\dfrac{7\pi}{6} + \dfrac{\pi}{2} = \dfrac{7\pi}{6} + \dfrac{3\pi}{6} = \dfrac{10\pi}{6} = \dfrac{5\pi}{3}$ $\left(\dfrac{5\pi}{3}, 2\right)$

Ending: $\dfrac{5\pi}{3} + \dfrac{\pi}{2} = \dfrac{10\pi}{6} + \dfrac{3\pi}{6} = \dfrac{13\pi}{6}$ $\left(\dfrac{13\pi}{6}, 4\right)$

The red graph is the graph of

$y = 2\sin\left(x - \dfrac{\pi}{6}\right) + 4$. We obtain

the blue graph by graphing the asymptotes where sine is equal to zero and sketching the graph in the opposite direction of the red graph. The blue graph is the

graph of $y = 2\csc\left(x - \dfrac{\pi}{6}\right) + 4$.

The dotted green line shows the vertical shift.

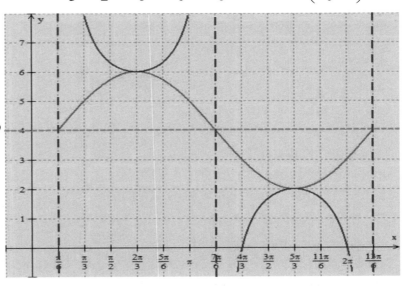

▶EXAMPLE 6

Sketch the graph of $y = 2\sec(2x - \pi) - 3$ over one period.

▶▶Solution

Begin by graphing the cosine function: $y = 2\cos(2x - \pi) - 3$.

Amplitude: $|2| = 2$ **Period:** $\dfrac{2\pi}{b} = \dfrac{2\pi}{2} = \pi$

Beginning and Ending:
$$0 \le 2x - \pi \le 2\pi$$
$$\frac{\pi}{2} \le x \le \frac{3\pi}{2}$$
    Beginning: $\dfrac{\pi}{2}$ **Ending:** $\dfrac{3\pi}{2}$

Phase Shift: $\dfrac{\pi}{2}$ right   **Vertical Shift:** Down 3   **Reflection:** None

Fourths:

$\dfrac{Period}{4} = \dfrac{\pi}{4}$

Beginning:	$\dfrac{\pi}{2}$	$\left(\dfrac{\pi}{2}, -1\right)$
1ˢᵗ Inflection Point:	$\dfrac{\pi}{2} + \dfrac{\pi}{4} = \dfrac{2\pi}{4} + \dfrac{\pi}{4} = \dfrac{3\pi}{4}$	$\left(\dfrac{3\pi}{4}, -3\right)$
Minimum:	$\dfrac{3\pi}{4} + \dfrac{\pi}{4} = \dfrac{4\pi}{4} = \pi$	$(\pi, -5)$
2ⁿᵈ Inflection Point:	$\pi + \dfrac{\pi}{4} = \dfrac{4\pi}{4} + \dfrac{\pi}{4} = \dfrac{5\pi}{4}$	$\left(\dfrac{5\pi}{4}, -3\right)$
Ending:	$\dfrac{5\pi}{4} + \dfrac{\pi}{4} = \dfrac{6\pi}{4} = \dfrac{3\pi}{2}$	$\left(\dfrac{3\pi}{2}, -1\right)$

The red graph is the graph of

$y = 2\cos(2x - \pi) - 3$. We obtain the blue graph by graphing the asymptotes where cosine is equal to zero and sketching the graph in the opposite direction of the red graph. The blue graph is the graph of $y = 2\sec(2x - \pi) - 3$. The dotted green line shows the vertical shift.

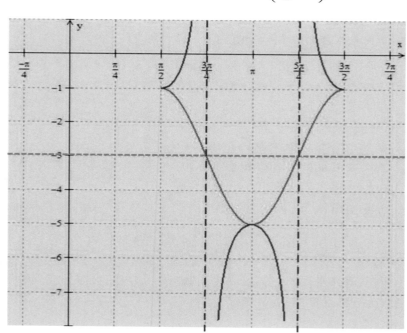

HOMEWORK

Objectives 1and 2

Graph the following cosecant and secant functions over one period. Label the fourths.

See Examples 1-6.

1. $y = -\csc 3x$

2. $y = -3\sec\left(2x - \dfrac{\pi}{2}\right)$

3. $y = \tfrac{1}{2}\sec(x + \pi)$

4. $y = \sec(x + \pi) + 1$

5. $y = 2\csc\left(2x + \dfrac{\pi}{2}\right) + 1$

6. $y = \tfrac{1}{2}\sec(3x - \pi)$

7. $y = 3\sec\left(\dfrac{1}{2}x + \dfrac{\pi}{8}\right)$

8. $y = 2\csc 4x$

9. $y = -\csc\left(2x + \dfrac{\pi}{6}\right)$

10. $y = -2\sec\left(x - \dfrac{\pi}{2}\right)$

11. $y = \tfrac{1}{2}\csc\left(x - \dfrac{\pi}{3}\right)$

12. $y = \sec(x + \pi) - 2$

13. $y = 3\csc\left(x - \dfrac{\pi}{4}\right)$

14. $y = \tfrac{1}{2}\csc x + 2$

15. $y = \csc\left(4x - \dfrac{\pi}{8}\right)$

16. $y = 3\sec 4x - 3$

ANSWERS

1.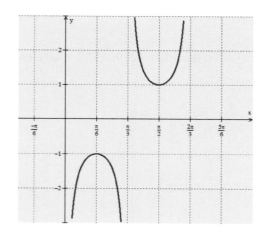

2.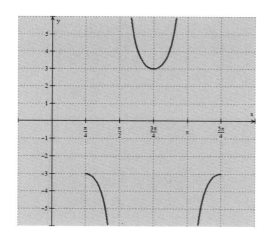

3.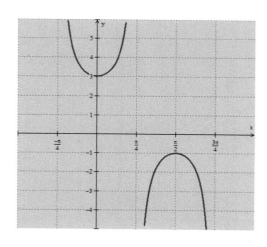

4.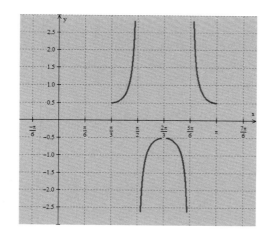

5.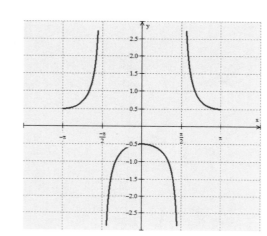

6.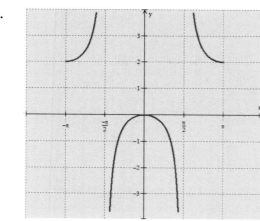

7.

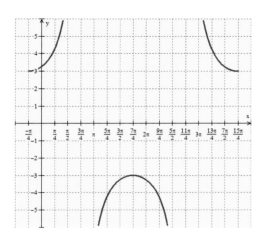

8.

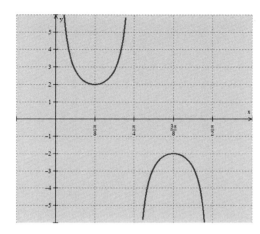

9.

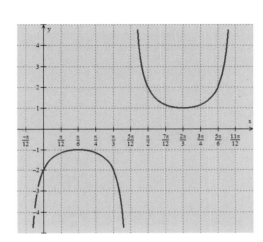

10.

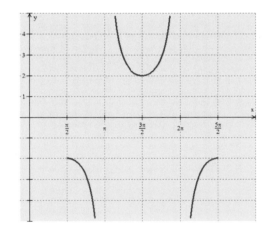

11.

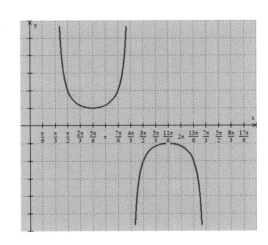

12.

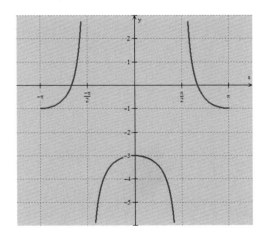

13.

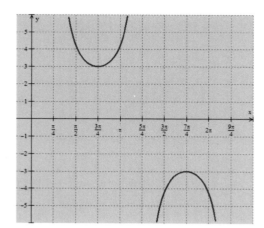

14.

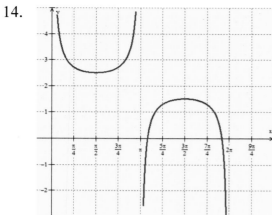

15.

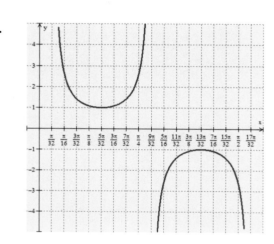

16.

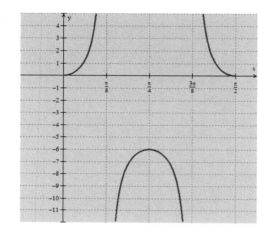

Section 11: Graph of the Tangent and Cotangent Functions

Learning Outcomes:

- The student will correctly memorize and apply trigonometric formulas, definitions, identities, and properties.
- The student will illustrate and examine the graphs of: the six trigonometric functions and their inverses, parametric equations, and polar equations.

Objectives: At the conclusion of this lesson you should be able to:

1.   Graph the tangent function.
2.   Graph the cotangent function.

We have graphed four of the six trigonometric functions in the previous three sections. In this section we will graph the remaining two functions, tangent and cotangent. The graphs of tangent and cotangent functions look very different from those of sine, cosine, cosecant, and secant functions.

Graphing Tangent

Since $\tan x = \dfrac{\sin x}{\cos x}$, the graph of tangent will be undefined when the denominator cosine is equal to 0. Cosine is equal to 0 for all x-values in the following set:

$$\left\{ \ldots -\frac{5\pi}{2}, -\frac{3\pi}{2}, -\frac{\pi}{2}, \frac{\pi}{2}, \frac{3\pi}{2}, \frac{5\pi}{2}, \ldots \right\}$$ and therefore the tangent function is undefined and will have

vertical asymptotes at these values. The domain of $y = \tan x$ is all real numbers except $x = \dfrac{\pi}{2} + k\pi$, for any integer k.

The vertical asymptotes for $y = \tan x$ are $x = \dfrac{\pi}{2} + k\pi$, for any integer k. Consider the graph of $y = \tan x$ over the interval $\left(-\dfrac{3\pi}{2}, \dfrac{3\pi}{2} \right)$ as shown in **Figure 11.1**. Notice the tangent graph repeats itself in intervals of π and we can conclude that the period of tangent is π.

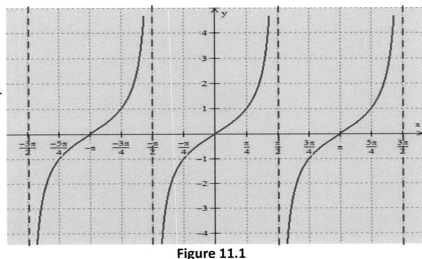

Figure 11.1

If we look at $y = \tan x$ over the interval $\left(-\dfrac{\pi}{2}, \dfrac{\pi}{2}\right)$ we have the following table:

x	$-\dfrac{\pi}{2}$	$-\dfrac{\pi}{3}$	$-\dfrac{\pi}{4}$	$-\dfrac{\pi}{6}$	0	$\dfrac{\pi}{6}$	$\dfrac{\pi}{4}$	$\dfrac{\pi}{3}$	$\dfrac{\pi}{2}$
$\tan x$	undefined	$-\sqrt{3}$	-1	$-\dfrac{\sqrt{3}}{3}$	0	$\dfrac{\sqrt{3}}{3}$	1	$\sqrt{3}$	undefined

If we plot the values from the table we have **Figure 11.2**. The graph of tangent over one period resembles that of the x^3 graph from College Algebra. Also notice that the graph of tangent has symmetry with respect to the origin which implies $\tan(-x) = -\tan x$ and that tangent is odd. When graphing the sine and cosine functions, we considered 5 key points that divided their graphs into fourths over one period. When graphing tangent we consider the beginning asymptote, the key points $\left(-\dfrac{\pi}{4}, -1\right)$, $(0,0)$, $\left(\dfrac{\pi}{4}, 1\right)$ and the ending asymptote which divide the graph of the tangent function into fourths.

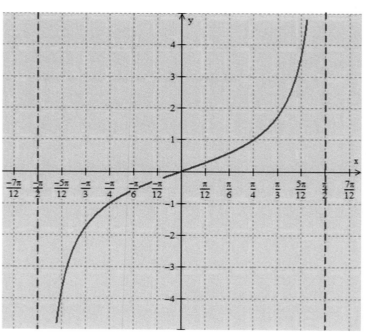

Figure 11.2

Now that we have graphed the basic tangent curve, we will study the effects of the constants **a, b, c,** and **d** in the general tangent equation: $y = a\tan(bx \pm c) \pm d$.

The Constant a

The a in the general tangent equation represents a vertical stretching if $|a| > 1$ or vertical shrinkage if $|a| < 1$. If a is negative, $a < 0$, the tangent graph will have a reflection on the x-axis as shown in **Figure 11.3**. The **red graph** is the graph of $y = \tan x$ and the **blue graph** is the graph of $y = -\tan x$. Notice that the graph of $y = \tan x$ is increasing and the graph of $y = -\tan x$ is decreasing.

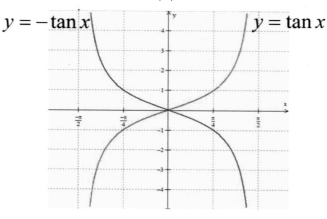

Figure 11.3

In **Figure 11.4** the graphs of $y = 2\tan x$ (**red graph**), $y = \tan x$ (**blue graph**), and $y = \frac{1}{2}\tan x$ (**green graph**). The basic tangent graph is the **blue graph,** $y = \tan x$. Notice the **red graph,** $y = 2\tan x$ has a vertical stretching of 2 and the **green graph,** $y = \frac{1}{2}\tan x$ has a vertical shrinkage of $\frac{1}{2}$.

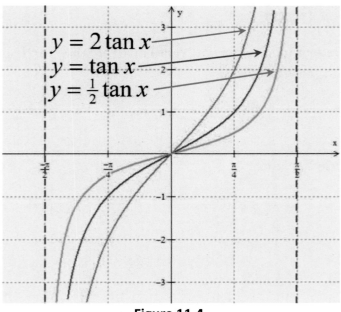

The Constant b

The period of $y = \tan x$ is π when $b = 1$:
$$-\frac{\pi}{2} < x < \frac{\pi}{2}.$$

Figure 11.4

The period of $y = \tan 2x$ is $\dfrac{\pi}{2}$ and the period of $y = \tan\frac{1}{2}x$ is 2π as shown in **Figure 11.5**.

$$-\frac{\pi}{2} < 2x < \frac{\pi}{2}$$

$$\frac{1}{2}\left(-\frac{\pi}{2}\right) < \frac{1}{2}(2x) < \frac{1}{2}\left(\frac{\pi}{2}\right)$$

$$-\frac{\pi}{4} < x < \frac{\pi}{4}$$

$$\text{Period} = \frac{\pi}{2}$$

$$-\frac{\pi}{2} < \frac{1}{2}x < \frac{\pi}{2}$$

$$2\left(-\frac{\pi}{2}\right) < 2\left(\frac{1}{2}x\right) < 2\left(\frac{\pi}{2}\right)$$

$$-\pi < x < \pi$$

$$\text{Period} = 2\pi$$

Figure 11.5

Since the period is π when $b = 1$, we can find the period of other tangent functions by using $Period = \dfrac{\pi}{b}$, $p = \dfrac{\pi}{b}$. If $b > 1$ the tangent graph has a horizontal shrinkage and if $b < 1$ the tangent graph has a horizontal stretching.

▶EXAMPLE 1

Determine the period, vertical stretching/vertical shrinkage and reflection of the following tangent functions.

a.) $y = 3\tan 6x$ b.) $y = -2\tan\frac{1}{4}x$ c.) $y = \frac{1}{2}\tan 3x$

▶▶**Solution**

a.) $y = 3\tan 6x$ $p = \dfrac{\pi}{6}$ The period is $\dfrac{\pi}{6}$.

Stretching of 3

No Reflection

b.) $y = -2\tan\frac{1}{4}x$ $p = \dfrac{\pi}{\frac{1}{4}} = \pi \cdot 4 = 4\pi$ The period is 4π.

Stretching of 2

Reflection on the x-axis

c.) $y = \frac{1}{2}\tan 3x$ $p = \dfrac{\pi}{3}$ The period is $\dfrac{\pi}{3}$.

Shrinkage of $\dfrac{1}{2}$.

No Reflection

The Constant c

The constant c is used in determining the phase shift. Rewriting $y = a\tan(bx \pm c) \pm d$ by factoring out the b, we have $y = a\tan\left[b\left(x \pm \dfrac{c}{b}\right)\right] \pm d$ where $\dfrac{c}{b}$ is the phase shift for the point where $x = 0$. When $\dfrac{c}{b} > 0$ there is a phase shift to the left $\dfrac{c}{b}$ and when $\dfrac{c}{b} < 0$ there is a phase shift to the right $\dfrac{c}{b}$.

The Constant d

The constant d tells us the vertical shift of the graph. If $d > 0$ we have a vertical shift up of d, if $d < 0$ we have vertical shift down of d.

► EXAMPLE 2

Sketch the graph of $y = 2\tan\left(3x - \dfrac{\pi}{6}\right)$ **over one period.**

►► Solution

Vertical Stretching: $|2| = 2$ **Period:** $\dfrac{\pi}{b} = \dfrac{\pi}{3}$

Beginning and Ending: $-\dfrac{\pi}{2} < 3x - \dfrac{\pi}{6} < \dfrac{\pi}{2}$ **Beginning Asymptote:** $x = -\dfrac{\pi}{9}$

$$-\dfrac{\pi}{9} < x < \dfrac{2\pi}{9}$$ **Ending Asymptote:** $x = \dfrac{2\pi}{9}$

Phase Shift: $\dfrac{\pi}{18}$ right **Vertical Shift:** None **Reflection:** None

Fourths: **Beginning Asymptote:** $-\dfrac{\pi}{9}$ $x = -\dfrac{\pi}{9}$

$\dfrac{Period}{4} = \dfrac{\dfrac{\pi}{3}}{4} = \dfrac{\pi}{12}$

1st Key Point: $-\dfrac{\pi}{9} + \dfrac{\pi}{12} = -\dfrac{4\pi}{36} + \dfrac{3\pi}{36} = -\dfrac{\pi}{36}$ $\left(-\dfrac{\pi}{36}, -2\right)$

2nd Key Point: $-\dfrac{\pi}{36} + \dfrac{\pi}{12} = -\dfrac{\pi}{36} + \dfrac{3\pi}{36} = \dfrac{2\pi}{36} = \dfrac{\pi}{18}$ $\left(\dfrac{\pi}{18}, 0\right)$

3rd Key Point: $\dfrac{\pi}{18} + \dfrac{\pi}{12} = \dfrac{2\pi}{36} + \dfrac{3\pi}{36} = \dfrac{5\pi}{36}$ $\left(\dfrac{5\pi}{36}, 2\right)$

Ending Asymptote: $\dfrac{5\pi}{36} + \dfrac{\pi}{12} = \dfrac{5\pi}{36} + \dfrac{3\pi}{36} = \dfrac{8\pi}{36} = \dfrac{2\pi}{9}$ $x = \dfrac{2\pi}{9}$

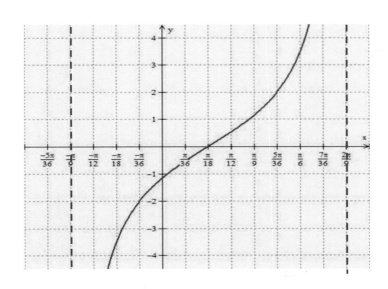

▶EXAMPLE 3

Graph $y = -\tan\left(x + \dfrac{\pi}{4}\right)$ over one period. Label the fourths.

▶▶Solution

Vertical Stretching/Shrinkage: None   **Period:** $\dfrac{\pi}{b} = \dfrac{\pi}{1} = \pi$

Beginning and Ending: $-\dfrac{\pi}{2} < x + \dfrac{\pi}{4} < \dfrac{\pi}{2}$ Beginning Asymptote: $x = -\dfrac{3\pi}{4}$

$$-\dfrac{3\pi}{4} < x < \dfrac{\pi}{4}$$ Ending Asymptote: $x = \dfrac{\pi}{4}$

Phase Shift: $\dfrac{\pi}{4}$ left   **Vertical Shift:** None   **Reflection:** x-axis

Fourths: Beginning Asymptote: $-\dfrac{3\pi}{4}$ $x = -\dfrac{3\pi}{4}$

$\dfrac{Period}{4} = \dfrac{\pi}{4}$ 1ˢᵗ Key Point: $-\dfrac{3\pi}{4} + \dfrac{\pi}{4} = -\dfrac{2\pi}{4} = -\dfrac{\pi}{2}$ $\left(-\dfrac{\pi}{2}, 1\right)$

          2ⁿᵈ Key Point: $-\dfrac{\pi}{2} + \dfrac{\pi}{4} = -\dfrac{2\pi}{4} + \dfrac{\pi}{4} = -\dfrac{\pi}{4}$ $\left(-\dfrac{\pi}{4}, 0\right)$

          3ʳᵈ Key Point: $-\dfrac{\pi}{4} + \dfrac{\pi}{4} = 0$ $(0, -1)$

          Ending Asymptote: $0 + \dfrac{\pi}{4} = \dfrac{\pi}{4}$ $x = \dfrac{\pi}{4}$

The graph of $y = \tan\left(x + \dfrac{\pi}{4}\right)$ is

the **red graph** and the blue graph

is the graph of $y = -\tan\left(x + \dfrac{\pi}{4}\right)$.

Notice the blue graph is the
reflection on the red graph.

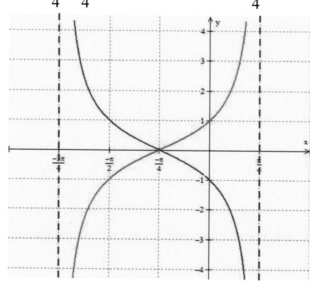

▶ EXAMPLE 4

Graph $y = \frac{1}{2}\tan\left(2x - \frac{\pi}{8}\right) + 2$ **over one period. Label the fourths.**

▶▶ Solution

Vertical Shrinkage: $\left|\frac{1}{2}\right| = \frac{1}{2}$

Period: $\frac{\pi}{b} = \frac{\pi}{2}$

Beginning and Ending:

$$-\frac{\pi}{2} < 2x - \frac{\pi}{8} < \frac{\pi}{2}$$

$$-\frac{3\pi}{16} < x < \frac{5\pi}{16}$$

Beginning Asymptote: $x = -\frac{3\pi}{16}$

Ending Asymptote: $x = \frac{5\pi}{16}$

Phase Shift: $\frac{\pi}{16}$ right   **Vertical Shift:** Up 2   **Reflection:** None

Fourths:

$$\frac{Period}{4} = \frac{\frac{\pi}{2}}{4} = \frac{\pi}{2} \cdot \frac{1}{4} = \frac{\pi}{8}$$

Beginning Asymptote: $-\frac{3\pi}{16}$ $x = -\frac{3\pi}{16}$

1st Key Point: $-\frac{3\pi}{16} + \frac{\pi}{8} = -\frac{3\pi}{16} + \frac{2\pi}{16} = -\frac{\pi}{16}$ $\left(-\frac{\pi}{16}, -0.5\right)$

2nd Key Point: $-\frac{\pi}{16} + \frac{\pi}{8} = -\frac{\pi}{16} + \frac{2\pi}{16} = \frac{\pi}{16}$ $\left(\frac{\pi}{16}, 0\right)$

3rd Key Point: $\frac{\pi}{16} + \frac{\pi}{8} = \frac{\pi}{16} + \frac{2\pi}{16} = \frac{3\pi}{16}$ $\left(\frac{3\pi}{16}, 0.5\right)$

Ending Asymptote: $\frac{3\pi}{16} + \frac{\pi}{8} = \frac{3\pi}{16} + \frac{2\pi}{16} = \frac{5\pi}{16}$ $x = \frac{5\pi}{16}$

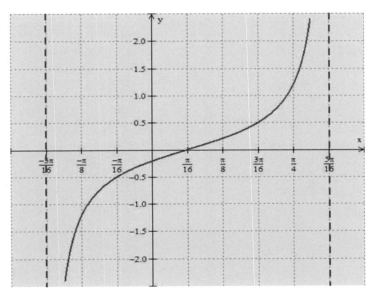

Graphing Cotangent

Since $\cot x = \dfrac{\cos x}{\sin x}$, the graph of cotangent will be undefined when the denominator sine is equal to 0. Sine is equal to 0 for all x-values in the following set: $\{\ldots -3\pi, -2\pi, -\pi, 0, \pi, 2\pi, 3\pi, \ldots\}$ and therefore the cotangent function is undefined and will have vertical asymptotes at these values. The domain of $y = \cot x$ is all real numbers except $x = k\pi$, for any integer k. The vertical asymptotes for $y = \cot x$ are $x = k\pi$, for any integer k. Consider the graph of $y = \cot x$ over the interval $(-\pi, 2\pi)$ as shown in **Figure 11.6**. Notice the cotangent graph repeats itself in intervals of π and we can conclude that the period of cotangent is π.

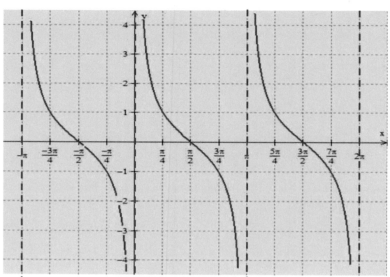

Figure 11.6

If we look at $y = \cot x$ over the interval $(0, \pi)$ we have the following table:

x	0	$\dfrac{\pi}{6}$	$\dfrac{\pi}{4}$	$\dfrac{\pi}{3}$	$\dfrac{\pi}{2}$	$\dfrac{2\pi}{3}$	$\dfrac{3\pi}{4}$	$\dfrac{5\pi}{6}$	π
$\cot x$	*undefined*	$\sqrt{3}$	1	$\dfrac{\sqrt{3}}{3}$	0	$-\dfrac{\sqrt{3}}{3}$	-1	$-\sqrt{3}$	*undefined*

Plotting the values from the table we have **Figure 11.7**. When graphing cotangent we consider the beginning asymptote, the key points $\left(\dfrac{\pi}{4}, 1\right)$, $\left(\dfrac{\pi}{2}, 0\right)$, $\left(\dfrac{3\pi}{4}, -1\right)$ and the ending asymptote which divide the graph of the cotangent function into fourths.

Now that we have graphed the basic cotangent curve, we will study the effects of the constants **a**, **b**, **c**, and **d** in the general cotangent equation:

$$y = a\cot(bx \pm c) \pm d.$$

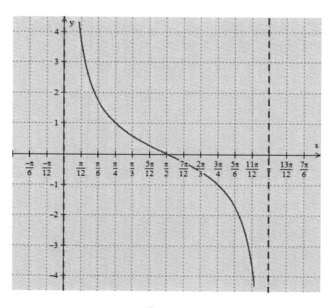

Figure 11.6

The Constant a

The a in the general cotangent equation represents a vertical stretching if $|a| > 1$ or vertical shrinkage if $|a| < 1$. If a is negative, $a < 0$, the cotangent graph will have a reflection on the x-axis as shown in **Figure 11.7**. The **pink graph** is the graph of $y = \cot x$ and the **green graph** is the graph of $y = -\cot x$. Notice that the graph of $y = -\cot x$ is increasing and the graph of $y = \cot x$ is decreasing.

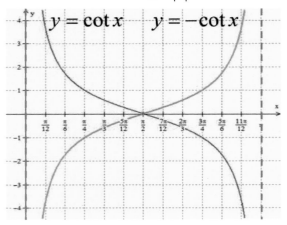

Figure 11.7

In **Figure 11.8** we see the graphs of $y = 2 \cot x$ (gold graph), $y = \cot x$ (blue graph), and $y = \frac{1}{2} \cot x$ (pink graph). The basic tangent graph is the **blue graph**, $y = \cot x$. Notice the gold graph, $y = 2 \cot x$ has a vertical stretching of 2 and the pink graph, $y = \frac{1}{2} \cot x$ has a vertical shrinkage of $\frac{1}{2}$.

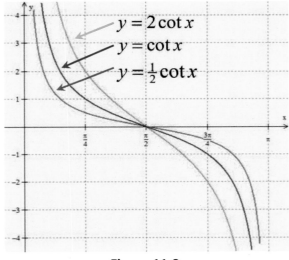

Figure 11.8

The Constant b

The period of $y = \cot x$ is π when $b = 1$: $0 < x < \pi$. The period of $y = \cot 2x$ is $\dfrac{\pi}{2}$ and the period of $y = \cot \frac{1}{2} x$ is 2π as shown in **Figure 11.9**.

$$0 < 2x < \pi \qquad\qquad 0 < \frac{1}{2}x < \pi$$

$$\frac{1}{2}(0) < \frac{1}{2}(2x) < \frac{1}{2}(\pi) \qquad 2(0) < 2\left(\frac{1}{2}x\right) < 2(\pi)$$

$$0 < x < \frac{\pi}{2} \qquad\qquad 0 < x < 2\pi$$

$$\text{Period} = \frac{\pi}{2} \qquad\qquad \text{Period} = 2\pi$$

Figure 11.9

Since the period is π when $b = 1$, we can find the period of other cotangent functions by using $Period = \dfrac{\pi}{b}$, $p = \dfrac{\pi}{b}$. If $b > 1$ the cotangent graph has a horizontal shrinkage and if $b < 1$ the cotangent graph has a horizontal stretching.

► EXAMPLE 5

Determine the period, vertical stretching/vertical shrinkage and reflection of the following cotangent functions.

a.) $y = -2\cot 4x$ b.) $y = 4\cot \tfrac{1}{3} x$ c.) $y = \tfrac{1}{3}\cot 5x$

►► Solution

a.) $y = -2\cot 4x$ $p = \dfrac{\pi}{4}$ The period is $\dfrac{\pi}{4}$.

             Stretching of 2
             Reflection on the x-axis

b.) $y = 4\cot \tfrac{1}{3} x$ $p = \dfrac{\pi}{\frac{1}{3}} = \pi \cdot 3 = 3\pi$ The period is 3π.

             Stretching of 4
             No Reflection

c.) $y = \tfrac{1}{3}\cot 5x$ $p = \dfrac{\pi}{5}$ The period is $\dfrac{\pi}{5}$.

             Shrinkage of $\dfrac{1}{3}$.
             No Reflection

The Constant c

The constant c is used in determining the phase shift. Rewriting $y = a\cot(bx \pm c) \pm d$ by factoring out the b, we have $y = a\cot\left[b\left(x \pm \dfrac{c}{b}\right)\right] \pm d$ where $\dfrac{c}{b}$ is the phase shift. When $\dfrac{c}{b} > 0$ there is a phase shift to the left $\dfrac{c}{b}$ and when $\dfrac{c}{b} < 0$ there is a phase shift to the right $\dfrac{c}{b}$.

The Constant d

The constant d tells us the vertical shift of the graph. If $d > 0$ we have a vertical shift up of d, if $d < 0$ we have vertical shift down of d.

► EXAMPLE 6

Sketch the graph of $y = -\dfrac{1}{2}\cot\left(2x + \dfrac{\pi}{4}\right)$ **over one period.**

►►Solution

Vertical Shrinkage: $\left|-\dfrac{1}{2}\right| = \dfrac{1}{2}$ **Period:** $\dfrac{\pi}{b} = \dfrac{\pi}{2}$

Beginning and Ending: $0 < 2x + \dfrac{\pi}{4} < \pi$ Beginning Asymptote: $x = -\dfrac{\pi}{8}$

$$-\dfrac{\pi}{8} < x < \dfrac{3\pi}{8}$$ Ending Asymptote: $x = \dfrac{3\pi}{8}$

Phase Shift: $\dfrac{\pi}{8}$ left   **Vertical Shift:** None   **Reflection:** x-axis

Fourths:

$$\dfrac{Period}{4} = \dfrac{\dfrac{\pi}{2}}{4} = \dfrac{\pi}{8}$$

Beginning Asymptote: $-\dfrac{\pi}{8}$ $x = -\dfrac{\pi}{8}$

1st Key Point: $-\dfrac{\pi}{8} + \dfrac{\pi}{8} = 0$ $(0, -0.5)$

2nd Key Point: $0 + \dfrac{\pi}{8} = \dfrac{\pi}{8}$ $\left(\dfrac{\pi}{8}, 0\right)$

3rd Key Point: $\dfrac{\pi}{8} + \dfrac{\pi}{8} = \dfrac{2\pi}{8} = \dfrac{\pi}{4}$ $\left(\dfrac{\pi}{4}, 0.5\right)$

Ending Asymptote: $\dfrac{\pi}{4} + \dfrac{\pi}{8} = \dfrac{2\pi}{8} + \dfrac{\pi}{8} = \dfrac{3\pi}{8}$ $x = \dfrac{3\pi}{8}$

The **blue graph** is the graph of $y = \dfrac{1}{2}\cot\left(2x + \dfrac{\pi}{4}\right)$ and the **green graph** is the reflected graph $y = -\dfrac{1}{2}\cot\left(2x + \dfrac{\pi}{4}\right)$.

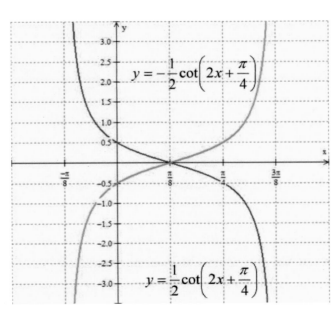

►EXAMPLE 7

Sketch the graph of $y = 2\cot\left(x - \dfrac{\pi}{6}\right) - 3$ over one period.

►►Solution

Vertical Stretching: $|2| = 2$ **Period:** $\dfrac{\pi}{b} = \dfrac{\pi}{1} = \pi$

Beginning and Ending: $0 < x - \dfrac{\pi}{6} < \pi$ **Beginning Asymptote:** $x = \dfrac{\pi}{6}$

$\dfrac{\pi}{6} < x < \dfrac{7\pi}{6}$ **Ending Asymptote:** $x = \dfrac{7\pi}{6}$

Phase Shift: $\dfrac{\pi}{6}$ right   **Vertical Shift:** Down 3   **Reflection:** None

Fourths:

$\dfrac{Period}{4} = \dfrac{\pi}{4}$

Beginning Asymptote: $\dfrac{\pi}{6}$ $x = \dfrac{\pi}{6}$

1ˢᵗ Key Point: $\dfrac{\pi}{6} + \dfrac{\pi}{4} = \dfrac{2\pi}{12} + \dfrac{3\pi}{4} = \dfrac{5\pi}{12}$ $\left(\dfrac{5\pi}{12}, -1\right)$

2ⁿᵈ Key Point: $\dfrac{5\pi}{12} + \dfrac{\pi}{4} = \dfrac{5\pi}{12} + \dfrac{3\pi}{4} = \dfrac{8\pi}{12} = \dfrac{2\pi}{3}$ $\left(\dfrac{2\pi}{3}, -3\right)$

3ʳᵈ Key Point: $\dfrac{2\pi}{3} + \dfrac{\pi}{4} = \dfrac{8\pi}{12} + \dfrac{3\pi}{4} = \dfrac{11\pi}{12}$ $\left(\dfrac{11\pi}{12}, -5\right)$

Ending Asymptote: $\dfrac{11\pi}{12} + \dfrac{\pi}{4} = \dfrac{11\pi}{12} + \dfrac{3\pi}{4} = \dfrac{14\pi}{12} = \dfrac{7\pi}{6}$ $x = \dfrac{7\pi}{6}$

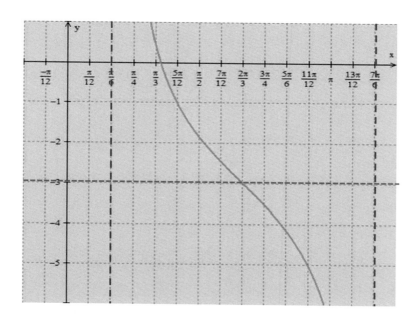

Section Wrap Up

$$y = a\tan(bx \pm c) \pm d$$

$a < 0$ Reflection on the x-axis

$|a| < 1$ Vertical Shrinkage of a

$|a| > 1$ Vertical Stretching of a

$Period = \dfrac{\pi}{b}$ $Fourths = \dfrac{Period}{4}$

$Phase\ Shift = \dfrac{c}{b}$, $x = 0$

Beginning and Ending Asymptotes

$$-\dfrac{\pi}{2} < bx \pm c < \dfrac{\pi}{2}$$

Vertical Shift: $\pm d$

Domain: $x \neq \dfrac{\pi}{2} + k\pi$, for any integer k.

Range: $(-\infty, \infty)$

Odd Function: $\tan(-x) = -\tan x$

$$y = a\cot(bx \pm c) \pm d$$

$a < 0$ Reflection on the x-axis

$|a| < 1$ Vertical Shrinkage of a

$|a| > 1$ Vertical Stretching of a

$Period = \dfrac{\pi}{b}$ $Fourths = \dfrac{Period}{4}$

$Phase\ Shift = \dfrac{c}{b}$

Beginning and Ending Asymptotes

$$0 < bx \pm c < \pi$$

Vertical Shift: $\pm d$

Domain: $x \neq k\pi$, for any integer k.

Range: $(-\infty, \infty)$

Odd Function: $\cot(-x) = -\cot x$

HOMEWORK

Objectives 1and 2

Graph the following tangent and cotangent functions over one period. Label the fourths.

See Examples 1-7.

1. $y = -\cot 2x$

2. $y = -2\tan\left(x - \dfrac{\pi}{2}\right)$

3. $y = \frac{1}{2}\tan(2x + \pi)$

4. $y = 2\cot(x - \pi) + 2$

5. $y = \tan\left(2x + \dfrac{\pi}{2}\right) + 1$

6. $y = \frac{1}{2}\cot(3x - \pi)$

7. $y = 3\cot\left(\dfrac{1}{2}x + \dfrac{\pi}{8}\right)$

8. $y = 2\tan 4x$

9. $y = -2\cot\left(2x + \dfrac{\pi}{6}\right)$

10. $y = -\tan(x + \pi)$

11. $y = 3\cot\left(x - \dfrac{\pi}{3}\right)$

12. $y = \tan\left(x + \dfrac{\pi}{2}\right) - 2$

13. $y = \cot\left(2x - \dfrac{\pi}{4}\right)$

14. $y = \frac{1}{2}\tan x + 2$

15. $y = \tan\left(4x - \dfrac{\pi}{8}\right)$

16. $y = 3\cot 2x - 3$

ANSWERS

1.
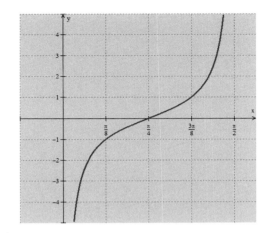

2.
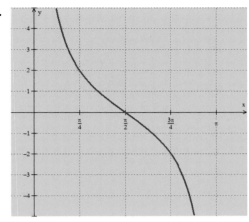

3.
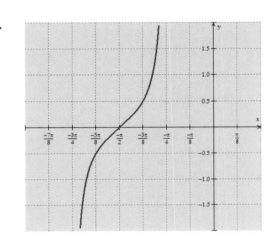

4.
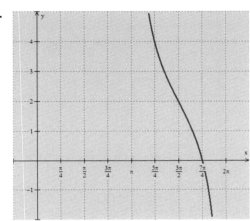

5.
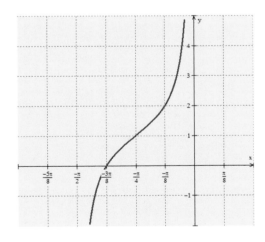

6.
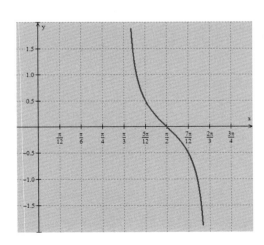

7.

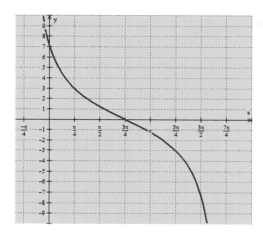

8.

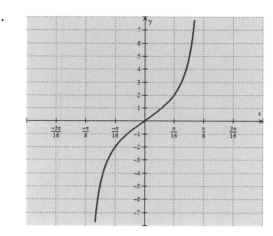

9.

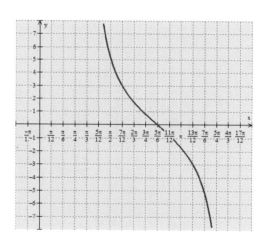

10.

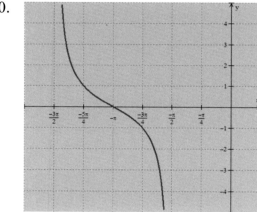

11.

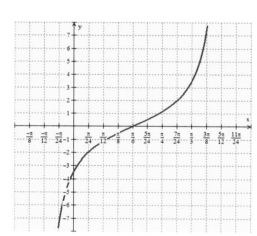

12.

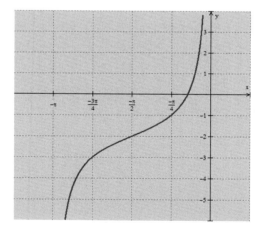

13.

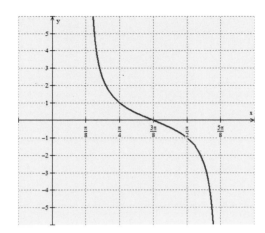

14.

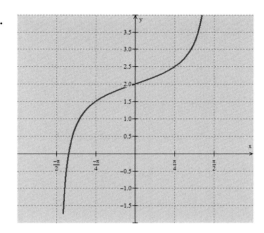

15.

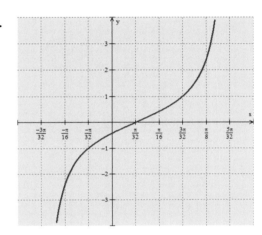

16.

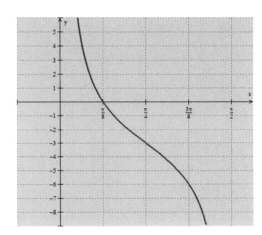

Section 12:   Inverse Trigonometric Functions

Learning Outcomes:

- The student will correctly memorize and apply trigonometric formulas, definitions, identities, and properties.
- The student will illustrate and examine the graphs of: the six trigonometric functions and their inverses, parametric equations, and polar equations.

Objectives: At the conclusion of this lesson you should be able to:

1.   Evaluate inverse trigonometric functions.
2.   Evaluate the compositions of trigonometric functions.

Recall the definition of a function: for each value in the domain, there is assigned exactly one range value. For a function to have an inverse, the function has to be one-to-one. A one-to-one function is unique in that for each value in the domain, there is assigned exactly one range value and for each value in the range, there is assigned exactly one domain value. In other words every element in the domain corresponds to only one element in the range and each element in the range corresponds to only one element in the domain. A graph of a one-to-one function has to pass the horizontal line test. If we look at the graph of sine, **Figure 12.1** we notice that sine does not pass the horizontal line test and is not one-to-one.

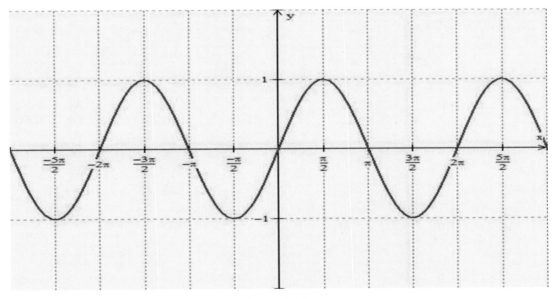

Figure 12.1

If we restrict the domain of sine to $\left[-\dfrac{\pi}{2},\dfrac{\pi}{2}\right]$ as shown in the sine graph in **Figure 12.2**, the sine

function is one-to-one.   We choose this interval since it is centrally located and the sine function attains all possible range values in the interval. Notice in the graph that the range is still $[-1,1]$. An equation for the inverse of $y=\sin x$,

$-\dfrac{\pi}{2}\le x\le\dfrac{\pi}{2}$ is obtained by interchanging x and y. The implicit form of the inverse function is $x=\sin y$, and the explicit form called the inverse sine of x is $\sin^{-1}x=y$ or $\arcsin x=y$, $-1\le x\le1$. The graph of the inverse sine, $\sin^{-1}x=y$, is shown in **Figure 12.3**. The domain and the range have been interchanged.

For $y=\sin x$ the domain is $\left[-\dfrac{\pi}{2},\dfrac{\pi}{2}\right]$ and the

range is $[-1,1]$ and for $\sin^{-1}x=y$ the domain is $[-1,1]$ and the range is $\left[-\dfrac{\pi}{2},\dfrac{\pi}{2}\right]$. The point

$\left(-1,-\dfrac{\pi}{2}\right)$ is on the graph of $\arcsin x=y$,

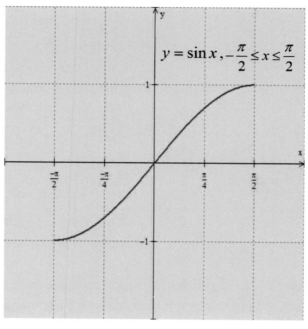

$$y = \sin x,\ -\frac{\pi}{2}\le x\le\frac{\pi}{2}$$

Figure 12.2

therefore $\arcsin(-1)=-\dfrac{\pi}{2}$. Alternatively we

can think of $\arcsin(-1)=y$ as "y is the number

(angle) in $\left[-\dfrac{\pi}{2},\dfrac{\pi}{2}\right]$ whose sine is -1."   We can

rewrite the equation $\arcsin(-1)=y$ as

$\sin y=-1$. Because $\sin\left(-\dfrac{\pi}{2}\right)=-1$ and

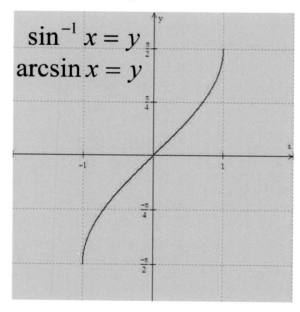

$$\sin^{-1}x = y$$
$$\arcsin x = y$$

Figure 12.3

$-\dfrac{\pi}{2}$ is in the range of the inverse sine function, $y=-\dfrac{\pi}{2}$.   Think of $\sin^{-1}x=y$ or

$\arcsin x=y$ as "y is the number (angle) in the interval $\left[-\dfrac{\pi}{2},\dfrac{\pi}{2}\right]$ whose sine is x."

Inverse Sine Function

$\sin^{-1} x = y$ or $\arcsin x = y$ means that $x = \sin y$ where $-1 \le x \le 1$ and $-\dfrac{\pi}{2} \le y \le \dfrac{\pi}{2}$

▶ EXAMPLE 1

Find the value of each real number y if it exists.

a.) $y = \arcsin \dfrac{1}{2}$ b.) $y = \sin^{-1}\left(-\dfrac{\sqrt{3}}{2}\right)$ c.) $y = \sin^{-1}(1)$

▶▶ Solution

a.)   Recognize that y is the angle in the interval $\left[-\dfrac{\pi}{2}, \dfrac{\pi}{2}\right]$ whose sine is $\dfrac{1}{2}$. Where on the

unit circle in the interval $\left[-\dfrac{\pi}{2}, \dfrac{\pi}{2}\right]$ does $\sin y = \dfrac{1}{2}$? Recall that the ordered pair at $\dfrac{\pi}{6}$ is

$\left(\dfrac{\sqrt{3}}{2}, \dfrac{1}{2}\right)$ where $\cos\dfrac{\pi}{6} = \dfrac{\sqrt{3}}{2}$ and $\sin\dfrac{\pi}{6} = \dfrac{1}{2}$. The angle y we are looking for is $\dfrac{\pi}{6}$, so

$\dfrac{\pi}{6} = \arcsin\dfrac{1}{2}$

b.)   Recognize that y is the angle in the interval $\left[-\dfrac{\pi}{2}, \dfrac{\pi}{2}\right]$ whose sine is $-\dfrac{\sqrt{3}}{2}$. Where on the

unit circle in the interval $\left[-\dfrac{\pi}{2}, \dfrac{\pi}{2}\right]$ does $\sin y = -\dfrac{\sqrt{3}}{2}$? Recall that the ordered pair at $\dfrac{\pi}{3}$ is

$\left(\dfrac{1}{2}, \dfrac{\sqrt{3}}{2}\right)$ where $\cos\dfrac{\pi}{3} = \dfrac{1}{2}$ and $\sin\dfrac{\pi}{3} = \dfrac{\sqrt{3}}{2}$. The angle in Quadrant I where $\dfrac{\sqrt{3}}{2}$ is

positive is $\dfrac{\pi}{3}$ and $-\dfrac{\sqrt{3}}{2}$ at $-\dfrac{\pi}{3}$ is in Quadrant IV. Therefore, $-\dfrac{\pi}{3} = \sin^{-1}\left(-\dfrac{\sqrt{3}}{2}\right)$.

c.)   Recognize that y is the angle in the interval $\left[-\dfrac{\pi}{2}, \dfrac{\pi}{2}\right]$ whose sine is 1. Where on the

unit circle in the interval $\left[-\dfrac{\pi}{2}, \dfrac{\pi}{2}\right]$ does $\sin y = 1$? Recall that the ordered pair at $\dfrac{\pi}{2}$ is

$(0,1)$ where $\cos\dfrac{\pi}{2} = 0$ and $\sin\dfrac{\pi}{2} = 1$. Therefore, $\dfrac{\pi}{2} = \sin^{-1}(1)$.

If we look at the graph of cosine, **Figure 12.4** we notice that cosine, like sine, does not pass the horizontal line test and is not one-to-one.

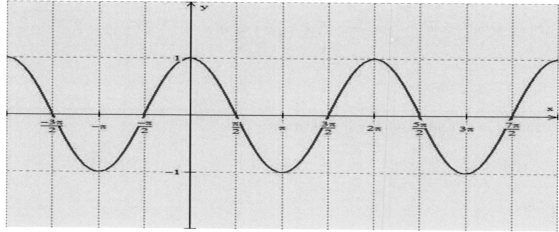

Figure 12.4

If we restrict the domain of cosine to $[0, \pi]$ as shown in the cosine graph in **Figure 12.5**, the cosine function is one-to-one on the interval $[0, \pi]$. We choose this interval because the cosine function attains all possible range values in the interval. Notice in the graph that the range is still $[-1,1]$. An equation for the inverse of $y = \cos x$, $0 \le x \le \pi$ is obtained by interchanging x and y. The implicit form of the inverse function is $x = \cos y$, and the explicit form called the inverse cosine of x is

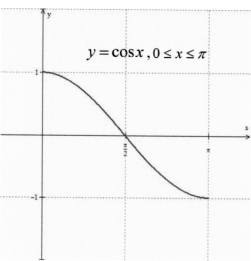

$y = \cos x, 0 \le x \le \pi$

Figure 12.5

$\cos^{-1} x = y$ or $\arccos x = y$, $-1 \le x \le 1$.

The graph of the inverse cosine, $\cos^{-1} x = y$, is shown in **Figure 12.6**. The domain and the range have been interchanged. For $y = \cos x$ the domain is $[0, \pi]$ and the range is $[-1,1]$ and for $\cos^{-1} x = y$ the domain is $[-1,1]$ and the range is $[0, \pi]$. As with inverse sine, think of $\cos^{-1} x = y$ or $\arccos x = y$ as "y is the number (angle) in the interval $[0, \pi]$ whose cosine is x."

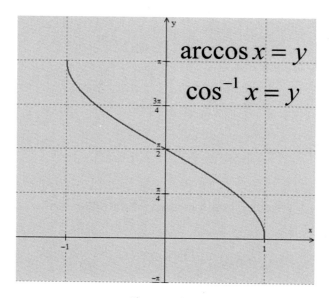

$\arccos x = y$

$\cos^{-1} x = y$

Figure 12.6

Inverse Cosine Function

$\cos^{-1} x = y$ or $\arccos x = y$ means that $x = \cos y, -1 \le x \le 1$ and $0 \le y \le \pi$

► EXAMPLE 2

Find the value of each real number y if it exists.

a.) $y = \arccos \dfrac{\sqrt{2}}{2}$ b.) $y = \cos^{-1}\left(-\dfrac{1}{2}\right)$ c.) $y = \cos^{-1}(-1)$

►►Solution

a.)   Recognize that y is the angle in the interval $[0, \pi]$ whose cosine is $\dfrac{\sqrt{2}}{2}$. Where on the

unit circle in the interval $[0, \pi]$ does $\cos y = \dfrac{\sqrt{2}}{2}$? Recall that the ordered pair at $\dfrac{\pi}{4}$ is

$\left(\dfrac{\sqrt{2}}{2}, \dfrac{\sqrt{2}}{2}\right)$ where $\cos \dfrac{\pi}{4} = \dfrac{\sqrt{2}}{2}$. The angle y we are looking for is $\dfrac{\pi}{4}$. Therefore,

$\dfrac{\pi}{4} = \arccos \dfrac{\sqrt{2}}{2}$

b.)   Recognize that y is the angle in the interval $[0, \pi]$ whose cosine is $-\dfrac{1}{2}$. Where on the

unit circle in the interval $[0, \pi]$ does $\cos y = -\dfrac{1}{2}$? Recall that the ordered pair at $\dfrac{\pi}{3}$ is

$\left(\dfrac{1}{2}, \dfrac{\sqrt{3}}{2}\right)$ where $\cos \dfrac{\pi}{3} = \dfrac{1}{2}$. The angle in Quadrant I where $\dfrac{1}{2}$ is positive is $\dfrac{\pi}{3}$ and $-\dfrac{1}{2}$ at

$\dfrac{2\pi}{3}$ in Quadrant II. Therefore, $\dfrac{2\pi}{3} = \cos^{-1}\left(-\dfrac{1}{2}\right)$.

c.)   Recognize that y is the angle in the interval $[0, \pi]$ whose cosine is -1. Where on the unit

circle in the interval $[0, \pi]$ does $\cos y = -1$? Recall that the ordered pair at π is $(-1, 0)$

where $\cos \pi = -1$. Therefore, $\pi = \cos^{-1}(1)$.

For tangent, we will restrict the domain to $\left(-\dfrac{\pi}{2}, \dfrac{\pi}{2}\right)$ in order to obtain a one-to-one function. The corresponding range is $(-\infty, \infty)$ as shown in **Figure 12.7**. An equation for the inverse of $y = \tan x$, $-\dfrac{\pi}{2} < x < \dfrac{\pi}{2}$ is obtained by interchanging x and y. The implicit form of the inverse function is $x = \tan y$, and the explicit form called the inverse tangent of x is $\tan^{-1} x = y$ or $\arctan x = y$, $-\infty < x < \infty$.

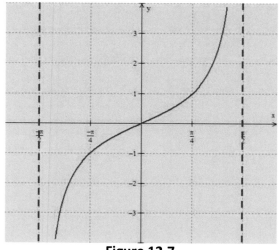

Figure 12.7

The graph of the inverse tangent, $\tan^{-1} x = y$, is shown in **Figure 12.8**. The domain and the range have been interchanged. For $y = \tan x$ the domain is $\left(-\dfrac{\pi}{2}, \dfrac{\pi}{2}\right)$ and the range is $(-\infty, \infty)$ and for $\tan^{-1} x = y$ the domain is $(-\infty, \infty)$ and the range is $\left(-\dfrac{\pi}{2}, \dfrac{\pi}{2}\right)$. As with inverse sine and cosine, think of $\tan^{-1} x = y$ or $\arctan x = y$ as "y is the number (angle) in the interval $\left(-\dfrac{\pi}{2}, \dfrac{\pi}{2}\right)$ whose tangent is x."

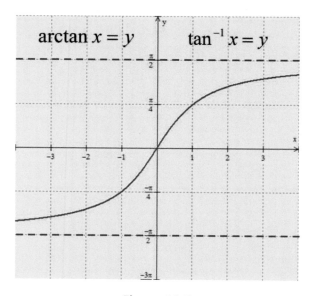

Figure 12.8

Inverse Tangent Function

$\tan^{-1} x = y$ or $\arctan x = y$ means that $x = \tan y$ where $-\infty < x < \infty$ and $-\dfrac{\pi}{2} < y < \dfrac{\pi}{2}$

► EXAMPLE 3

Find the value of each real number y if it exists.

a.) $y = \arctan 1$ b.) $y = \tan^{-1}\left(-\dfrac{\sqrt{3}}{3}\right)$ c.) $y = \tan^{-1} 0$

►►Solution

a.)   Where on the unit circle in the interval $\left(-\dfrac{\pi}{2}, \dfrac{\pi}{2}\right)$ does $\tan y = 1$? Recall that the ordered

pair at $\dfrac{\pi}{4}$ is $\left(\dfrac{\sqrt{2}}{2}, \dfrac{\sqrt{2}}{2}\right)$ where $\tan\dfrac{\pi}{4} = 1$. The angle y we are looking for is $\dfrac{\pi}{4}$.

Therefore, $\dfrac{\pi}{4} = \arctan 1$

b.)   Where on the unit circle in the interval $\left(-\dfrac{\pi}{2}, \dfrac{\pi}{2}\right)$ does $\tan y = -\dfrac{\sqrt{3}}{3}$? Recall that the

ordered pair at $\dfrac{\pi}{6}$ is $\left(\dfrac{\sqrt{3}}{2}, \dfrac{1}{2}\right)$ where $\tan\dfrac{\pi}{6} = \dfrac{\sqrt{3}}{3}$. The angle in Quadrant I is $\dfrac{\pi}{6}$ and

the angle in Quadrant IV is $-\dfrac{\pi}{6}$. Therefore, $-\dfrac{\pi}{6} = \tan^{-1}\left(-\dfrac{\sqrt{3}}{3}\right)$.

c.)   Where on the unit circle in the interval $\left(-\dfrac{\pi}{2}, \dfrac{\pi}{2}\right)$ does $\tan y = 0$? Recall that the ordered

pair at 0 is $(1,0)$ where $\tan \pi = 0$. Therefore, $0 = \tan^{-1} 0$.

The other three inverse trigonometric functions are defined similarly. We restrict the domain in order to obtain a one-to-one function. The inverse cotangent, cosecant, and secant functions are defined in the following chart.

Inverse Cotangent, Cosecant, and Secant Functions

$\cot^{-1} x = y$ or $arc\cot x = y$ means that $x = \cot y$ where $-\infty < x < \infty$ and $0 < y < \pi$

$\csc^{-1} x = y$ or $arc\csc x = y$ means that $x = \csc y$ where $|x| \geq 1$ and $-\dfrac{\pi}{2} \leq y \leq \dfrac{\pi}{2}, y \neq 0$

$\sec^{-1} x = y$ or $arc\sec x = y$ means that $x = \sec y$ where $|x| \geq 1$ and $0 \leq y \leq \pi, y \neq \dfrac{\pi}{2}$

Recall from College Algebra that $f\left(f^{-1}(x)\right)=x$ for all x in the domain of f^{-1} and that $f^{-1}(f(x))=x$ for all x in the domain of f. In terms of the sine, cosine, and tangent functions and their inverse, these properties are of the same form.

Inverse Function Properties for Sine, Cosine, and Tangent

$\sin\left(\sin^{-1}x\right)=x$ where $-1\le x\le 1$ $\sin^{-1}(\sin x)=x$ where $-\dfrac{\pi}{2}\le x\le\dfrac{\pi}{2}$

$\cos\left(\cos^{-1}x\right)=x$ where $-1\le x\le 1$ $\cos^{-1}(\cos x)=x$ where $0\le x\le\pi$

$\tan\left(\tan^{-1}x\right)=x$ where $-\infty<x<\infty$ $\tan^{-1}(\tan x)=x$ where $-\dfrac{\pi}{2}<x<\dfrac{\pi}{2}$

► EXAMPLE 4

Find the exact value of each composition using the properties of inverse trigonometric functions.

a.) $\sin\left(\sin^{-1}\dfrac{\sqrt{2}}{2}\right)$ b.) $\arccos\left(\cos\dfrac{\pi}{4}\right)$ c.) $\tan\left(\arctan\dfrac{\pi}{4}\right)$

►► Solution

a.) $\sin\left(\sin^{-1}\dfrac{\sqrt{2}}{2}\right)=\dfrac{\sqrt{2}}{2}$ since $\dfrac{\sqrt{2}}{2}$ is in $[-1,1]$

b.) $\arccos\left(\cos\dfrac{\pi}{4}\right)=\dfrac{\pi}{4}$ since $\dfrac{\pi}{4}$ is in $[0,\pi]$

c.) $\tan\left(\arctan\dfrac{\pi}{4}\right)=\dfrac{\pi}{4}$ since $\dfrac{\pi}{4}$ is in $\left(-\dfrac{\pi}{2},\dfrac{\pi}{2}\right)$

► EXAMPLE 5

Find the exact value of each expression.

a.) $\sin\left(\cos^{-1}\left(-\dfrac{\sqrt{2}}{2}\right)\right)$

b.) $\tan\left(\arcsin\dfrac{\sqrt{3}}{2}\right)$

c.) $\cos\left(\arctan\left(-\dfrac{\sqrt{3}}{3}\right)\right)$

►►Solution

a.) First we need to evaluate $\cos^{-1}\left(-\dfrac{\sqrt{2}}{2}\right)$. Where on the unit circle in the interval $[0,\pi]$

does $\cos^{-1}\left(-\dfrac{\sqrt{2}}{2}\right)$? Recall that the ordered pair at $\dfrac{\pi}{4}$ is $\left(\dfrac{\sqrt{2}}{2},\dfrac{\sqrt{2}}{2}\right)$ where $\cos\dfrac{\pi}{4}=\dfrac{\sqrt{2}}{2}$.

The angle we are looking for is $\dfrac{3\pi}{4}$ because $\cos\dfrac{3\pi}{4}=-\dfrac{\sqrt{2}}{2}$. Substituting $\dfrac{3\pi}{4}$ for

$\cos^{-1}\left(-\dfrac{\sqrt{2}}{2}\right)$ gives us $\sin\left(\dfrac{3\pi}{4}\right)$. The sine at $\dfrac{3\pi}{4}$ is $\dfrac{\sqrt{2}}{2}$ so, $\sin\left(\cos^{-1}\left(-\dfrac{\sqrt{2}}{2}\right)\right)=\dfrac{\sqrt{2}}{2}$.

b.) Where on the unit circle in the interval $\left[-\dfrac{\pi}{2},\dfrac{\pi}{2}\right]$ does $\sin y=\dfrac{\sqrt{3}}{2}$? Recall that the

ordered pair at $\dfrac{\pi}{3}$ is $\left(\dfrac{1}{2},\dfrac{\sqrt{3}}{2}\right)$ where $\sin\dfrac{\pi}{3}=\dfrac{\sqrt{3}}{2}$. Substituting $\dfrac{\pi}{3}$ for $\arcsin\dfrac{\sqrt{3}}{2}$ gives us

$\tan\left(\dfrac{\pi}{3}\right)$. The tangent at $\dfrac{\pi}{3}$ is $\sqrt{3}$ so, $\tan\left(\arcsin\dfrac{\sqrt{3}}{2}\right)=\tan\left(\dfrac{\pi}{3}\right)=\sqrt{3}$.

c.) Where on the unit circle in the interval $\left(-\dfrac{\pi}{2},\dfrac{\pi}{2}\right)$ does $\tan y=-\dfrac{\sqrt{3}}{3}$? Recall that the

ordered pair at $\dfrac{\pi}{6}$ is $\left(\dfrac{\sqrt{3}}{2},\dfrac{1}{2}\right)$ where $\tan\dfrac{\pi}{6}=\dfrac{\sqrt{3}}{3}$. $\dfrac{\pi}{6}$ is the angle in Quadrant I and $-\dfrac{\pi}{6}$

is in Quadrant IV and is the value for y we are looking for. Substituting $-\dfrac{\pi}{6}$ for

$\arctan\left(-\dfrac{\sqrt{3}}{3}\right)$ gives us $\cos\left(\arctan\left(-\dfrac{\sqrt{3}}{3}\right)\right)=\cos\left(-\dfrac{\pi}{6}\right)$. The cosine at $-\dfrac{\pi}{6}$ is $\dfrac{\sqrt{3}}{2}$ so,

$\cos\left(\arctan\left(-\dfrac{\sqrt{3}}{3}\right)\right)=\dfrac{\sqrt{3}}{2}$.

▶**EXAMPLE 6**

Find the exact value of each expression.

a.) $\quad \sin^{-1}\left(\cos\left(\dfrac{5\pi}{6}\right)\right)$

b.) $\quad \tan^{-1}\left(\sin\dfrac{\pi}{2}\right)$

c.) $\quad \arccos\left(\sin\left(\dfrac{5\pi}{3}\right)\right)$

▶▶**Solution**

a.) First we need to evaluate $\cos\left(\dfrac{5\pi}{6}\right)$. The $\cos\left(\dfrac{5\pi}{6}\right) = -\dfrac{\sqrt{3}}{2}$ and we substitute this into our

expression and we have $\sin^{-1}\left(\cos\left(\dfrac{5\pi}{6}\right)\right) = \sin^{-1}\left(-\dfrac{\sqrt{3}}{2}\right)$. Where on the unit circle in the

interval $\left[-\dfrac{\pi}{2}, \dfrac{\pi}{2}\right]$ does $\sin y = -\dfrac{\sqrt{3}}{2}$? Sine is equal to $-\dfrac{\sqrt{3}}{2}$ in the fourth quadrant at $-\dfrac{\pi}{3}$

so, $\sin^{-1}\left(\cos\left(\dfrac{5\pi}{6}\right)\right) = \sin^{-1}\left(-\dfrac{\sqrt{3}}{2}\right) = -\dfrac{\pi}{3}$.

b.) First we need to evaluate $\sin\left(\dfrac{\pi}{2}\right)$. The $\sin\left(\dfrac{\pi}{2}\right) = 1$ and we substitute this into our

expression and we have $\tan^{-1}\left(\sin\dfrac{\pi}{2}\right) = \tan^{-1}(1)$. Where on the unit circle in the

interval $\left(-\dfrac{\pi}{2}, \dfrac{\pi}{2}\right)$ does $\tan y = 1$? Tangent is equal to 1 in the first quadrant at $\dfrac{\pi}{4}$ so,

$\tan^{-1}\left(\sin\dfrac{\pi}{2}\right) = \tan^{-1}(1) = \dfrac{\pi}{4}$.

c.) First we need to evaluate $\sin\left(\dfrac{5\pi}{3}\right)$. The $\sin\left(\dfrac{5\pi}{3}\right) = -\dfrac{\sqrt{3}}{2}$ and we substitute this into our

expression and we have $\arccos\left(\sin\left(\dfrac{5\pi}{3}\right)\right) = \arccos\left(-\dfrac{\sqrt{3}}{2}\right)$. Where on the unit circle in

the interval $[0, \pi]$ does $\cos y = -\dfrac{\sqrt{3}}{2}$? Cosine is equal to $-\dfrac{\sqrt{3}}{2}$ in the second quadrant at

$\dfrac{5\pi}{6}$ so, $\arccos\left(\sin\left(\dfrac{5\pi}{3}\right)\right) = \arccos\left(-\dfrac{\sqrt{3}}{2}\right) = \dfrac{5\pi}{6}$.

►EXAMPLE 7

Find the exact value of the expression: $\sin\left(\arccos\left(-\dfrac{4}{5}\right)\right)$.

►►Solution

We will let θ equal $\arccos\left(-\dfrac{4}{5}\right)$ to obtain $\arccos\left(-\dfrac{4}{5}\right)=\theta$ which implies $\cos\theta=-\dfrac{4}{5}$. Since

cosine is negative in Quadrant II in the interval $[0,\pi]$, we will draw a triangle in Quadrant II and

label what we know. Using the Pythagorean Theorem we find

the third side to equal 3. Since we let $\arccos\left(-\dfrac{4}{5}\right)=\theta$ we can

substitute θ into our expression and have the following:

$\sin\left(\arccos\left(-\dfrac{4}{5}\right)\right)=\sin\theta$. Looking at our sketch we find that

$\sin\theta=\dfrac{3}{5}$ so $\sin\left(\arccos\left(-\dfrac{4}{5}\right)\right)=\sin\theta=\dfrac{3}{5}$.

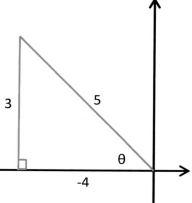

►EXAMPLE 8

Find the exact value of the expression: $\sec\left(\arctan\left(-\dfrac{5}{12}\right)\right)$.

►►Solution

We will let θ equal $\arctan\left(-\dfrac{5}{12}\right)$ to obtain $\arctan\left(-\dfrac{5}{12}\right)=\theta$ which implies $\tan\theta=-\dfrac{5}{12}$.

Since tangent is negative in Quadrant IV in the interval $\left(-\dfrac{\pi}{2},\dfrac{\pi}{2}\right)$, we will draw a triangle in

Quadrant IV and label what we know. Using the Pythagorean Theorem we find the hypotenuse

to equal 13. Since we let $\arctan\left(-\dfrac{5}{12}\right)=\theta$ we can

substitute θ into our expression and have the following:

$\sec\left(\arctan\left(-\dfrac{5}{12}\right)\right)=\sec\theta$. Looking at our sketch we find

that $\sec\theta=\dfrac{13}{12}$ so $\sec\left(\arctan\left(-\dfrac{5}{12}\right)\right)=\sec\theta=\dfrac{13}{12}$.

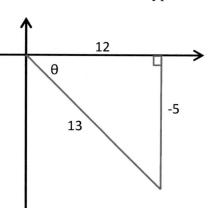

▶ EXAMPLE 9

Find an equivalent algebraic expression for $\cos\left(\arcsin\dfrac{5}{x}\right).$

▶▶ Solution

We will let θ equal $\arcsin\left(\dfrac{5}{x}\right)$ to obtain $\arcsin\left(\dfrac{5}{x}\right)=\theta$ which implies $\sin\theta=\dfrac{5}{x}$. We will draw

a triangle and label what we know. Using the Pythagorean Theorem we find the other leg to

equal $\sqrt{x^2-25}$. Since we let $\arcsin\left(\dfrac{5}{x}\right)=\theta$ we can

substitute θ into our expression and have the following:

$\cos\left(\arcsin\dfrac{5}{x}\right)=\cos\theta$. Looking at our sketch we find that

$\cos\theta=\dfrac{\sqrt{x^2-25}}{x}$ so $\cos\left(\arcsin\dfrac{5}{x}\right)=\cos\theta=\dfrac{\sqrt{x^2-25}}{x}$.

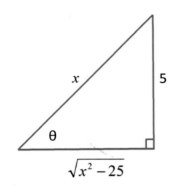

HOMEWORK

Objective 1

Find the exact value of the following expressions.

See Examples 1-3.

1. $\arcsin\left(-\dfrac{\sqrt{2}}{2}\right)$

2. $\cos^{-1}\left(-\dfrac{\sqrt{3}}{2}\right)$

3. $\arctan(1)$

4. $\arccos\left(\dfrac{1}{2}\right)$

5. $\sin^{-1}\left(-\dfrac{1}{2}\right)$

6. $\arcsin(0)$

7. $\cos^{-1}(-1)$

8. $\sin^{-1}\left(\dfrac{\sqrt{3}}{2}\right)$

9. $\arccos(-1)$

10. $\tan^{-1}\left(-\sqrt{3}\right)$

Objective 2

Find the exact value of the following compositions.

See Examples 4- 6.

11. $\tan\left(\arccos\left(\dfrac{1}{2}\right)\right)$

12. $\sin^{-1}\left(\cos\dfrac{2\pi}{3}\right)$

13. $\cos\left(\arcsin\left(\dfrac{\sqrt{2}}{2}\right)\right)$

14. $\arccos\left(\sin\left(\dfrac{7\pi}{6}\right)\right)$

15. $\sin^{-1}(\cos\pi)$

16. $\sin^{-1}\left(\sin\left(\dfrac{\pi}{6}\right)\right)$

17. $\sec\left(\arccos\dfrac{1}{2}\right)$

18. $\cos^{-1}\left(\sin\left(\dfrac{5\pi}{6}\right)\right)$

19. $\cos\left(\arccos\dfrac{1}{2}\right)$

20. $\sin^{-1}\left(\cos\left(\dfrac{5\pi}{6}\right)\right)$

21. $\csc\left(\arcsin\left(-\dfrac{1}{2}\right)\right)$

22. $\arcsin\left(\sin\left(\dfrac{7\pi}{6}\right)\right)$

Objective 2

Find the exact value of the following compositions.

See Examples 7&8.

23. $\cos\left(\arctan\left(-\dfrac{4}{3}\right)\right)$

24. $\tan\left(\arcsin\left(\dfrac{7}{25}\right)\right)$

25. $\cos\left(\arcsin\left(-\dfrac{12}{37}\right)\right)$

26. $\sin\left(\arccos\left(\dfrac{5}{13}\right)\right)$

27. $\sec\left(\arctan\left(\dfrac{9}{40}\right)\right)$

28. $\cot\left(\arcsin\left(-\dfrac{20}{29}\right)\right)$

Objective 2

Find the exact value of the following compositions.

See Example 9.

29. $\cos\left(\arctan\left(\dfrac{2}{x}\right)\right)$

30. $\tan\left(\arcsin\left(2x\right)\right)$

31. $\cos\left(\arcsin\left(\dfrac{x}{3}\right)\right)$

32. $\sin\left(\arccos\left(3x\right)\right)$

ANSWERS

1. $-\dfrac{\pi}{4}$

2. $\dfrac{5\pi}{6}$

3. $\dfrac{\pi}{4}$

4. $\dfrac{\pi}{3}$

5. $-\dfrac{\pi}{6}$

6. 0

7. π

8. $\dfrac{\pi}{3}$

9. π

10. $-\dfrac{\pi}{3}$

11. $\sqrt{3}$

12. $-\dfrac{\pi}{6}$

13. $\dfrac{\sqrt{2}}{2}$

14. $\dfrac{2\pi}{3}$

15. $-\dfrac{\pi}{2}$

16. $\dfrac{\pi}{6}$

17. 2

18. $\dfrac{\pi}{3}$

19. $\dfrac{1}{2}$

20. $-\dfrac{\pi}{3}$

21. -2

22. $-\dfrac{\pi}{6}$

23. $\dfrac{3}{5}$

24. $\dfrac{7}{24}$

25. $\dfrac{35}{37}$

26. $\dfrac{12}{13}$

27. $\dfrac{41}{40}$

28. $-\dfrac{21}{20}$

29. $\dfrac{x}{\sqrt{x^2+4}}$

30. $\dfrac{2x}{\sqrt{1-4x^2}}$

31. $\dfrac{\sqrt{9-x^2}}{3}$

32. $\sqrt{1-9x^2}$

Section 13:   Fundamental Identities

Learning Outcomes:

- The student will correctly memorize and apply trigonometric formulas, definitions, identities, and properties.

Objectives: At the conclusion of this lesson you should be able to:

1. Rewrite trigonometric expressions.
2. Verify trigonometric identities.
3. Use identities to evaluate trigonometric functions.

An identity is an equation that is satisfied by every number for which both sides are defined. We use identities to simplify expressions and determine whether expressions are equivalent. It is encouraged that you memorize the following basic trigonometric identities.

Reciprocal Identities

$$\sin\theta = \frac{1}{\csc\theta} \qquad \cos\theta = \frac{1}{\sec\theta} \qquad \tan\theta = \frac{1}{\cot\theta}$$

$$\csc\theta = \frac{1}{\sin\theta} \qquad \sec\theta = \frac{1}{\cos\theta} \qquad \cot\theta = \frac{1}{\tan\theta}$$

Even/Odd Identities

$$\cos(-\theta) = \cos\theta$$

$$\sin(-\theta) = -\sin\theta$$

$$\tan(-\theta) = -\tan\theta$$

Ratio/Quotient Identities

$$\tan\theta = \frac{\sin\theta}{\cos\theta} \qquad \tan\theta = \frac{\sec\theta}{\csc\theta}$$

$$\cot\theta = \frac{\cos\theta}{\sin\theta} \qquad \cot\theta = \frac{\csc\theta}{\sec\theta}$$

Pythagorean Identities

$$\sin^2\theta + \cos^2\theta = 1$$

$$\tan^2\theta + 1 = \sec^2\theta$$

$$1 + \cot^2\theta = \csc^2\theta$$

$$\boxed{\begin{array}{c}
\textit{Cofunction Identities} \\[0.5em]
\cos(90° - \theta) = \sin\theta \quad \sec(90° - \theta) = \csc\theta \quad \tan(90° - \theta) = \cot\theta \\[0.5em]
\cos\left(\dfrac{\pi}{2} - \theta\right) = \sin\theta \quad \sec\left(\dfrac{\pi}{2} - \theta\right) = \csc\theta \quad \tan\left(\dfrac{\pi}{2} - \theta\right) = \cot\theta \\[1.5em]
\sin(90° - \theta) = \cos\theta \quad \csc(90° - \theta) = \sec\theta \quad \cot(90° - \theta) = \tan\theta \\[0.5em]
\sin\left(\dfrac{\pi}{2} - \theta\right) = \cos\theta \quad \csc\left(\dfrac{\pi}{2} - \theta\right) = \sec\theta \quad \cot\left(\dfrac{\pi}{2} - \theta\right) = \tan\theta
\end{array}}$$

The basic trigonometric identities can be manipulated to achieve alternate forms because equality has already been established.

For example: The Pythagorean Identity $sin^2\theta + cos^2\theta = 1$ can be written as $sin^2\theta = 1 - cos^2\theta$ or $cos^2\theta = 1 - sin^2\theta$.

One of the skills required for calculus and advanced work in mathematics is the ability to use identities to write alternate forms of trigonometric expressions. There are no standard steps in rewriting a trigonometric expression in a similar form and some trigonometric expressions can be written more than one way. When we rewrite algebraic expressions we use techniques of factoring, distributing, common denominators, substitution, and other formulas. We use these same techniques and the basic trigonometric identities listed above to rewrite trigonometric expressions.

▶EXAMPLE 1

Use the fundamental identities to rewrite the expression: $\cos\theta \tan\theta$.

▶▶Solution

$\cos\theta \tan\theta$

$\dfrac{\cancel{\cos\theta}}{1} \cdot \dfrac{\sin\theta}{\cancel{\cos\theta}}$

$\sin\theta$

Use the Ratio/Quotient Identities. Substitute $\dfrac{\sin\theta}{\cos\theta}$ for $\tan\theta$ and reduce the common factor $\cos\theta$.

►EXAMPLE 2

Use the fundamental and cofunction identities to rewrite the expression: $\sin\left(\dfrac{\pi}{2} - x\right)\csc x$.

►►Solution

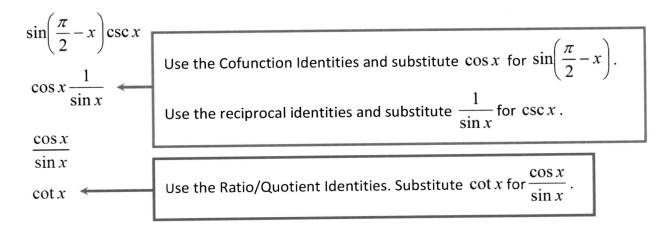

$\sin\left(\dfrac{\pi}{2} - x\right)\csc x$

$\cos x \dfrac{1}{\sin x}$ ← Use the Cofunction Identities and substitute $\cos x$ for $\sin\left(\dfrac{\pi}{2} - x\right)$.

Use the reciprocal identities and substitute $\dfrac{1}{\sin x}$ for $\csc x$.

$\dfrac{\cos x}{\sin x}$

$\cot x$ ← Use the Ratio/Quotient Identities. Substitute $\cot x$ for $\dfrac{\cos x}{\sin x}$.

►EXAMPLE 3

Use the fundamental identities to rewrite the expression: $\dfrac{1 - \sin^2 \beta}{\csc^2 \beta - 1}$.

►►Solution

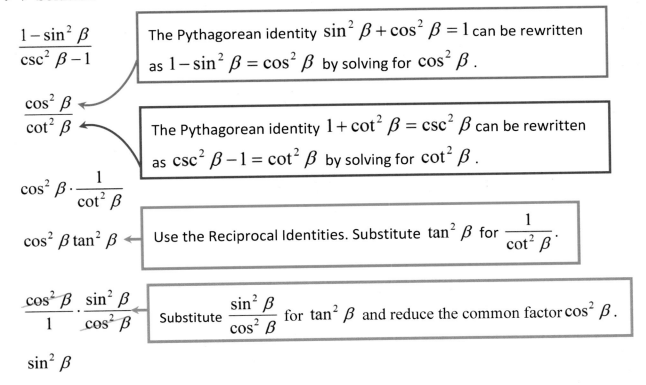

$\dfrac{1 - \sin^2 \beta}{\csc^2 \beta - 1}$ The Pythagorean identity $\sin^2 \beta + \cos^2 \beta = 1$ can be rewritten as $1 - \sin^2 \beta = \cos^2 \beta$ by solving for $\cos^2 \beta$.

$\dfrac{\cos^2 \beta}{\cot^2 \beta}$ The Pythagorean identity $1 + \cot^2 \beta = \csc^2 \beta$ can be rewritten as $\csc^2 \beta - 1 = \cot^2 \beta$ by solving for $\cot^2 \beta$.

$\cos^2 \beta \cdot \dfrac{1}{\cot^2 \beta}$

$\cos^2 \beta \tan^2 \beta$ ← Use the Reciprocal Identities. Substitute $\tan^2 \beta$ for $\dfrac{1}{\cot^2 \beta}$.

$\dfrac{\cos^2 \beta}{1} \cdot \dfrac{\sin^2 \beta}{\cos^2 \beta}$ Substitute $\dfrac{\sin^2 \beta}{\cos^2 \beta}$ for $\tan^2 \beta$ and reduce the common factor $\cos^2 \beta$.

$\sin^2 \beta$

In Example 4 you will see that there is more than one way to simplify a trigonometric expression.

▶ EXAMPLE 4

Use the fundamental identities to rewrite the expression: $\cos\theta\left(1+\tan^2\theta\right)$.

▶▶ Solution 1

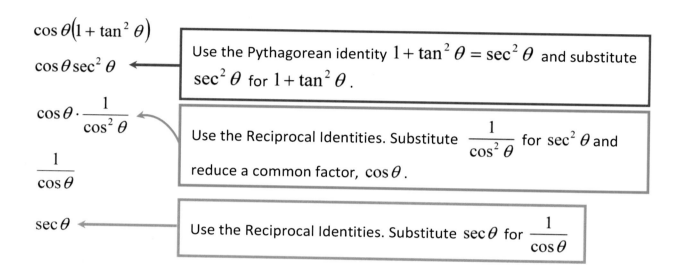

▶▶ Solution 2

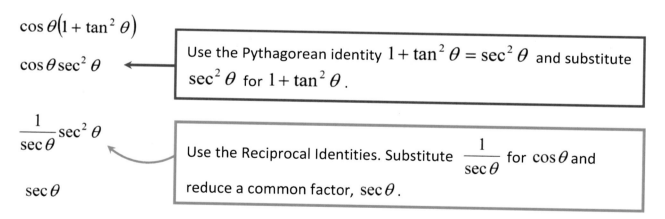

Now that we have simplified trigonometric expressions we are ready to verify trigonometric identities. Our primary strategy is to choose an expression on one side of the identity and to manipulate it and transform it so that it is identical to the expression on the other side. Follow the following guidelines to help you in verifying identities.

Guidelines for Verifying Identities

1. Memorize the basic trigonometric identities, which can be manipulated.

2. Work one side of the identity. It is often better to work with the more complex side. We cannot treat an identity as an equation that is true which means we cannot apply the properties of equality. In other words you can only manipulate one side of the identity to make it identical to the other side.

3. Look for opportunities to distribute, factor, combine by getting a common denominator, separate a fraction into multiple fractions, square a binomial, substitute an equivalent identity, etc.

4. If the preceding guidelines do not help, try converting all terms to sines and cosines.

5. Do not give up. If something you try does not work, try something else. Each path you take, whether it works or not, gives insights.

Verifying an identity using distribution.

►EXAMPLE 5

Verify that the following identity is true: $\cot\theta + 1 = \csc\theta(\cos\theta + \sin\theta)$.

►►Solution

$\cot\theta + 1 = \csc\theta(\cos\theta + \sin\theta)$

$\quad = \dfrac{1}{\sin\theta}(\cos\theta + \sin\theta)$

> We are going to work the right hand side of the equation. We will substitute $\dfrac{1}{\sin\theta}$ for $\csc\theta$.

$\quad = \dfrac{1}{\sin\theta}\cos\theta + \dfrac{1}{\sin\theta}\sin\theta$

> Use the distributive property and distribute $\dfrac{1}{\sin\theta}$.

$\quad = \dfrac{\cos\theta}{\sin\theta} + 1$

> Use the Reciprocal Identities. Substitute $\cot\theta$ for $\dfrac{\cos\theta}{\sin\theta}$.
>
> Notice the right hand side is the same as the left hand side; we have verified the identity is true.

$\quad = \cot\theta + 1$

Verifying an identity using factoring.

►EXAMPLE 5

Verify that the following identity is true: $\cos^2 \theta - \cos^4 \theta = \sin^2 \theta - \sin^4 \theta$.

►►Solution

$$\cos^2 \theta - \cos^4 \theta = \sin^2 \theta - \sin^4 \theta$$

> Notice that both sides look similar. We are going to work the left hand side of the equation.

$$\cos^2 \theta (1 - \cos^2 \theta) =$$

> Factor out $\cos^2 \theta$.

$$(1 - \sin^2 \theta)\sin^2 \theta =$$

> We have everything in terms of cosine and we need everything in terms of sine. Using the Pythagorean identities, we can substitute $1 - \sin^2 \theta$ for $\cos^2 \theta$ and $\sin^2 \theta$ for $1 - \cos^2 \theta$.

$$\sin^2 \theta - \sin^4 \theta =$$

> Distribute $\sin^2 \theta$ and we have the right hand side of the equation. The identity is verified.

►EXAMPLE 6

Verify that the following identity is true: $\cos^4 \theta - \sin^4 \theta = 2\cos^2 \theta - 1$.

►►Solution

$$\cos^4 \theta - \sin^4 \theta = 2\cos^2 \theta - 1$$

> We are going to work the left hand side of the equation.

$$\underbrace{}_{1}$$

$$(\cos^2 \theta + \sin^2 \theta)(\cos^2 \theta - \sin^2 \theta) =$$

> Notice that $\cos^4 \theta - \sin^4 \theta$ is a difference between two perfect squares and can be factored as shown.

$$1 \cdot (\cos^2 \theta - \sin^2 \theta) =$$

> Since $\cos^2 \theta + \sin^2 \theta = 1$, we will substitute 1 for $\cos^2 \theta + \sin^2 \theta$.

$$\cos^2 \theta - (1 - \cos^2 \theta) =$$

> Substitute $1 - \cos^2 \theta$ for $\sin^2 \theta$.

$$\cos^2 \theta - 1 + \cos^2 \theta =$$

> Simplify and combine like terms.

$$2\cos^2 \theta - 1 =$$

> We have verified the identity.

Verifying an identity by getting a common denominator.

▶ EXAMPLE 7

Verify that the following identity is true: $2\sec\theta = \dfrac{1+\sin\theta}{\cos\theta} + \dfrac{\cos\theta}{1+\sin\theta}.$

▶▶ Solution

$2\sec\theta = \dfrac{1+\sin\theta}{\cos\theta} + \dfrac{\cos\theta}{1+\sin\theta}$ | We are going to work the right hand side of the equation.

$= \dfrac{1+\sin\theta}{\cos\theta}\cdot\dfrac{1+\sin\theta}{1+\sin\theta} + \dfrac{\cos\theta}{1+\sin\theta}\cdot\dfrac{\cos\theta}{\cos\theta}$ | We need to combine the two terms on the right hand side of the equation by getting a common denominator.

$= \dfrac{\overbrace{1+\sin\theta+\sin\theta+\sin^2\theta+\cos^2\theta}^{1}}{\cos\theta(1+\sin\theta)}$ | F.O.I.L. the numerator in the first fraction and multiply the numerator in the second fraction.

$= \dfrac{1+2\sin\theta+1}{\cos\theta(1+\sin\theta)}$ | Simplify and substitute 1 for $\cos^2\theta+\sin^2\theta$.

$= \dfrac{2+2\sin\theta}{\cos\theta(1+\sin\theta)}$ | Simplify.

$= \dfrac{2(1+\sin\theta)}{\cos\theta(1+\sin\theta)}$ | Factor out the 2 and reduce the common factors.

$= \dfrac{2}{\cos\theta}$ | Rewrite.

$= 2\cdot\dfrac{1}{\cos\theta}$ | Substitute $\sec\theta$ for $\dfrac{1}{\cos\theta}$.

$= 2\sec\theta$ | We have verified that the identity is true.

▶**EXAMPLE 8**

Verify that the following identity is true: $\dfrac{1}{1-\sin x}+\dfrac{1}{1+\sin x}=2\sec^{2}x.$

▶▶**Solution**

$$\dfrac{1}{1-\sin x}+\dfrac{1}{1+\sin x}=2\sec^{2}x$$

> We are going to work the left hand side of the equation.

$$\dfrac{1}{1-\sin x}\cdot\dfrac{1+\sin x}{1+\sin x}+\dfrac{1}{1+\sin x}\cdot\dfrac{1-\sin x}{1-\sin x}=$$

> We need to combine the two terms on the left hand side of the equation by getting a common denominator.

$$\dfrac{1+\sin x+1-\sin x}{1+\sin x-\sin x-\sin^{2}x}=$$

> Multiply the numerators and F.O.I.L. the denominator.

$$\dfrac{2}{1-\sin^{2}x}=$$

> Simplify the numerator and the denominator.

$$\dfrac{2}{\cos^{2}x}=$$

> Substitute $\cos^{2}x$ for $1-\sin^{2}x$.

$$2\cdot\dfrac{1}{\cos^{2}x}=$$

> Substitute $\sec^{2}x$ for $\dfrac{1}{\cos^{2}x}$.

$$2\sec^{2}x=$$

> We have verified that the identity is true.

Verifying an identity by separating a fraction.

▶ EXAMPLE 9

Verify that the following identity is true: $\dfrac{1}{\sec x \tan x} = \csc x - \sin x$.

▶▶ Solution

$\dfrac{1}{\sec x \tan x} = \csc x - \sin x$

> We are going to work the left hand side of the equation.

$\dfrac{1}{\sec x} \cdot \dfrac{1}{\tan x} =$

> We are going to separate the fraction into the multiplication of two fractions.

$\cos x \cdot \cot x =$

> Substitute $\cos x$ for $\dfrac{1}{\sec x}$ and $\cot x$ for $\dfrac{1}{\tan x}$.

$\cos x \cdot \dfrac{\cos x}{\sin x} =$

> Substitute $\dfrac{\cos x}{\sin x}$ for $\cot x$.

$\dfrac{\cos^2 x}{\sin x} =$

> Multiply.

$\dfrac{1 - \sin^2 x}{\sin x} =$

> Substitute $1 - \sin^2 x$ for $\cos^2 x$.

$\dfrac{1}{\sin x} - \dfrac{\sin^2 x}{\sin x} =$

> Separate the fraction into two fractions. Substitute $\csc x$ for $\dfrac{1}{\sin x}$ and reduce $\dfrac{\sin^2 x}{\sin x} = \sin x$.

$\csc x - \sin x =$

> We have verified that the identity is true.

Verifying an identity by squaring a binomial.

▶ EXAMPLE 10

Verify that the following identity is true: $1 + 2\sin\alpha\cos\alpha = (\cos\alpha + \sin\alpha)^2$.

▶▶ Solution

$1 + 2\sin\alpha\cos\alpha = (\cos\alpha + \sin\alpha)^2$ | We are going to work the right hand side of the equation.

$= (\cos\alpha + \sin\alpha)(\cos\alpha + \sin\alpha)$ | Rewrite the left hand side and F.O.I.L.

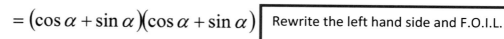

$= \cos^2\alpha + \cos\alpha\sin\alpha + \cos\alpha\sin\alpha + \sin^2\alpha$ | Simplify and substitute 1 for $\cos^2\alpha + \sin^2\alpha$.

$= 1 + 2\sin\alpha\cos\alpha$ | We have verified that the identity is true.

Verifying an identity by substituting.

▶ EXAMPLE 11

Verify that the following identity is true: $\dfrac{1}{\sin x} - \dfrac{1}{\cos x} = \csc x - \sec x$.

▶▶ Solution

$\dfrac{1}{\sin x} - \dfrac{1}{\cos x} = \csc x - \sec x$ | We are going to work the left hand side of the equation.

$\csc x - \sec x =$ | Substitute $\csc x$ for $\dfrac{1}{\sin x}$ and $\sec x$ for $\dfrac{1}{\cos x}$. We have verified that the identity is true.

► EXAMPLE 12

Verify that the following identity is true: $1+\csc^2 \beta - \cot^2 \beta = 2$.

►►Solution

$$1+\csc^2 \beta - \cot^2 \beta = 2$$

> We are going to work the left hand side of the equation.

$$1+\overbrace{\csc^2 \beta - \cot^2 \beta}^{1} =$$

> $\csc^2 \beta - \cot^2 \beta = 1$ is an alternate form of the Pythagorean identity $1+\cot^2 \beta = \csc^2 \beta$.

$$1+1 =$$

> Simplify.

$$2 =$$

> We have verified the identity.

► EXAMPLE 13

Verify that the following identity is true: $\tan^2(-\theta) - \dfrac{\sin(-\theta)}{\sin \theta} = \sec^2 \theta$.

►►Solution

$$\tan^2(-\theta) - \frac{\sin(-\theta)}{\sin \theta} = \sec^2 \theta$$

> We are going to work the left hand side of the equation.

$$(-\tan \theta)^2 - \frac{-\sin \theta}{\sin \theta} =$$

> Sine and tangent are odd functions so we will use $\sin(-\theta) = -\sin \theta$ and $\tan(-\theta) = -\tan \theta$.

$$\tan^2 \theta + 1 =$$

> Simplify.

$$\sec^2 \theta =$$

> Substitute $\sec^2 \theta$ for $\tan^2 \theta + 1$ and we have verified the identity.

▶ EXAMPLE 14

Verify that the following identity is true: $\sec\theta - \cos\theta = \dfrac{\sin\theta}{\cot\theta}$.

▶▶ Solution

$$\sec\theta - \cos\theta = \frac{\sin\theta}{\cot\theta}$$

> We are going to work the right hand side of the equation.

$$= \frac{\sin\theta}{1} \cdot \frac{1}{\cot\theta}$$

> Rewrite the fraction and substitute $\tan\theta$ for $\dfrac{1}{\cot\theta}$.

$$= \frac{\sin\theta}{1} \cdot \tan\theta$$

> Substitute $\tan\theta$ for $\dfrac{1}{\cot\theta}$.

$$= \frac{\sin\theta}{1} \cdot \frac{\sin\theta}{\cos\theta}$$

> Substitute $\dfrac{\sin\theta}{\cos\theta}$ for $\tan\theta$.

$$= \frac{\sin^2\theta}{\cos\theta}$$

> Multiply.

$$= \frac{1 - \cos^2\theta}{\cos\theta}$$

> Substitute $1 - \cos^2\theta$ for $\sin^2\theta$.

$$= \frac{1}{\cos\theta} - \frac{\cos^2\theta}{\cos\theta}$$

> Rewrite the fraction as two fractions.

$$= \frac{1}{\cos\theta} - \frac{\cos^2\theta}{\cos\theta}$$

> Substitute $\sec\theta$ for $\dfrac{1}{\cos\theta}$ and simplify $\dfrac{\cos^2\theta}{\cos\theta} = \cos\theta$.

$$= \sec\theta - \cos\theta$$

> We have verified that the identity is true.

A common use of trigonometric identities is to use given values of trigonometric functions to evaluate other trigonometric functions. Because all of the trigonometric functions are related by identities we can find their values for an angle if we know the value of any one of them and the quadrant of the angle.

▶ **EXAMPLE 15**

Use identities to find the value of the other 5 trigonometric functions given $\sin\theta = -\dfrac{5}{13}$ **and**

θ **is in Quadrant III.**

▶▶ **Solution**

$\sin\theta = -\dfrac{5}{13}$, θ is in Quadrant III

$\sin^2\theta + \cos^2\theta = 1$ | We can use the Pythagorean Identity $\sin^2\theta + \cos^2\theta = 1$ to find $\cos\theta$. |

$\left(-\dfrac{5}{13}\right)^2 + \cos^2\theta = 1$ | Substitute $-\dfrac{5}{13}$ for $\sin\theta$. |

$\dfrac{25}{169} + \cos^2\theta = 1$ | Isolate $\cos^2\theta$. |

Use the Reciprocal Identities to find the remaining trigonometric function values.

$\cos^2\theta = 1 - \dfrac{25}{169}$

$\cos^2\theta = \dfrac{169}{169} - \dfrac{25}{169}$ | Get a common denominator. |

| $\sin\theta = -\dfrac{5}{13}$ | $\csc\theta = -\dfrac{13}{5}$ |

$\cos^2\theta = \dfrac{144}{169}$

| $\cos\theta = -\dfrac{12}{13}$ | $\sec\theta = -\dfrac{13}{12}$ |

$\sqrt{\cos^2\theta} = \pm\sqrt{\dfrac{144}{169}}$ | Take the square root of both sides. |

| $\tan\theta = \dfrac{5}{12}$ | $\cot\theta = \dfrac{12}{5}$ |

$\cos\theta = -\dfrac{12}{13}$ | Cosine is negative in Quadrant III. |

$\tan\theta = \dfrac{-\dfrac{5}{13}}{-\dfrac{12}{13}}$ | Use the quotient identity $\tan\theta = \dfrac{\sin\theta}{\cos\theta}$ to find tangent. |

$\tan\theta = \dfrac{5}{13}\cdot\dfrac{13}{12} = \dfrac{5}{12}$

HOMEWORK

Objective 1

Use the fundamental and cofunction identities to rewrite the following trigonometric expressions. There is more than one correct form of each answer.

See Examples 1-4.

1. $\sin^2 \theta \left(1 + \tan^2 \theta\right)$

2. $\cot \beta \sec \beta$

3. $\dfrac{1 - \sin^2 \theta}{\csc^2 \theta - 1}$

4. $\dfrac{\cos^2 x}{1 - \sin^2 x}$

5. $\sec^2 \alpha \left(1 - \sin^2 \alpha\right)$

6. $\tan x \csc x$

7. $\sin \beta \sec \beta$

8. $\dfrac{\csc x}{1 + \cot^2 x}$

9. $\dfrac{\tan \alpha}{\sec \alpha}$

10. $\cot\left(\dfrac{\pi}{2} - x\right)\cos x$

Objective 2

Verify that the following equations are trigonometric identities. There is more than one correct form of each answer.

See Examples 5 -14.

11. $\sin t \csc t = 1$

12. $\left(1 + \sin t\right)\left(1 - \sin t\right) = \cos^2 t$

13. $\cos^2 \beta - \sin^2 \beta = 1 - 2\sin^2 \beta$

14. $\left(1 + \cos(-\theta)\right)\left(1 - \cos(-\theta)\right) = \sin^2 \theta$

15. $\dfrac{\cot x}{\csc x} = \cos x$

16. $\dfrac{1 - \sin^2 \alpha}{\cos \alpha} = \cos \alpha$

17. $\cot \beta + \tan \beta = \sec \beta \csc \beta$

18. $\dfrac{\sin^2 \theta}{\cos \theta} = \sec \theta - \cos \theta$

19. $\dfrac{\cos x + 1}{\tan^2 x} = \dfrac{\cos x}{\sec x - 1}$

20. $\left(1 + \cot^2 \alpha\right)\left(1 + \sin^2 \alpha\right) = 2 + \cot^2 \alpha$

21. $\dfrac{\cot x + \tan x}{\csc x} = \sec x$

22. $\dfrac{1}{\cos^2 \theta} + \dfrac{1}{\sin^2 \theta} = \csc^2 \theta \sec^2 \theta$

23. $\dfrac{\sin^4 \beta - \cos^4 \beta}{\cos^2 \beta} = \sec^2 \beta - 2$

24. $\dfrac{\sin^4 \beta - \cos^4 \beta}{\sin^3 \beta + \cos^3 \beta} = \dfrac{\sin \beta - \cos \beta}{1 - \sin \beta \cos \beta}$

25. $\dfrac{\cot \beta}{\cot \beta + \tan \beta} = 1 - \sin^2 \beta$

26. $\dfrac{\sin^4 \beta - \cos^4 \beta}{\sin^3 \beta - \cos^3 \beta} = \dfrac{\sin \beta + \cos \beta}{1 + \sin \beta \cos \beta}$

27. $\dfrac{\cos x}{1 + \cos x} - \dfrac{\cos x}{1 - \cos x} = -2\cot^2 x$

28. $\dfrac{\cos\left(\dfrac{\pi}{2} - \theta\right)}{\sin\left(\dfrac{\pi}{2} - \theta\right)} = \tan \theta$

29. $\dfrac{\sin x \cos x}{\cos^2 x - \sin^2 x} = \dfrac{\tan x}{1 - \tan^2 x}$

30. $3\sin^2 \alpha + 4\cos^2 \alpha = 3 + \cos^2 \alpha$

31. $\tan \theta + \dfrac{\cos \theta}{1 + \sin \theta} = \sec \theta$

32. $\sec^4 x - \sec^2 x = \tan^2 x + \tan^4 x$

33. $\sec \theta - \cos \theta = \sin \theta \tan \theta$

34. $\left(1 - \sin x\right)^2 + \cos^2 x = 2\left(1 - \sin x\right)$

35. $\dfrac{1}{\sin \beta \cos \beta} - \dfrac{1}{\cot \beta} = \cot \beta$

36. $\dfrac{1 - \cos x}{\sin x} = \csc x - \cot x$

Objective 3

Use identities to find the value of the other 5 trigonometric functions.

See Example 15.

37. $\sin \beta = -\dfrac{20}{29}$, β *in Quadrant IV*

38. $\cos \beta = -\dfrac{15}{17}$, β *in Quadrant II*

39. $\tan \beta = \dfrac{8}{15}$, β *in Quadrant III*

40. $\sec \beta = \dfrac{25}{24}$, β *in Quadrant IV*

ANSWERS

Answers may vary.

1. $tan^2\theta$ 2. $csc\beta$ 3. $sin^2\theta$ 4.   1

5. 1   6. $secx$ 7. $tan\beta$ 8. $sinx$

9. $sin\alpha$ 10. $sinx$

Answers may vary.

11. $sintcsct = 1$

$sint \cdot \dfrac{1}{sint} =$

$1 = 1$

12. $(1 + sint)(1 - sint) = cos^2t$

$1 - sint + sint - sin^2t =$

$1 - sin^2t =$

$cos^2t = cos^2t$

13. $cos^2\beta - sin^2\beta = 1 - 2sin^2\beta$

$1 - sin^2\beta - sin^2\beta =$

$1 - 2sin^2\beta = 1 - 2sin^2\beta$

14. $\left(1 + cos(-\theta)\right)\left(1 - cos(-\theta)\right) = sin^2\theta$

$(1 + cos\theta)(1 - cos\theta) =$

$1 - cos\theta + cos\theta - cos^2\theta =$

$1 - cos^2\theta =$

$sin^2\theta = sin^2\theta$

15. $\dfrac{cotx}{cscx} = cosx$

$cotx \cdot \dfrac{1}{cscx} =$

$\dfrac{cosx}{sinx} \cdot sinx =$

$cosx = cosx$

16. $\dfrac{1 - sin^2\alpha}{cos\alpha} = cos\alpha$

$\dfrac{cos^2\alpha}{cos\alpha} =$

$cos\alpha = cos\alpha$

17. $cot\beta + tan\beta = sec\beta csc\beta$

$\dfrac{cos\beta}{sin\beta} + \dfrac{sin\beta}{cos\beta} =$

$\dfrac{cos^2\beta + sin^2\beta}{cos\beta sin\beta} =$

$\dfrac{1}{cos\beta sin\beta} =$

$sec\beta csc\beta = sec\beta csc\beta$

18. $\dfrac{sin^2\theta}{cos\theta} = sec\theta - cos\theta$

$\dfrac{1 - cos^2\theta}{cos\theta} =$

$\dfrac{1}{cos\theta} - \dfrac{cos^2\theta}{cos\theta} =$

$sec\theta - cos\theta = sec\theta - cos\theta$

19. $\dfrac{cosx + 1}{tan^2x} = \dfrac{cosx}{secx - 1}$

$= \dfrac{cosx}{secx - 1} \cdot \dfrac{secx + 1}{secx + 1}$

$= \dfrac{cosxsecx + cosx}{sec^2x + secx - secx - 1}$

$= \dfrac{cosx \cdot \dfrac{1}{cosx} + cosx}{sec^2x - 1}$

$= \dfrac{1 + cosx}{tan^2x}$

$\dfrac{cosx + 1}{tan^2x} = \dfrac{cosx + 1}{tan^2x}$

20. $(1 + cot^2\alpha)(1 + sin^2\alpha) = 2 + cot^2\alpha$

$1 + sin^2\alpha + cot^2\alpha + cot^2\alpha sin^2\alpha =$

$1 + sin^2\alpha + cot^2\alpha + \dfrac{cos^2\alpha}{sin^2\alpha} \cdot sin^2\alpha =$

$1 + sin^2\alpha + cot^2\alpha + cos^2\alpha =$

$1 + 1 + cot^2\alpha =$

$2 + cot^2\alpha = 2 + cot^2\alpha$

22. $\dfrac{1}{cos^2\theta} + \dfrac{1}{sin^2\theta} = csc^2\theta sec^2\theta$

$\dfrac{sin^2\theta + cos^2\theta}{sin^2\theta cos^2\theta} =$

$\dfrac{1}{sin^2\theta cos^2\theta} =$

$csc^2\theta sec^2\theta = csc^2\theta sec^2\theta$

21. $\dfrac{cotx + tanx}{cscx} = secx$

$\dfrac{\dfrac{cosx}{sinx} + \dfrac{sinx}{cosx}}{\dfrac{1}{sinx}} \cdot \dfrac{sinxcosx}{sinxcosx} =$

$\dfrac{cos^2x + sin^2x}{cosx} =$

$\dfrac{1}{cosx} =$

$secx = secx$

23. $\dfrac{sin^4\beta - cos^4\beta}{cos^2\beta} = sec^2\beta - 2$

$\dfrac{(sin^2\beta - cos^2\beta)(sin^2\beta + cos^2\beta)}{cos^2\beta} =$

$\dfrac{1 - cos^2\beta - cos^2\beta}{cos^2\beta} =$

$\dfrac{1 - 2cos^2\beta}{cos^2\beta} =$

$\dfrac{1}{cos^2\beta} - \dfrac{2cos^2\beta}{cos^2\beta} =$

$sec^2\beta - 2 =$

24.

$$\frac{\sin^4\beta - \cos^4\beta}{\sin^3\beta + \cos^3\beta} = \frac{\sin\beta - \cos\beta}{1 - \sin\beta\cos\beta}$$

$$\frac{(\sin^2\beta - \cos^2\beta)(\sin^2\beta + \cos^2\beta)}{(\sin\beta + \cos\beta)(\sin^2\beta - \sin\beta\cos\beta + \cos^2\beta)} =$$

$$\frac{(\sin\beta - \cos\beta)(\sin\beta + \cos\beta)(1)}{(\sin\beta + \cos\beta)(1 - \sin\beta\cos\beta)} =$$

$$\frac{\sin\beta - \cos\beta}{1 - \sin\beta\cos\beta} = \frac{\sin\beta - \cos\beta}{1 - \sin\beta\cos\beta}$$

25.

$$\frac{\cot\beta}{\cot\beta + \tan\beta} = 1 - \sin^2\beta$$

$$\frac{\dfrac{\cos\beta}{\sin\beta}}{\dfrac{\cos\beta}{\sin\beta} + \dfrac{\sin\beta}{\cos\beta}} \cdot \frac{\cos\beta\sin\beta}{\cos\beta\sin\beta} =$$

$$\frac{\cos^2\beta}{\cos^2\beta + \sin^2\beta} =$$

$$\frac{\cos^2\beta}{1} =$$

$$1 - \sin^2\beta = 1 - \sin^2\beta$$

26.

$$\frac{\sin^4\beta - \cos^4\beta}{\sin^3\beta - \cos^3\beta} = \frac{\sin\beta + \cos\beta}{1 + \sin\beta\cos\beta}$$

$$\frac{(\sin^2\beta - \cos^2\beta)(\sin^2\beta + \cos^2\beta)}{(\sin\beta - \cos\beta)(\sin^2\beta + \sin\beta\cos\beta + \cos^2\beta)} =$$

$$\frac{(\sin\beta - \cos\beta)(\sin\beta + \cos\beta)(1)}{(\sin\beta - \cos\beta)(1 + \sin\beta\cos\beta)} =$$

$$\frac{\sin\beta + \cos\beta}{1 + \sin\beta\cos\beta} = \frac{\sin\beta + \cos\beta}{1 + \sin\beta\cos\beta}$$

27.
$$\frac{\cos x}{1+\cos x}-\frac{\cos x}{1-\cos x}=-2\cot^2 x$$

$$\frac{\cos x(1-\cos x)-\cos x(1+\cos x)}{(1+\cos x)(1-\cos x)}=$$

$$\frac{\cos x-\cos^2 x-\cos x-\cos^2 x}{1-\cos x+\cos x-\cos^2 x}=$$

$$\frac{-2\cos^2 x}{1-\cos^2 x}=$$

$$\frac{-2\cos^2 x}{\sin^2 x}=$$

$$-2\cot^2 x=-2\cot^2 x$$

28.
$$\frac{\cos\left(\frac{\pi}{2}-\theta\right)}{\sin\left(\frac{\pi}{2}-\theta\right)}=\tan\theta$$

$$\frac{\sin\theta}{\cos\theta}=$$

$$\tan\theta=\tan\theta$$

29.
$$\frac{\sin x\cos x}{\cos^2 x-\sin^2 x}=\frac{\tan x}{1-\tan^2 x}$$

$$=\frac{\dfrac{\sin x}{\cos x}}{1-\dfrac{\sin^2 x}{\cos^2 x}}\cdot\frac{\cos^2 x}{\cos^2 x}$$

$$\frac{\sin x\cos x}{\cos^2 x-\sin^2 x}=\frac{\sin x\cos x}{\cos^2 x-\sin^2 x}$$

30.
$$3\sin^2\alpha+4\cos^2\alpha=3+\cos^2\alpha$$

$$3(1-\cos^2\alpha)+4\cos^2\alpha=$$

$$3-3\cos^2\alpha+4\cos^2\alpha=$$

$$3+\cos^2\alpha=3+\cos^2\alpha$$

31.
$$\tan\theta+\frac{\cos\theta}{1+\sin\theta}=\sec\theta$$

$$\frac{\sin\theta}{\cos\theta}+\frac{\cos\theta}{1+\sin\theta}=$$

$$\frac{\sin\theta(1+\sin\theta)+\cos\theta\cos\theta}{\cos\theta(1+\sin\theta)}=$$

$$\frac{\sin\theta+\sin^2\theta+\cos^2\theta}{\cos\theta(1+\sin\theta)}=$$

$$\frac{\sin\theta+1}{\cos\theta(1+\sin\theta)}=$$

$$\frac{1}{\cos\theta}=$$

$$\sec\theta=\sec\theta$$

32.
$$\sec^4 x-\sec^2 x=\tan^2 x+\tan^4 x$$

$$\sec^2 x(\sec^2 x-1)=$$

$$(\tan^2 x+1)\tan^2 x=$$

$$\tan^2 x+\tan^4 x=\tan^2 x+\tan^4 x$$

33. $\sec\theta - \cos\theta = \sin\theta\tan\theta$

$$= \sin\theta \cdot \frac{\sin\theta}{\cos\theta}$$

$$= \frac{\sin^2\theta}{\cos\theta}$$

$$= \frac{1 - \cos^2\theta}{\cos\theta}$$

$$= \frac{1}{\cos\theta} - \frac{\cos^2\theta}{\cos\theta}$$

$$\sec\theta - \cos\theta = \sec\theta - \cos\theta$$

34. $(1 - \sin x)^2 + \cos^2 x = 2(1 - \sin x)$

$$1 - \sin x - \sin x + \sin^2 x + \cos^2 x =$$

$$1 - 2\sin x + 1 =$$

$$2 - 2\sin x =$$

$$2(1 - \sin x) = 2(1 - \sin x)$$

35.

$$\frac{1}{\sin\beta\cos\beta} - \frac{1}{\cot\beta} = \cot\beta$$

$$\frac{1}{\sin\beta\cos\beta} - \tan\beta =$$

$$\frac{1}{\sin\beta\cos\beta} - \frac{\sin\beta}{\cos\beta} =$$

$$\frac{1}{\sin\beta\cos\beta} - \frac{\sin\beta}{\cos\beta} \cdot \frac{\sin\beta}{\sin\beta} =$$

$$\frac{1 - \sin^2\beta}{\sin\beta\cos\beta} =$$

$$\frac{\cos^2\beta}{\sin\beta\cos\beta} =$$

$$\frac{\cos\beta}{\sin\beta} =$$

$$\cot\beta = \cot\beta$$

36. $\dfrac{1 - \cos x}{\sin x} = \csc x - \cot x$

$$\frac{1}{\sin x} - \frac{\cos x}{\sin x} =$$

$$\csc x - \cot x = \csc x - \cot x$$

37.
$$\sin\beta = -\frac{20}{29} \qquad \csc\beta = -\frac{29}{20}$$

$$\cos\beta = \frac{21}{29} \qquad \sec\beta = \frac{29}{21}$$

$$\tan\beta = -\frac{20}{21} \qquad \cot\beta = -\frac{21}{20}$$

38.
$$\sin\beta = \frac{8}{17} \qquad \csc\beta = \frac{17}{8}$$

$$\cos\beta = -\frac{15}{17} \qquad \sec\beta = -\frac{17}{15}$$

$$\tan\beta = -\frac{8}{15} \qquad \cot\beta = -\frac{15}{8}$$

39.
$$\sin\beta = -\frac{8}{17} \qquad \csc\beta = -\frac{17}{8}$$

$$\cos\beta = -\frac{15}{17} \qquad \sec\beta = -\frac{17}{15}$$

$$\tan\beta = \frac{8}{15} \qquad \cot\beta = \frac{15}{8}$$

40.
$$\sin\beta = -\frac{7}{25} \qquad \csc\beta = -\frac{25}{7}$$

$$\cos\beta = \frac{24}{25} \qquad \sec\beta = \frac{25}{24}$$

$$\tan\beta = -\frac{7}{24} \qquad \cot\beta = -\frac{24}{7}$$

Section 14: Sum and Difference, Double Angle, and Half Angle Identities

Learning Outcomes:

- The student will correctly memorize and apply trigonometric formulas, definitions, identities, and properties.

Objectives: At the conclusion of this lesson you should be able to:

1.   Apply the sum and difference identities.
2.   Apply the double angle identities.
3.   Apply the half-angle identities.
4.   Verify identities involving the sum and difference identities, double angle identities, and half-angle identities.

Sum and Difference Identities

In this section we will continue our work with identities by obtaining identities that involve the sum or difference of two angles. On the unit circle consider the point A on the terminal side of angle θ and point B on the terminal side of angle β. The radius is one on the unit circle and therefore the point at A is $(\cos\theta, \sin\theta)$ and the point at B is $(\cos\beta, \sin\beta)$ as shown in **Figure 14.1**. We will use the distance formula to find the length of the segment $\overline{AB}$.

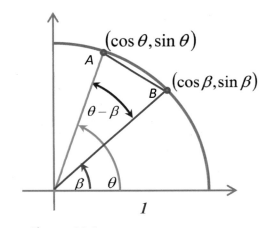

Figure 14.1

$$\overline{AB} = \sqrt{(\cos\theta - \cos\beta)^2 + (\sin\theta - \sin\beta)^2}$$

$$\overline{AB} = \sqrt{\cos^2\theta - 2\cos\theta\cos\beta + \cos^2\beta + \sin^2\theta - 2\sin\theta\sin\beta + \sin^2\beta}$$

$$\overline{AB} = \sqrt{-2\cos\theta\cos\beta - 2\sin\theta\sin\beta + (\cos^2\theta + \sin^2\theta) + (\cos^2\beta + \sin^2\beta)}$$

$$\overline{AB} = \sqrt{-2\cos\theta\cos\beta - 2\sin\theta\sin\beta + 1 + 1}$$

$$\overline{AB} = \sqrt{-2\cos\theta\cos\beta - 2\sin\theta\sin\beta + 2}$$

If we place the angle $\theta - \beta$ in standard position as shown in **Figure 14.2** the point B has coordinates of $(1,0)$ and the point A has coordinates of $(\cos(\theta - \beta), \sin(\theta - \beta))$.

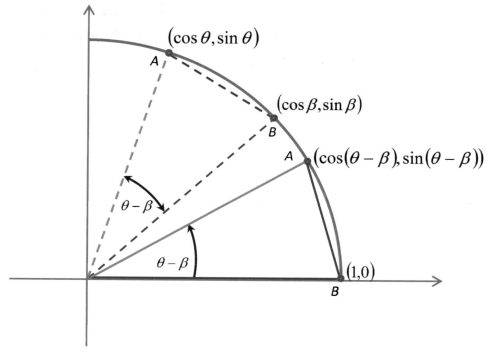

Figure 14.2

Notice that the length of the segment $\overline{AB}$ will remain the same. If we use the distance formula again to find the length of the segment $\overline{AB}$, we have the following.

$$\overline{AB} = \sqrt{(\cos(\theta-\beta)-1)^2 + (\sin(\theta-\beta)-0)^2}$$

$$\overline{AB} = \sqrt{\cos^2(\theta-\beta)-2\cos(\theta-\beta)+1+\sin^2(\theta-\beta)}$$

$$\overline{AB} = \sqrt{-2\cos(\theta-\beta)+1+\left(\cos^2(\theta-\beta)+\sin^2(\theta-\beta)\right)}$$

$$\overline{AB} = \sqrt{-2\cos(\theta-\beta)+2}$$

Since both expressions represent the length of the segment $\overline{AB}$ we can set them equal to each other and solve for $\cos(\theta-\beta)$.

$$\sqrt{-2\cos(\theta-\beta)+2} = \sqrt{-2\cos\theta\cos\beta-2\sin\theta\sin\beta+2}$$

$$\left(\sqrt{-2\cos(\theta-\beta)+2}\right)^2 = \left(\sqrt{-2\cos\theta\cos\beta-2\sin\theta\sin\beta+2}\right)^2$$

$$-2\cos(\theta-\beta)+2 = -2\cos\theta\cos\beta-2\sin\theta\sin\beta+2$$

$$-2\cos(\theta-\beta) = -2\cos\theta\cos\beta-2\sin\theta\sin\beta$$

$$\frac{-2\cos(\theta-\beta)}{-2} = \frac{-2\cos\theta\cos\beta}{-2} - \frac{2\sin\theta\sin\beta}{-2}$$

$$\cos(\theta-\beta) = \cos\theta\cos\beta+\sin\theta\sin\beta$$

The result $\cos(\theta - \beta) = \cos\theta\cos\beta + \sin\theta\sin\beta$ is called the difference identity for cosine. We can find the sum identity for cosine by substituting $-\beta$ for β and simplifying to obtain $\cos(\theta + \beta) = \cos\theta\cos\beta - \sin\theta\sin\beta$.

$$\cos(\theta - \beta) = \cos\theta\cos\beta + \sin\theta\sin\beta$$

$$\cos(\theta - (-\beta)) = \cos\theta\cos(-\beta) + \sin\theta\sin(-\beta)$$

$$\cos(\theta + \beta) = \cos\theta\cos\beta - \sin\theta\sin\beta$$

To find the sine sum and sine difference identities, we will use the cofunction identity $\sin(\theta) = \cos\left(\dfrac{\pi}{2} - \theta\right)$.

$$\sin(\theta + \beta) = \cos\left(\dfrac{\pi}{2} - (\theta + \beta)\right) \qquad \text{Use the sine cofunction Identity.}$$

$$\sin(\theta + \beta) = \cos\left(\left(\dfrac{\pi}{2} - \theta\right) - \beta\right) \qquad \text{Regroup using the Associative Property.}$$

$$\sin(\theta + \beta) = \cos\left(\dfrac{\pi}{2} - \theta\right)\cos\beta + \sin\left(\dfrac{\pi}{2} - \theta\right)\sin\beta \qquad \text{Use the cosine difference Identity.}$$

$$\sin(\theta + \beta) = \sin\theta\cos\beta + \cos\theta\sin\beta \qquad \text{Use the sine and cosine cofunction Identities.}$$

The sine sum identity is $\sin(\theta + \beta) = \sin\theta\cos\beta + \cos\theta\sin\beta$. To find the sine difference identity, substitute $-\beta$ for β and simplify to obtain $\sin(\theta - \beta) = \sin\theta\cos\beta - \cos\theta\sin\beta$.

$$\sin(\theta + \beta) = \sin\theta\cos\beta + \cos\theta\sin\beta$$

$$\sin(\theta + (-\beta)) = \sin\theta\cos(-\beta) + \cos\theta\sin(-\beta)$$

$$\sin(\theta - \beta) = \sin\theta\cos\beta - \cos\theta\sin\beta$$

We can use the ratio/quotient identity $\tan\theta = \dfrac{\sin\theta}{\cos\theta}$ to find the sum identity for tangent

$$\tan(\theta + \beta) = \dfrac{\sin(\theta + \beta)}{\cos(\theta + \beta)}.$$

$$\tan(\theta + \beta) = \frac{\sin(\theta + \beta)}{\cos(\theta + \beta)}$$

$$\tan(\theta + \beta) = \frac{\sin \theta \cos \beta + \cos \theta \sin \beta}{\cos \theta \cos \beta - \sin \theta \sin \beta}$$ *Apply the sine sum identity and cosine sum identity.*

$$\tan(\theta + \beta) = \frac{\sin \theta \cos \beta + \cos \theta \sin \beta}{\cos \theta \cos \beta - \sin \theta \sin \beta} \cdot \frac{\dfrac{1}{\cos \theta \cos \beta}}{\dfrac{1}{\cos \theta \cos \beta}}$$ *Multiply the numerator and denominator by an equivalency of one.*

$$\tan(\theta + \beta) = \frac{\dfrac{\sin \theta \cos \beta}{\cos \theta \cos \beta} + \dfrac{\cos \theta \sin \beta}{\cos \theta \cos \beta}}{\dfrac{\cos \theta \cos \beta}{\cos \theta \cos \beta} - \dfrac{\sin \theta \sin \beta}{\cos \theta \cos \beta}}$$ *Simplify and use the ratio/quotient identity for tangent.*

$$\tan(\theta + \beta) = \frac{\tan \theta + \tan \beta}{1 - \tan \theta \tan \beta}$$

The tangent sum identity is $\tan(\theta + \beta) = \dfrac{\tan \theta + \tan \beta}{1 - \tan \theta \tan \beta}$. To find the tangent difference identity,

substitute $-\beta$ for β and simplify to obtain $\tan(\theta - \beta) = \dfrac{\tan \theta - \tan \beta}{1 + \tan \theta \tan \beta}$.

$$\tan(\theta + \beta) = \frac{\tan \theta + \tan \beta}{1 - \tan \theta \tan \beta}$$

$$\tan(\theta + (-\beta)) = \frac{\tan \theta + \tan(-\beta)}{1 - \tan \theta \tan(-\beta)}$$

$$\tan(\theta - \beta) = \frac{\tan \theta - \tan \beta}{1 + \tan \theta \tan \beta}$$

We have derived the sum and difference identities for sine, cosine and tangent. One use of these identities is to obtain the exact value of the sine, cosine, and/or tangent for nonstandard angles on the unit circle.

Sum and Difference Identities

$$\cos(\theta + \beta) = \cos \theta \cos \beta - \sin \theta \sin \beta$$

$$\cos(\theta - \beta) = \cos \theta \cos \beta + \sin \theta \sin \beta$$

$$\sin(\theta + \beta) = \sin \theta \cos \beta + \cos \theta \sin \beta$$

$$\sin(\theta - \beta) = \sin \theta \cos \beta - \cos \theta \sin \beta$$

$$\tan(\theta + \beta) = \frac{\tan \theta + \tan \beta}{1 - \tan \theta \tan \beta}$$

$$\tan(\theta - \beta) = \frac{\tan \theta - \tan \beta}{1 + \tan \theta \tan \beta}$$

► EXAMPLE 1

Find the exact value of $\cos 15°$.

►►Solution

$\cos 15°$ ◄— 15° is not a standard angle on the unit circle. We can obtain 15° using $45° - 30°$ or $60° - 45°$. We will use the cosine difference identity and $45° - 30°$.

$$\cos 15° = \cos(45° - 30°) = \cos 45° \cos 30° + \sin 45° \sin 30°$$

$$\cos 15° = \cos(45° - 30°) = \left(\frac{\sqrt{2}}{2}\right)\left(\frac{\sqrt{3}}{2}\right) + \left(\frac{\sqrt{2}}{2}\right)\left(\frac{1}{2}\right)$$ Substitute the values for sine and cosine.

$$\cos 15° = \cos(45° - 30°) = \frac{\sqrt{6}}{4} + \frac{\sqrt{2}}{4}$$ Simplify.

$$\cos 15° = \frac{\sqrt{6} + \sqrt{2}}{4}$$ This is the exact value of $\cos 15°$.

▶ **EXAMPLE 2**

Find the exact value of $\sin \dfrac{7\pi}{12}$.

▶▶ **Solution**

$\sin \dfrac{7\pi}{12}$ ⟵ $\dfrac{7\pi}{12}$ is not a standard angle on the unit circle. We need to rewrite the standard angles with a denominator of 12: $\dfrac{\pi}{6} = \dfrac{2\pi}{12}$, $\dfrac{\pi}{4} = \dfrac{3\pi}{12}$, and $\dfrac{\pi}{3} = \dfrac{4\pi}{12}$. We will use $\dfrac{\pi}{4} = \dfrac{3\pi}{12}$ and $\dfrac{\pi}{3} = \dfrac{4\pi}{12}$ because they add to equal $\dfrac{7\pi}{12}$.

$$\sin \frac{7\pi}{12} = \sin\left(\frac{\pi}{4} + \frac{\pi}{3}\right) = \sin\frac{\pi}{4}\cos\frac{\pi}{3} + \cos\frac{\pi}{4}\sin\frac{\pi}{3}$$

$$\sin \frac{7\pi}{12} = \sin\left(\frac{\pi}{4} + \frac{\pi}{3}\right) = \left(\frac{\sqrt{2}}{2}\right)\left(\frac{1}{2}\right) + \left(\frac{\sqrt{2}}{2}\right)\left(\frac{\sqrt{3}}{2}\right)$$ Substitute the values for sine and cosine.

$$\sin \frac{7\pi}{12} = \sin\left(\frac{\pi}{4} + \frac{\pi}{3}\right) = \frac{\sqrt{2}}{4} + \frac{\sqrt{6}}{4}$$ Simplify.

$$\sin \frac{7\pi}{12} = \frac{\sqrt{6} + \sqrt{2}}{4}$$ This is the exact value of $\sin \dfrac{7\pi}{12}$.

► EXAMPLE 3

Find the exact value of $\tan 75°$.

►► Solution

$\tan 75°$ ← | 75° is not a standard angle on the unit circle. We can obtain 75° using $30° + 45°$. |

$$\tan 75° = \tan(30° + 45°) = \frac{\tan 30° + \tan 45°}{1 - \tan 30° \tan 45°}$$

$$\tan 75° = \tan(30° + 45°) = \frac{\frac{\sqrt{3}}{3} + 1}{1 - \left(\frac{\sqrt{3}}{3}\right)(1)}$$ | Substitute the values for sine and cosine. |

$$\tan 75° = \tan(30° + 45°) = \frac{\frac{\sqrt{3}}{3} + 1}{1 - \frac{\sqrt{3}}{3}} \cdot \frac{3}{3}$$ | We need to eliminate the complex fraction by multiplying by the least common denominator. |

$$\tan 75° = \tan(30° + 45°) = \frac{\sqrt{3} + 3}{3 - \sqrt{3}}$$

$$\tan 75° = \frac{\sqrt{3} + 3}{3 - \sqrt{3}}$$ | This is the exact value of $\tan 75°$. |

▶ EXAMPLE 4

Given: $\cos\theta = -\dfrac{15}{17}$ and $\sin\beta = \dfrac{4}{5}$, **θ is in quadrant III and β is in quadrant II, find the following.**

a.) $\sin(\theta+\beta)$ **b.)** $\cos(\theta+\beta)$ **c.)**   What quadrant is $(\theta+\beta)$ located in?

▶▶ Solution

Since θ is in is Quadrant III, we will draw a triangle in Quadrant III.

Since β is in is Quadrant II, we will draw a triangle in Quadrant II.

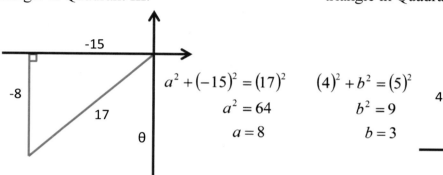

$$a^2 + (-15)^2 = (17)^2$$
$$a^2 = 64$$
$$a = 8$$

$$\sin\theta = -\frac{8}{17}, \quad \cos\theta = -\frac{15}{17}$$

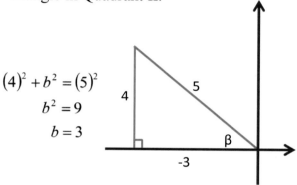

$$(4)^2 + b^2 = (5)^2$$
$$b^2 = 9$$
$$b = 3$$

$$\sin\beta = \frac{4}{5}, \quad \cos\beta = -\frac{3}{5}$$

a.) $\sin(\theta+\beta) = \sin\theta\cos\beta + \cos\theta\sin\beta$

$$\sin(\theta+\beta) = \left(-\frac{8}{17}\right)\left(-\frac{3}{5}\right) + \left(-\frac{15}{17}\right)\left(\frac{4}{5}\right)$$

$$\sin(\theta+\beta) = \frac{24}{85} - \frac{60}{85}$$

$$\sin(\theta+\beta) = -\frac{36}{85}$$

b.) $\cos(\theta+\beta) = \cos\theta\cos\beta - \sin\theta\sin\beta$

$$\cos(\theta+\beta) = \left(-\frac{15}{17}\right)\left(-\frac{3}{5}\right) - \left(-\frac{8}{17}\right)\left(\frac{4}{5}\right)$$

$$\cos(\theta+\beta) = \frac{45}{85} + \frac{32}{85}$$

$$\cos(\theta+\beta) = \frac{77}{85}$$

c.)   Sine is negative in Quadrants III and IV and cosine is positive in Quadrants I and IV therefore $(\theta+\beta)$ must be in Quadrant IV.

► **EXAMPLE 5**

Find the exact value of the given expression.

a.) $\sin 23° \cos 37° + \cos 23° \sin 37°$

b.) $\cos 53° \cos 37° - \sin 53° \sin 37°$

►►**Solution**

a.) $\sin 23° \cos 37° + \cos 23° \sin 37°$

> Notice the resemblance to sine sum identity:
> $\sin(\theta + \beta) = \sin \theta \cos \beta + \cos \theta \sin \beta$.

$\sin 23° \cos 37° + \cos 23° \sin 37° = \sin(23° + 37°)$ | Apply the sine sum identity.

$\sin 23° \cos 37° + \cos 23° \sin 37° = \sin(23° + 37°) = \sin 60°$ | Evaluate the sine of 60°.

$\sin 23° \cos 37° + \cos 23° \sin 37° = \sin(23° + 37°) = \sin 60° = \dfrac{\sqrt{3}}{2}$ | The sine at 60° is $\dfrac{\sqrt{3}}{2}$.

$\sin 23° \cos 37° + \cos 23° \sin 37° = \dfrac{\sqrt{3}}{2}$

b.) $\cos 53° \cos 37° - \sin 53° \sin 37°$

> Notice the resemblance to the cosine sum identity:
> $\cos(\theta + \beta) = \cos \theta \cos \beta - \sin \theta \sin \beta$.

$\cos 53° \cos 37° - \sin 53° \sin 37° = \cos(53° + 37°)$ | Apply the cosine sum identity.

$\cos 53° \cos 37° - \sin 53° \sin 37° = \cos(53° + 37°) = \cos 90°$ | Evaluate the cosine of 90°.

$\cos 53° \cos 37° - \sin 53° \sin 37° = \cos(53° + 37°) = \cos 90° = 0$ | The cosine at 90° is 0.

$\cos 53° \cos 37° - \sin 53° \sin 37° = 0$

Double Angle Identities

We can use the sum and difference identities to derive the double angle identities. The double angle identity for cosine has three forms.

$$\sin(2\theta) = \sin(\theta + \theta) = \sin\theta\cos\theta + \cos\theta\sin\theta$$

$$\sin(2\theta) = 2\sin\theta\cos\theta$$

$$\cos(2\theta) = \cos(\theta + \theta) = \cos\theta\cos\theta - \sin\theta\sin\theta$$

$$\cos(2\theta) = \cos^2\theta - \sin^2\theta$$

$$\cos(2\theta) = \cos^2\theta - \sin^2\theta \qquad\qquad \cos(2\theta) = \cos^2\theta - \sin^2\theta$$

$$\cos(2\theta) = \cos^2\theta - (1 - \cos^2\theta) \qquad \cos(2\theta) = (1 - \sin^2\theta) - \sin^2\theta$$

$$\cos(2\theta) = \cos^2\theta - 1 + \cos^2\theta \qquad\quad \cos(2\theta) = 1 - \sin^2\theta - \sin^2\theta$$

$$\cos(2\theta) = 2\cos^2\theta - 1 \qquad\qquad\qquad \cos(2\theta) = 1 - 2\sin^2\theta$$

$$\tan 2\theta = \tan(\theta + \theta) = \frac{\tan\theta + \tan\theta}{1 - \tan\theta\tan\theta}$$

$$\tan 2\theta = \frac{2\tan\theta}{1 - \tan^2\theta}$$

Double Angle Identities

$$\sin(2\theta) = 2\sin\theta\cos\theta$$

$$\cos(2\theta) = \cos^2\theta - \sin^2\theta \quad \cos(2\theta) = 2\cos^2\theta - 1 \quad \cos(2\theta) = 1 - 2\sin^2\theta$$

$$\tan 2\theta = \frac{2\tan\theta}{1 - \tan^2\theta}$$

▶ **EXAMPLE 6**

Find the exact value of $\sin 2\theta$, $\cos 2\theta$ **and** $\tan 2\theta$ **for** $\sin \theta = -\dfrac{21}{29}$ **and** θ **is in Quadrant IV.**

▶▶ **Solution**

$\sin \theta = -\dfrac{21}{29}$, θ is in Quadrant IV

In order to find $\sin 2\theta$, we need to know both $\sin \theta$ and $\cos \theta$. We will draw a triangle in the fourth quadrant and label what we know. Using the Pythagorean Theorem we can find the measurement of the missing side.

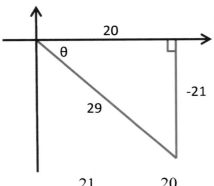

$$a^2 + (-21)^2 = (29)^2$$
$$a^2 = (29)^2 - (-21)^2$$
$$\sqrt{a} = \sqrt{841 - 441}$$
$$a = \sqrt{400}$$
$$a = 20$$

$$\sin \theta = -\frac{21}{29}, \quad \cos \theta = \frac{20}{29}, \quad \tan \theta = -\frac{21}{20}$$

$$\sin 2\theta = 2 \sin \theta \cos \theta$$
$$\sin 2\theta = 2\left(-\frac{21}{29}\right)\left(\frac{20}{29}\right)$$
$$\sin 2\theta = -\frac{840}{841}$$

$$\cos 2\theta = \cos^2 \theta - \sin^2 \theta$$
$$\cos 2\theta = \left(\frac{20}{29}\right)^2 - \left(-\frac{21}{29}\right)^2$$
$$\cos 2\theta = \frac{400}{841} - \frac{441}{841}$$
$$\cos 2\theta = -\frac{41}{841}$$

$$\tan 2\theta = \frac{2 \tan \theta}{1 - \tan^2 \theta}$$
$$\tan 2\theta = \frac{2\left(-\dfrac{21}{20}\right)}{1 - \left(-\dfrac{21}{20}\right)^2}$$
$$\tan 2\theta = \frac{-\dfrac{21}{10}}{1 - \dfrac{441}{400}} \cdot \frac{400}{400}$$
$$\tan 2\theta = \frac{-840}{400 - 441}$$
$$\tan 2\theta = \frac{840}{41}$$

► **EXAMPLE 7**

Use a double angle identity to find the exact value for the expressions.

a.) $\sin 15^\circ \cos 15^\circ$

b.) $\dfrac{2 \tan \dfrac{\pi}{8}}{1 - \tan^2 \dfrac{\pi}{8}}$

►► **Solution**

a.) $\sin 15^\circ \cos 15^\circ$ resembles the form of $\sin 2\theta = 2 \sin \theta \cos \theta$.

$\sin 15^\circ \cos 15^\circ$ | We need a 2 in front of $\sin 15^\circ \cos 15^\circ$ for it to equal $\sin 2\theta$. |

$\dfrac{1}{2}\left(2 \sin 15^\circ \cos 15^\circ\right)$ | We cannot change the value of the expression. If we put the 2, we have to put $\dfrac{1}{2}$ in front. 2 times $\dfrac{1}{2}$ is equivalent to one and we are not changing the value of the expression. |

$\dfrac{1}{2}\left(\sin(2 \cdot 15^\circ)\right)$ | Using the double angle identity for sine, we can rewrite the expression. |

$\dfrac{1}{2}\left(\sin 30^\circ\right)$

$\dfrac{1}{2}\left(\dfrac{1}{2}\right)$ | Substitute the value for $\sin 30^\circ$ which is $\dfrac{1}{2}$. |

$\dfrac{1}{4}$ | Simplify. $\sin 15^\circ \cos 15^\circ = \dfrac{1}{4}$ |

b.) $\dfrac{2 \tan \dfrac{\pi}{8}}{1 - \tan^2 \dfrac{\pi}{8}}$ resembles the form of $\tan 2\theta = \dfrac{2 \tan \theta}{1 - \tan^2 \theta}$.

$$\dfrac{2 \tan \dfrac{\pi}{8}}{1 - \tan^2 \dfrac{\pi}{8}} = \tan\left(2 \cdot \dfrac{\pi}{8}\right) = \tan \dfrac{\pi}{4} = 1$$

Half-Angle Identities

We will use the double angle identity $\cos(2\theta) = 1 - 2\sin^2\theta$ to derive the half-angle identity for sine.

$$\cos(2\theta) = 1 - 2\sin^2\theta \qquad \boxed{\text{We will begin by solving for } \sin^2\theta.}$$

$$2\sin^2\theta + \cos(2\theta) = 1$$

$$2\sin^2\theta = 1 - \cos(2\theta)$$

$$\sin^2\theta = \frac{1 - \cos(2\theta)}{2}$$

$$\sin^2\frac{\theta}{2} = \frac{1 - \cos\left(2 \cdot \frac{\theta}{2}\right)}{2} \qquad \boxed{\text{Substitute } \frac{\theta}{2} \text{ for } \theta.}$$

$$\sin^2\frac{\theta}{2} = \frac{1 - \cos\theta}{2}$$

$$\sqrt{\sin^2\frac{\theta}{2}} = \pm\sqrt{\frac{1 - \cos\theta}{2}} \qquad \boxed{\text{Solve for sine by taking the square root of both sides.}}$$

$$\sin\frac{\theta}{2} = \pm\sqrt{\frac{1 - \cos\theta}{2}} \qquad \boxed{\text{This is the half-angle for sine.}}$$

We will use the double angle identity $\cos(2\theta) = 2\cos^2\theta - 1$ to derive the half-angle identity for cosine.

$$\cos(2\theta) = 2\cos^2\theta - 1 \qquad \boxed{\text{We will begin by solving for } \cos^2\theta.}$$

$$2\cos^2\theta = 1 + \cos(2\theta)$$

$$\frac{2\cos^2\theta}{2} = \frac{1 + \cos(2\theta)}{2}$$

$$\cos^2\theta = \frac{1 + \cos(2\theta)}{2} \qquad \boxed{\text{Substitute } \frac{\theta}{2} \text{ for } \theta \text{ and simplify.}}$$

$$\cos^2\frac{\theta}{2} = \frac{1 + \cos\theta}{2}$$

$$\sqrt{\cos^2\frac{\theta}{2}} = \pm\sqrt{\frac{1 + \cos\theta}{2}} \qquad \boxed{\text{Solve for sine by taking the square root of both sides.}}$$

$$\cos\frac{\theta}{2} = \pm\sqrt{\frac{1 + \cos\theta}{2}} \qquad \boxed{\text{This is the half-angle for sine.}}$$

We will use the ratio/quotient identity $\tan\theta = \dfrac{\sin\theta}{\cos\theta}$ to derive a half-angle identity for tangent.

$$\tan\frac{\theta}{2} = \frac{\sin\dfrac{\theta}{2}}{\cos\dfrac{\theta}{2}} = \frac{\pm\sqrt{\dfrac{1-\cos\theta}{2}}}{\sqrt{\dfrac{1+\cos\theta}{2}}}$$

$$\tan\frac{\theta}{2} = \pm\sqrt{\frac{\dfrac{1-\cos\theta}{2}}{\dfrac{1+\cos\theta}{2}}}$$

$$\tan\frac{\theta}{2} = \pm\sqrt{\frac{1-\cos\theta}{2}\cdot\frac{2}{1+\cos\theta}}$$

$$\tan\frac{\theta}{2} = \pm\sqrt{\frac{1-\cos\theta}{1+\cos\theta}}$$

There are two other half-angle identities for tangent. Here is one form.

$$\tan\frac{\theta}{2} = \frac{\sin\dfrac{\theta}{2}}{\cos\dfrac{\theta}{2}}\cdot\frac{2\cos\dfrac{\theta}{2}}{2\cos\dfrac{\theta}{2}} = \frac{2\sin\dfrac{\theta}{2}\cos\dfrac{\theta}{2}}{2\cos^2\dfrac{\theta}{2}}$$

$$\tan\frac{\theta}{2} = \frac{2\sin\dfrac{\theta}{2}\cos\dfrac{\theta}{2}}{2\cos^2\dfrac{\theta}{2}} = \frac{\sin\left(2\cdot\dfrac{\theta}{2}\right)}{2\left(\pm\sqrt{\dfrac{1+\cos\theta}{2}}\right)^2}$$

$$\tan\frac{\theta}{2} = \frac{\sin\left(2\cdot\dfrac{\theta}{2}\right)}{2\left(\pm\sqrt{\dfrac{1+\cos\theta}{2}}\right)^2} = \frac{\sin\theta}{2\left(\dfrac{1+\cos\theta}{2}\right)} = \frac{\sin\theta}{1+\cos\theta}$$

Here is another form of the half-angle identity for tangent.

$$\tan\frac{\theta}{2} = \frac{\sin\dfrac{\theta}{2}}{\cos\dfrac{\theta}{2}} \cdot \frac{2\sin\dfrac{\theta}{2}}{2\sin\dfrac{\theta}{2}} = \frac{2\sin^2\dfrac{\theta}{2}}{2\sin\dfrac{\theta}{2}\cos\dfrac{\theta}{2}}$$

$$\tan\frac{\theta}{2} = \frac{2\sin^2\dfrac{\theta}{2}}{2\sin\dfrac{\theta}{2}\cos\dfrac{\theta}{2}} = \frac{2\left(\pm\sqrt{\dfrac{1-\cos\theta}{2}}\right)^2}{\sin\left(2\cdot\dfrac{\theta}{2}\right)}$$

$$\tan\frac{\theta}{2} = \frac{2\left(\pm\sqrt{\dfrac{1-\cos\theta}{2}}\right)^2}{\sin\left(2\cdot\dfrac{\theta}{2}\right)} = \frac{2\left(\dfrac{1-\cos\theta}{2}\right)}{\sin\theta} = \frac{1-\cos\theta}{\sin\theta}$$

Half-Angle Identities

$$\sin\frac{\theta}{2} = \pm\sqrt{\frac{1-\cos\theta}{2}}$$

$$\cos\frac{\theta}{2} = \pm\sqrt{\frac{1+\cos\theta}{2}}$$

$$\tan\frac{\theta}{2} = \pm\sqrt{\frac{1-\cos\theta}{1+\cos\theta}} \qquad \tan\frac{\theta}{2} = \frac{\sin\theta}{1+\cos\theta} \qquad \tan\frac{\theta}{2} = \frac{1-\cos\theta}{\sin\theta}$$

▶ **EXAMPLE 8**

Use a half- angle identity to find the exact value for the expressions.

a.) $\cos 15°$

b.) $\sin 112.5°$

▶▶ **Solution**

a.)   Since 15° is in *Quadrant I*, $\cos 15°$ is positive.

$$\cos 15° = \sqrt{\frac{1 + \cos(2 \cdot 15°)}{2}}$$

$$\cos 15° = \sqrt{\frac{1 + \cos 30°}{2}}$$

$$\cos 15° = \sqrt{\frac{1 + \frac{\sqrt{3}}{2}}{2}}$$

$$\cos 15° = \sqrt{\frac{1 + \frac{\sqrt{3}}{2}}{2} \cdot \frac{2}{2}}$$

$$\cos 15° = \sqrt{\frac{2 + \sqrt{3}}{4}}$$

$$\cos 15° = \frac{\sqrt{2 + \sqrt{3}}}{2}$$

b.)   Since 112.5° is in *Quadrant II* and sine is positive in *Quadrant II*, $\sin 112.5°$ is positive.

$$\sin 112.5° = \sqrt{\frac{1 + \cos(2 \cdot 112.5°)}{2}}$$

$$\sin 112.5° = \sqrt{\frac{1 + \cos 225°}{2}}$$

$$\sin 112.5° = \sqrt{\frac{1 + \frac{\sqrt{2}}{2}}{2}}$$

$$\sin 112.5° = \sqrt{\frac{1 + \frac{\sqrt{2}}{2}}{2} \cdot \frac{2}{2}}$$

$$\sin 112.5° = \sqrt{\frac{2 + \sqrt{2}}{4}}$$

$$\sin 112.5° = \frac{\sqrt{2 + \sqrt{2}}}{2}$$

► **EXAMPLE 9**

Use all three tangent half-angle identities to find the exact value for $\tan\dfrac{\pi}{8}$.

►► **Solution**

$$\tan\frac{\pi}{8} = \sqrt{\frac{1-\cos\left(2\cdot\frac{\pi}{8}\right)}{1+\cos\left(2\cdot\frac{\pi}{8}\right)}} \qquad \tan\frac{\pi}{8} = \frac{\sin\left(2\cdot\frac{\pi}{8}\right)}{1+\cos\left(2\cdot\frac{\pi}{8}\right)} \qquad \tan\frac{\pi}{8} = \frac{1-\cos\left(2\cdot\frac{\pi}{8}\right)}{\sin\left(2\cdot\frac{\pi}{8}\right)}$$

$$\tan\frac{\pi}{8} = \sqrt{\frac{1-\cos\frac{\pi}{4}}{1+\cos\frac{\pi}{4}}} \qquad \tan\frac{\pi}{8} = \frac{\sin\frac{\pi}{4}}{1+\cos\frac{\pi}{4}} \qquad \tan\frac{\pi}{8} = \frac{1-\cos\frac{\pi}{4}}{\sin\frac{\pi}{4}}$$

$$\tan\frac{\pi}{8} = \sqrt{\frac{1-\frac{\sqrt{2}}{2}}{1+\frac{\sqrt{2}}{2}}\cdot\frac{2}{2}} \qquad \tan\frac{\pi}{8} = \frac{\frac{\sqrt{2}}{2}}{1+\frac{\sqrt{2}}{2}}\cdot\frac{2}{2} \qquad \tan\frac{\pi}{8} = \frac{1-\frac{\sqrt{2}}{2}}{\frac{\sqrt{2}}{2}}\cdot\frac{2}{2}$$

$$\tan\frac{\pi}{8} = \sqrt{\frac{2-\sqrt{2}}{2+\sqrt{2}}} \qquad \tan\frac{\pi}{8} = \frac{\sqrt{2}}{2+\sqrt{2}}\cdot\frac{2-\sqrt{2}}{2-\sqrt{2}} \qquad \tan\frac{\pi}{8} = \frac{2-\sqrt{2}}{\sqrt{2}}\cdot\frac{\sqrt{2}}{\sqrt{2}}$$

$$\tan\frac{\pi}{8} = \sqrt{\frac{2-\sqrt{2}}{2+\sqrt{2}}\cdot\frac{2-\sqrt{2}}{2-\sqrt{2}}} \qquad \tan\frac{\pi}{8} = \frac{2\sqrt{2}-2}{4-2\sqrt{2}+2\sqrt{2}-2} \qquad \tan\frac{\pi}{8} = \frac{2\sqrt{2}-2}{2}$$

$$\tan\frac{\pi}{8} = \sqrt{\frac{\left(2-\sqrt{2}\right)^2}{4-2\sqrt{2}+2\sqrt{2}-2}} \qquad \tan\frac{\pi}{8} = \frac{2\sqrt{2}-2}{2} \qquad \tan\frac{\pi}{8} = \frac{2\left(\sqrt{2}-1\right)}{2}$$

$$\tan\frac{\pi}{8} = \sqrt{\frac{\left(2-\sqrt{2}\right)^2}{2}} \qquad \tan\frac{\pi}{8} = \frac{2\left(\sqrt{2}-1\right)}{2} \qquad \tan\frac{\pi}{8} = \sqrt{2}-1$$

$$\tan\frac{\pi}{8} = \frac{2-\sqrt{2}}{\sqrt{2}}\cdot\frac{\sqrt{2}}{\sqrt{2}}$$

$$\tan\frac{\pi}{8} = \frac{2\sqrt{2}-2}{2} \qquad \tan\frac{\pi}{8} = \sqrt{2}-1$$

$$\tan\frac{\pi}{8} = \frac{2\left(\sqrt{2}-1\right)}{2} = \sqrt{2}-1$$

▶ EXAMPLE 10

Find the exact value for $\sin\dfrac{\theta}{2}, \cos\dfrac{\theta}{2}$, **and** $\tan\dfrac{\theta}{2}$ **given** $\sin\theta = -\dfrac{33}{65}$, θ **is in Quadrant III.**

▶▶ Solution

In order to use the half-angle identities we need to know the value of $\cos\theta$. We will draw a triangle in Quadrant III and find the $\cos\theta$.

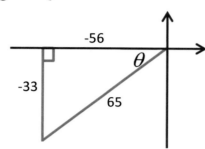

$$a^2 + (-33)^2 = (65)^2$$
$$a^2 = (65)^2 - (-33)^2$$
$$a^2 = 4225 - 1089$$
$$a^2 = 3136$$
$$\sqrt{a^2} = \sqrt{3136}$$
$$a = 56$$

$$\sin\theta = -\frac{33}{65}$$

$$\cos\theta = -\frac{56}{65}$$

$$\pi < \theta < \frac{3\pi}{2}$$

$$\frac{1}{2}\cdot\pi < \frac{1}{2}\cdot\theta < \frac{1}{2}\cdot\frac{3\pi}{2}$$

$$\frac{\pi}{2} < \frac{\theta}{2} < \frac{3\pi}{4}$$

Since θ is in Quadrant III, $\pi < \theta < \dfrac{3\pi}{2}$ we can find what quadrant $\dfrac{\theta}{2}$ is located in. Since $\dfrac{\theta}{2}$ is in Quadrant II, cosine and tangent are negative and sine is positive.

$$\sin\frac{\theta}{2} = \sqrt{\frac{1-\cos\theta}{2}}$$

$$\sin\frac{\theta}{2} = \sqrt{\frac{1-\left(-\dfrac{56}{65}\right)}{2}}$$

$$\sin\frac{\theta}{2} = \sqrt{\frac{1+\dfrac{56}{65}}{2}\cdot\frac{65}{65}}$$

$$\sin\frac{\theta}{2} = \sqrt{\frac{65+56}{130}}$$

$$\sin\frac{\theta}{2} = \sqrt{\frac{121}{130}}$$

$$\sin\frac{\theta}{2} = \frac{11}{\sqrt{130}}$$

$$\cos\frac{\theta}{2} = -\sqrt{\frac{1+\cos\theta}{2}}$$

$$\cos\frac{\theta}{2} = -\sqrt{\frac{1+\left(-\dfrac{56}{65}\right)}{2}}$$

$$\cos\frac{\theta}{2} = -\sqrt{\frac{1-\dfrac{56}{65}}{2}\cdot\frac{65}{65}}$$

$$\cos\frac{\theta}{2} = -\sqrt{\frac{65-56}{130}}$$

$$\cos\frac{\theta}{2} = -\sqrt{\frac{9}{130}}$$

$$\cos\frac{\theta}{2} = -\frac{3}{\sqrt{130}}$$

$$\tan\frac{\theta}{2} = \frac{\sin\dfrac{\theta}{2}}{\cos\dfrac{\theta}{2}}$$

$$\tan\frac{\theta}{2} = \frac{\dfrac{11}{\sqrt{130}}}{-\dfrac{3}{\sqrt{130}}}$$

$$\tan\frac{\theta}{2} = \frac{11}{\sqrt{130}}\cdot-\frac{\sqrt{130}}{3}$$

$$\tan\frac{\theta}{2} = -\frac{11}{3}$$

►EXAMPLE 11

Verify the following identity: $\dfrac{\cos(2\theta)}{1-\cos^2\theta} = \cot^2\theta - 1.$

►►Solution

$\dfrac{\cos(2\theta)}{1-\cos^2\theta} = \cot^2\theta - 1$

> We are going to work the left hand side of the equation.

$\dfrac{\cos^2\theta - \sin^2\theta}{\sin^2\theta} =$

> Substitute $\cos^2\theta - \sin^2\theta$ for the cosine double angle and substitute $1-\cos^2\theta$ for $\sin^2\theta$.

$\dfrac{\cos^2\theta}{\sin^2\theta} - \dfrac{\sin^2\theta}{\sin^2\theta} =$

> Separate the fraction into two fractions.

$\cot^2\theta - 1 =$

> Substitute and simplify. The identity is verified.

►EXAMPLE 12

Verify the following identity: $\cos^2\left(\dfrac{x}{2}\right) - \sin^2\left(\dfrac{x}{2}\right) = \cos x.$

►►Solution

$\cos^2\left(\dfrac{x}{2}\right) - \sin^2\left(\dfrac{x}{2}\right) = \cos x$

> We are going to work the left hand side of the equation.

$\left(\pm\sqrt{\dfrac{1+\cos x}{2}}\right)^2 - \left(\pm\sqrt{\dfrac{1-\cos x}{2}}\right)^2 =$

> Substitute for the cosine half-angle and the sine half-angle.

$\dfrac{1+\cos\theta}{2} - \dfrac{1-\cos x}{2} =$

> Simplify.

$\dfrac{1+\cos\theta - 1 + \cos x}{2} =$

> Combine the fractions.

$\dfrac{2\cos x}{2} =$

> Simplify.

$\cos x =$

> The identity is verified.

HOMEWORK

Objective 1

Use sum and difference identities to find the exact value.

See Examples 1-3.

1. $\cos\dfrac{5\pi}{12}$

2. $\sin\dfrac{\pi}{12}$

3. $\tan 75°$

4. $\cos 105°$

5. $\sin 105°$

6. $\tan\left(-\dfrac{\pi}{12}\right)$

7. $\cos(-15°)$

8. $\cos\dfrac{7\pi}{12}$

9. $\sin 75°$

10. $\tan 105°$

11. $\sin 15°$

12. $\tan\dfrac{5\pi}{12}$

Objective 1

Find the exact value of the given expressions.

See Example 4.

13.   Given: $\cos\theta = -\dfrac{12}{13}$ and $\sin\beta = -\dfrac{7}{25}$, θ is in quadrant III and β is in quadrant IV, find the following.

   a.) $\sin(\theta+\beta)$ b.) $\cos(\theta+\beta)$ c.)   What quadrant is $(\theta+\beta)$ located in?

14.   Given: $\tan\theta = -\dfrac{21}{20}$ and $\cos\beta = -\dfrac{33}{65}$, θ is in quadrant II and β is in quadrant III, find the following.

   a.) $\sin(\theta-\beta)$ b.) $\tan(\theta-\beta)$ c.)   What quadrant is $(\theta-\beta)$ located in?

15.   Given: $\sec\theta = \dfrac{37}{35}$ and $\csc\beta = -\dfrac{17}{8}$, θ is in quadrant IV and β is in quadrant III, find the following.

   a.) $\cos(\theta+\beta)$ b.) $\tan(\theta+\beta)$ c.)   What quadrant is $(\theta+\beta)$ located in?

Objective 1

Find the exact value of the given expressions.

See Example 5.

16. $\sin\left(\dfrac{11\pi}{24}\right)\cos\left(\dfrac{5\pi}{24}\right)-\cos\left(\dfrac{11\pi}{24}\right)\sin\left(\dfrac{5\pi}{24}\right)$

17. $\cos 183°\cos 153°+\sin 183°\sin 153°$

18. $\dfrac{\tan 10°+\tan 50°}{1-\tan 10°\tan 50°}$

19. $\dfrac{\tan\left(\dfrac{11\pi}{21}\right)-\tan\left(\dfrac{4\pi}{21}\right)}{1+\tan\left(\dfrac{11\pi}{21}\right)\tan\left(\dfrac{4\pi}{21}\right)}$

20. $\cos\left(\dfrac{7\pi}{36}\right)\cos\left(\dfrac{5\pi}{36}\right)-\sin\left(\dfrac{7\pi}{36}\right)\sin\left(\dfrac{5\pi}{36}\right)$

21. $\sin 137°\cos 47°-\cos 137°\sin 47°$

Objective 2

Find the exact value of $\sin 2\theta$, $\cos 2\theta$ and $\tan 2\theta$.

See Example 6.

22. $\sin\theta=-\dfrac{12}{13}$, θ *in QIII*

23. $\tan\theta=-\dfrac{9}{40}$, θ *in QIV*

24. $\cos\theta=-\dfrac{13}{85}$, θ *in QII*

25. $\sin\theta=\dfrac{45}{53}$, θ *in QII*

Objective 2

Use a double angle identity to find the exact value for the expressions.

See Example 7.

26. $\cos 75°\sin 75°$

27. $1-2\sin^2\dfrac{\pi}{8}$

28. $\cos^2 15°-\sin^2 15°$

29. $\dfrac{2\tan 22.5°}{1-\tan^2 22.5°}$

30. $2\sin 15°\cos 15°$

31. $\dfrac{2\tan\left(\dfrac{\pi}{12}\right)}{1-\tan^2\left(\dfrac{\pi}{12}\right)}$

Objective 3

Use a half- angle identity to find the exact value for the expressions.

See Examples 8 and 9.

32. $\cos \dfrac{5\pi}{12}$ 33. $\sin \dfrac{\pi}{12}$ 34. $\sin 75°$ 35. $\tan 15°$

36. $\cos \dfrac{7\pi}{12}$ 37. $\tan \dfrac{\pi}{8}$ 38. $\cos 67.5°$ 39. $\sin 22.5°$

Objective 3

Find the exact value for $\sin \dfrac{\theta}{2}$, $\cos \dfrac{\theta}{2}$, and $\tan \dfrac{\theta}{2}$.

See Example 10.

40. $\sin \theta = -\dfrac{20}{29}$, θ in Quadrant IV

41. $\cos \theta = -\dfrac{15}{17}$, θ in Quadrant II

42. $\tan \theta = \dfrac{15}{8}$, θ in Quadrant III

43. $\sec \theta = \dfrac{25}{24}$, θ in Quadrant IV

Objective 4

Verify the following identities.

See Examples 11 and 12.

44. $\dfrac{2\sin x \cos x}{\cos^2 x - \sin^2 x} = \tan(2x)$

45. $\dfrac{1 - 2\sin^2 x}{\cos^2 x - \sin^2 x} = 1$

46. $(\sin \theta + \cos \theta)^2 = 1 + \sin(2\theta)$

47. $\cot \beta - \tan \beta = \dfrac{2\cos(2\beta)}{\sin(2\beta)}$

48. $\sin 2y = \dfrac{2\tan y}{1 + \tan^2 y}$

49. $\cos 2\beta = \dfrac{1 - \tan^2 \beta}{1 + \tan^2 \beta}$

50. $2\cos^2 \dfrac{x}{2} \tan x = \tan x + \sin x$

51. $\csc \beta \sin 2\beta - \sec \beta = \cos 2\beta \sec \beta$

52. $2\cos^2 \left(\dfrac{\theta}{2} \right) - 1 = \cos \theta$

53. $1 + \tan^2 \alpha = 2\tan \alpha \csc 2\alpha$

ANSWERS

1. $\dfrac{\sqrt{6}-\sqrt{2}}{4}$

2. $\dfrac{\sqrt{6}-\sqrt{2}}{4}$

3. $\dfrac{\sqrt{3}+1}{\sqrt{3}-1}$ or $2+\sqrt{3}$

4. $\dfrac{\sqrt{2}-\sqrt{6}}{4}$

5. $\dfrac{\sqrt{6}+\sqrt{2}}{4}$

6. $\dfrac{1-\sqrt{3}}{1+\sqrt{3}}$ or $-2+\sqrt{3}$

7. $\dfrac{\sqrt{6}+\sqrt{2}}{4}$

8. $\dfrac{\sqrt{2}-\sqrt{6}}{4}$

9. $\dfrac{\sqrt{2}+\sqrt{6}}{4}$

10. $\dfrac{1-\sqrt{3}}{1+\sqrt{3}}$ or $-2+\sqrt{3}$

11. $\dfrac{\sqrt{6}-\sqrt{2}}{4}$

12. $\dfrac{\sqrt{3}+1}{\sqrt{3}-1}$ or $2+\sqrt{3}$

13. $-\dfrac{36}{325},-\dfrac{323}{325},QIII$

14. $-\dfrac{1813}{1885},\dfrac{1813}{516},QIII$

15. $-\dfrac{621}{629},\dfrac{100}{621},QIII$

16. $\dfrac{\sqrt{2}}{2}$

17. $\dfrac{\sqrt{3}}{2}$

18. $\sqrt{3}$

19. $\sqrt{3}$

20. $\dfrac{1}{2}$

21. 1

22. $sin2\theta=\dfrac{120}{169},cos2\theta=-\dfrac{119}{169},tan2\theta=-\dfrac{120}{119}$

23. $sin2\theta=-\dfrac{720}{1681},cos2\theta=\dfrac{1519}{1681},tan2\theta=-\dfrac{720}{1519}$

24. $sin2\theta=-\dfrac{2184}{7225},cos2\theta=-\dfrac{6887}{7225},tan2\theta=\dfrac{2184}{6887}$

25. $sin2\theta=\dfrac{-2520}{2809},cos2\theta=-\dfrac{1241}{2809},tan2\theta=\dfrac{2520}{1241}$

26. $\dfrac{1}{4}$

27. $\dfrac{\sqrt{2}}{2}$

28. $\dfrac{\sqrt{3}}{2}$

29. 1

30. $\dfrac{1}{2}$

31. $\dfrac{\sqrt{3}}{3}$

32. $\dfrac{\sqrt{2-\sqrt{3}}}{2}$

33. $\dfrac{\sqrt{2-\sqrt{3}}}{2}$

34. $\dfrac{\sqrt{2+\sqrt{3}}}{2}$

35. $2-\sqrt{3}$

36. $-\dfrac{\sqrt{2-\sqrt{3}}}{2}$

37. $\sqrt{2}-1$

38. $\dfrac{\sqrt{2-\sqrt{2}}}{2}$

39. $\dfrac{\sqrt{2-\sqrt{2}}}{2}$

40. $sin\dfrac{\theta}{2}=\dfrac{2}{\sqrt{29}},cos\dfrac{\theta}{2}=-\dfrac{5}{\sqrt{29}},tan\dfrac{\theta}{2}=-\dfrac{2}{5}$

41. $sin\dfrac{\theta}{2}=\dfrac{4}{\sqrt{17}},cos\dfrac{\theta}{2}=\dfrac{1}{\sqrt{17}},tan\dfrac{\theta}{2}=4$

42. $\sin\dfrac{\theta}{2} = \dfrac{5}{\sqrt{34}}, \cos\dfrac{\theta}{2} = -\dfrac{3}{\sqrt{34}}, \tan\dfrac{\theta}{2} = -\dfrac{5}{3}$

43. $\sin\dfrac{\theta}{2} = \dfrac{1}{5\sqrt{2}}, \cos\dfrac{\theta}{2} = -\dfrac{7}{5\sqrt{2}}, \tan\dfrac{\theta}{2} = -\dfrac{1}{7}$

44.
$$\frac{2\sin x\cos x}{\cos^2 x - \sin^2 x} = \tan(2x)$$

$$\frac{\sin(2x)}{\cos(2x)} =$$

$$\tan(2x) = \tan(2x)$$

45.
$$\frac{1 - 2\sin^2 x}{\cos^2 x - \sin^2 x} = 1$$

$$\frac{\cos(2x)}{\cos(2x)} =$$

$$1 = 1$$

46.
$$(\sin\theta + \cos\theta)^2 = 1 + \sin(2\theta)$$

$$\sin^2\theta + \sin\theta\cos\theta + \sin\theta\cos\theta + \cos^2\theta =$$

$$1 + 2\sin\theta\cos\theta =$$

$$1 + \sin(2\theta) = 1 + \sin(2\theta)$$

47. $\cot\beta - \tan\beta = \dfrac{2\cos(2\beta)}{\sin(2\beta)}$

$$= \frac{2(\cos^2\beta - \sin^2\beta)}{2\sin\beta\cos\beta}$$

$$= \frac{2\cos^2\beta - 2\sin^2\beta}{2\sin\beta\cos\beta}$$

$$= \frac{2\cos^2\beta}{2\sin\beta\cos\beta} - \frac{2\sin^2\beta}{2\sin\beta\cos\beta}$$

$$= \frac{\cos\beta}{\sin\beta} - \frac{\sin\beta}{\cos\beta}$$

$$\cot\beta - \tan\beta = \cot\beta - \tan\beta$$

48.
$$sin2y = \frac{2tany}{1 + tan^2y}$$

$$= \frac{2\frac{siny}{cosy}}{1 + \frac{sin^2y}{cos^2y}} \cdot \frac{cos^2y}{cos^2y}$$

$$= \frac{2sinycosy}{cos^2y + sin^2y}$$

$$= \frac{sin2y}{1}$$

$$sin2y = sin2y$$

49.
$$cos2\beta = \frac{1 - tan^2\beta}{1 + tan^2\beta}$$

$$= \frac{1 - \frac{sin^2\beta}{cos^2\beta}}{1 + \frac{sin^2\beta}{cos^2\beta}} \cdot \frac{cos^2\beta}{cos^2\beta}$$

$$= \frac{cos^2\beta - sin^2\beta}{cos^2\beta + sin^2\beta}$$

$$= \frac{cos2\beta}{1}$$

$$cos2\beta = cos2\beta$$

50.
$$2cos^2\frac{x}{2}tanx = tanx + sinx$$

$$2\left(\pm\sqrt{\frac{1 + cosx}{2}}\right)^2 tanx =$$

$$2\left(\frac{1 + cosx}{2}\right)tanx =$$

$$(1 + cosx)tanx =$$

$$tanx + cosxtanx =$$

$$tanx + cosx \cdot \frac{sinx}{cosx} =$$

$$tanx + sinx = tanx + sinx$$

51.
$$csc\beta sin2\beta - sec\beta = cos2\beta sec\beta$$

$$\frac{1}{sin\beta} \cdot 2sin\beta cos\beta - \frac{1}{cos\beta} =$$

$$2cos\beta - \frac{1}{cos\beta} =$$

$$\frac{2cos^2\beta - 1}{cos\beta} =$$

$$cos2\beta \cdot \frac{1}{cos\beta} =$$

$$cos2\beta sec\beta = cos2\beta sec\beta$$

52.
$$2cos^2\left(\frac{\theta}{2}\right) - 1 = cos\theta$$

$$2\left(\pm\sqrt{\frac{1 + cos\theta}{2}}\right)^2 - 1 =$$

$$2\left(\frac{1 + cos\theta}{2}\right) - 1 =$$

$$1 + cos\theta - 1 =$$

$$cos\theta = cos\theta$$

53.

$$1 + \tan^2\alpha = 2\tan\alpha\csc2\alpha$$

$$= 2 \cdot \frac{\sin\alpha}{\cos\alpha} \cdot \frac{1}{\sin2\alpha}$$

$$= 2 \cdot \frac{\sin\alpha}{\cos\alpha} \cdot \frac{1}{2\sin\alpha\cos\alpha}$$

$$= \frac{1}{\cos^2\alpha}$$

$$= \sec^2\alpha$$

$$1 + \tan^2\alpha = 1 + \tan^2\alpha$$

Section 15: Product-to-Sum, Sum-to-Product, Power Reduction Identities

Learning Outcomes:

- The student will correctly memorize and apply trigonometric formulas, definitions, identities, and properties.

Objectives: At the conclusion of this lesson you should be able to:

1. Apply the product-to-sum identities.
2. Apply the sum-to-product identities.
3. Apply the power reduction identities.

In this section we are going to look at a group of identities that allow us to restate a product of two trigonometric functions as a sum, and a sum of two trigonometric functions as a product. We will also look at power reducing identities.

Sum to product identities are used in the study of sound waves in music to convert sum forms to more convenient product forms.

Product-To-Sum Identities

The product-to-sum identities are used in calculus to convert product forms to more convenient sum forms. We can derive these identities by adding or subtracting sum and difference identities for sine and cosine. If we add the cosine sum and the cosine difference, we can derive the product-to-sum identity $\cos\theta\cos\beta$ as shown below.

$$\cos(\theta+\beta)+\cos(\theta-\beta)=(\cos\theta\cos\beta-\sin\theta\sin\beta)+(\cos\theta\cos\beta+\sin\theta\sin\beta)$$

$$\cos(\theta+\beta)+\cos(\theta-\beta)=\cos\theta\cos\beta-\sin\theta\sin\beta+\cos\theta\cos\beta+\sin\theta\sin\beta$$

$$\cos(\theta+\beta)+\cos(\theta-\beta)=2\cos\theta\cos\beta$$

$$\frac{\cos(\theta+\beta)+\cos(\theta-\beta)}{2}=\frac{2\cos\theta\cos\beta}{2}$$

$$\frac{1}{2}[\cos(\theta+\beta)+\cos(\theta-\beta)]=\cos\theta\cos\beta$$

The other product-to-sum identities can be derived the same way.

> ### Product-To-Sum Identities
>
> $$\cos\theta\cos\beta = \frac{1}{2}\left[\cos(\theta+\beta)+\cos(\theta-\beta)\right]$$
>
> $$\sin\theta\sin\beta = \frac{1}{2}\left[\cos(\theta-\beta)-\cos(\theta+\beta)\right]$$
>
> $$\sin\theta\cos\beta = \frac{1}{2}\left[\sin(\theta+\beta)+\sin(\theta-\beta)\right]$$
>
> $$\cos\theta\sin\beta = \frac{1}{2}\left[\sin(\theta+\beta)-\sin(\theta-\beta)\right]$$

► EXAMPLE 1

Write each product as a sum using the product-to-sum identities.

a.) $\cos(5\theta)\sin(3\theta)$

b.) $4\sin\left(\dfrac{5\theta}{9}\right)\sin\left(\dfrac{4\theta}{9}\right)$

►► Solution

a.) $\cos(5\theta)\sin(3\theta)$

$\cos(5\theta)\sin(3\theta) = \dfrac{1}{2}\left[\sin(5\theta+3\theta)-\sin(5\theta-3\theta)\right]$ Substitute into the product-to-sum identity.

$\cos(5\theta)\sin(3\theta) = \dfrac{1}{2}\left[\sin(8\theta)-\sin(2\theta)\right]$ Simplify.

b.) $4\sin\left(\dfrac{5\theta}{9}\right)\sin\left(\dfrac{4\theta}{9}\right)$

$4\sin\left(\dfrac{5\theta}{9}\right)\sin\left(\dfrac{4\theta}{9}\right) = 4\cdot\dfrac{1}{2}\left[\cos\left(\dfrac{5\theta}{9}-\dfrac{4\theta}{9}\right)-\cos\left(\dfrac{5\theta}{9}+\dfrac{4\theta}{9}\right)\right]$ Substitute into the product-to-sum identity.

$4\sin\left(\dfrac{5\theta}{9}\right)\sin\left(\dfrac{4\theta}{9}\right) = 2\left[\cos\left(\dfrac{\theta}{9}\right)-\cos(\theta)\right]$ Simplify.

► **EXAMPLE 2**

Find the exact value using product-to-sum identities.

a.) $\cos(15°)\cos(135°)$

b.) $4\sin\left(\dfrac{7\pi}{12}\right)\sin\left(-\dfrac{\pi}{12}\right)$

►►**Solution**

a.) $\cos(15°)\cos(135°)$

$$\cos(15°)\cos(135°) = \frac{1}{2}\left[\cos(150°) + \cos(-120°)\right]$$ Substitute into the product-to-sum identity.

$$\cos(15°)\cos(135°) = \frac{1}{2}\left[\cos(150°) + \cos(120°)\right]$$ Cosine is an even function, $\cos(-120°) = \cos(120°)$

$$\cos(15°)\cos(135°) = \frac{1}{2}\left[-\frac{\sqrt{3}}{2} + \left(-\frac{1}{2}\right)\right]$$ Substitute the values for $\cos 150°$ and $\cos 120°$

$$\cos(15°)\cos(135°) = \frac{1}{2}\left[\frac{-\sqrt{3}-1}{2}\right]$$ Simplify.

$$\cos(15°)\cos(135°) = \frac{-\sqrt{3}-1}{4}$$

b.) $4\sin\left(\dfrac{7\pi}{12}\right)\sin\left(-\dfrac{\pi}{12}\right)$

$$4\sin\left(\frac{7\pi}{12}\right)\sin\left(-\frac{\pi}{12}\right) = 4 \cdot \frac{1}{2}\left[\cos\left(\frac{7\pi}{12} - \left(-\frac{\pi}{12}\right)\right) - \cos\left(\frac{7\pi}{12} + \left(-\frac{\pi}{12}\right)\right)\right]$$ Substitute into the product-to-sum identity.

$$4\sin\left(\frac{7\pi}{12}\right)\sin\left(-\frac{\pi}{12}\right) = 2\left[\cos\left(\frac{8\pi}{12}\right) - \cos\left(\frac{6\pi}{12}\right)\right]$$ Simplify.

$$4\sin\left(\frac{7\pi}{12}\right)\sin\left(-\frac{\pi}{12}\right) = 2\left[\cos\left(\frac{2\pi}{3}\right) - \cos\left(\frac{\pi}{2}\right)\right]$$ Substitute the values for $\cos\left(\dfrac{2\pi}{3}\right)$ and $\cos\left(\dfrac{\pi}{2}\right)$

$$4\sin\left(\frac{7\pi}{12}\right)\sin\left(-\frac{\pi}{12}\right) = 2\left[-\frac{1}{2} - 0\right]$$ Simplify.

$$4\sin\left(\frac{7\pi}{12}\right)\sin\left(-\frac{\pi}{12}\right) = -1$$

Sum-To-Product Identities

Sometimes we need to write a sum of two trigonometric functions as a product. The sum-to-product identities are used to express sums and differences involving sines and cosines as products involving sines and cosines. To derive a sum-to-product identity, we will transform a product-to-sum identity. We will transform $\cos\theta\cos\beta = \dfrac{1}{2}\left[\cos(\theta+\beta)+\cos(\theta-\beta)\right]$ to a sum-to-product identity. We will let $2\theta = x+y$ and $2\beta = x-y$. Solving for θ and for β gives us $\theta = \dfrac{x+y}{2}$ and $\beta = \dfrac{x-y}{2}$. Substitute into our product-to-sum identity and simplify.

$$\cos\theta\cos\beta = \frac{1}{2}\left[\cos(\theta+\beta)+\cos(\theta-\beta)\right]$$

$$\cos\left(\frac{x+y}{2}\right)\cos\left(\frac{x-y}{2}\right) = \frac{1}{2}\left[\cos\left(\frac{x+y}{2}+\frac{x-y}{2}\right)+\cos\left(\frac{x+y}{2}-\frac{x-y}{2}\right)\right]$$

$$2\cdot\left[\cos\left(\frac{x+y}{2}\right)\cos\left(\frac{x-y}{2}\right)\right] = 2\cdot\left[\frac{1}{2}\left[\cos\left(\frac{x+y+x-y}{2}\right)+\cos\left(\frac{x+y-x+y}{2}\right)\right]\right]$$

$$2\cos\left(\frac{x+y}{2}\right)\cos\left(\frac{x-y}{2}\right) = \cos\left(\frac{2x}{2}\right)+\cos\left(\frac{2y}{2}\right)$$

$$2\cos\left(\frac{x+y}{2}\right)\cos\left(\frac{x-y}{2}\right) = \cos x + \cos y$$

The other three sum-to-product identities can be obtained by following similar procedures .

Sum-To-Product Identities

$$\cos x + \cos y = 2\cos\left(\frac{x+y}{2}\right)\cos\left(\frac{x-y}{2}\right)$$

$$\sin x + \sin y = 2\sin\left(\frac{x+y}{2}\right)\cos\left(\frac{x-y}{2}\right)$$

$$\cos x - \cos y = -2\sin\left(\frac{x+y}{2}\right)\sin\left(\frac{x-y}{2}\right)$$

$$\sin x - \sin y = 2\cos\left(\frac{x+y}{2}\right)\sin\left(\frac{x-y}{2}\right)$$

► EXAMPLE 3

Write each sum as a product using the sum-to-product identities.

a.) $\cos(5\theta)+\cos(3\theta)$

b.) $\sin\left(\dfrac{5\theta}{9}\right)-\sin\left(\dfrac{4\theta}{9}\right)$

►►Solution

a.) $\cos(5\theta)+\cos(3\theta)$

$\cos(5\theta)+\cos(3\theta)=2\cos\left(\dfrac{5\theta+3\theta}{2}\right)\cos\left(\dfrac{5\theta-3\theta}{2}\right)$ | Substitute into the sum-to-product identity.

$\cos(5\theta)+\cos(3\theta)=2\cos(4\theta)\cos(\theta)$ | Simplify.

b.) $\sin\left(\dfrac{5\theta}{9}\right)-\sin\left(\dfrac{4\theta}{9}\right)$

$\sin\left(\dfrac{5\theta}{9}\right)-\sin\left(\dfrac{4\theta}{9}\right)=2\cos\left(\dfrac{\dfrac{5\theta}{9}+\dfrac{4\theta}{9}}{2}\right)\sin\left(\dfrac{\dfrac{5\theta}{9}-\dfrac{4\theta}{9}}{2}\right)$ | Substitute into the product-to-sum identity.

$\sin\left(\dfrac{5\theta}{9}\right)-\sin\left(\dfrac{4\theta}{9}\right)=2\cos\left(\dfrac{\theta}{2}\right)\sin\left(\dfrac{\theta}{18}\right)$ | Simplify.

► EXAMPLE 4

Find the exact value using sum-to-product identities.

a.) $\sin(15°)+\sin(75°)$

b.) $\cos\left(\dfrac{7\pi}{12}\right)-\cos\left(-\dfrac{\pi}{12}\right)$

►►Solution

a.) $\sin(15°)+\sin(75°)=2\sin\left(\dfrac{15°+75°}{2}\right)\cos\left(\dfrac{15°-75°}{2}\right)$ | Substitute into the product-to-sum identity.

$\sin(15°)+\sin(75°)=2\sin\left(\dfrac{90°}{2}\right)\cos\left(-\dfrac{60°}{2}\right)$

$\sin(15°)+\sin(75°)=2\sin(45°)\cos(-30°)$ | Cosine is an even function, $\cos(-30°)=\cos(30°)$

$\sin(15°)+\sin(75°)=2\left(\dfrac{\sqrt{2}}{2}\right)\left(\dfrac{\sqrt{3}}{2}\right)$ | Substitute the values for $\sin 45°$ and $\cos 30°$.

$\sin(15°)+\sin(75°)=\dfrac{\sqrt{6}}{2}$ | Simplify.

b.)

$$\cos\left(\frac{7\pi}{12}\right) - \cos\left(-\frac{\pi}{12}\right) = -2\sin\left(\frac{\frac{7\pi}{12} + \left(-\frac{\pi}{12}\right)}{2}\right)\sin\left(\frac{\frac{7\pi}{12} - \left(-\frac{\pi}{12}\right)}{2}\right)$$

> Substitute into the product-to-sum identity.

$$\cos\left(\frac{7\pi}{12}\right) - \cos\left(-\frac{\pi}{12}\right) = -2\sin\left(\frac{\frac{6\pi}{12}}{2}\right)\sin\left(\frac{\frac{8\pi}{12}}{2}\right)$$

> Simplify.

$$\cos\left(\frac{7\pi}{12}\right) - \cos\left(-\frac{\pi}{12}\right) = -2\sin\left(\frac{\pi}{2}\cdot\frac{1}{2}\right)\sin\left(\frac{2\pi}{3}\cdot\frac{1}{2}\right)$$

> Simplify.

$$\cos\left(\frac{7\pi}{12}\right) - \cos\left(-\frac{\pi}{12}\right) = -2\sin\left(\frac{\pi}{4}\right)\sin\left(\frac{\pi}{3}\right)$$

> Substitute the values for $\sin\left(\frac{\pi}{4}\right)$ and $\sin\left(\frac{\pi}{3}\right)$.

$$\cos\left(\frac{7\pi}{12}\right) - \cos\left(-\frac{\pi}{12}\right) = -2\left(\frac{\sqrt{2}}{2}\right)\left(\frac{\sqrt{3}}{2}\right)$$

> Simplify.

$$\cos\left(\frac{7\pi}{12}\right) - \cos\left(-\frac{\pi}{12}\right) = -\frac{\sqrt{6}}{2}$$

Power Reduction Identities

We can rewrite even powers of trigonometric functions in terms of cosine to the power of one. When the power is four or higher, you may have to apply the power reduction identities more than once to eliminate all the exponents. The power reduction identities are very helpful in calculus and differential equations. We will use double angle identities to derive the power reduction identity for sine and cosine as shown below.

$$2\cos^2\theta - 1 = \cos 2\theta \qquad\qquad 1 - 2\sin^2\theta = \cos 2\theta$$

$$2\cos^2\theta = 1 + \cos 2\theta \qquad\qquad -2\sin^2\theta = -1 + \cos 2\theta$$

$$\frac{2\cos^2\theta}{2} = \frac{1 + \cos 2\theta}{2} \qquad\qquad \frac{-2\sin^2\theta}{-2} = \frac{-1 + \cos 2\theta}{-2}$$

$$\cos^2\theta = \frac{1 + \cos 2\theta}{2} \qquad\qquad \sin^2\theta = \frac{1 - \cos 2\theta}{2}$$

We can find the power reduction identity for tangent using identity $\tan^2\theta = \frac{\sin^2\theta}{\cos^2\theta}$.

$$\tan^2\theta = \frac{\sin^2\theta}{\cos^2\theta} \quad\longrightarrow\quad \tan^2\theta = \frac{\dfrac{1 - \cos 2\theta}{2}}{\dfrac{1 + \cos 2\theta}{2}} \quad\longrightarrow\quad \tan^2\theta = \frac{1 - \cos 2\theta}{1 + \cos 2\theta}$$

Power Reduction Identities

$$\cos^2 \theta = \frac{1 + \cos 2\theta}{2}$$

$$\sin^2 \theta = \frac{1 - \cos 2\theta}{2}$$

$$\tan^2 \theta = \frac{1 - \cos 2\theta}{1 + \cos 2\theta}$$

► EXAMPLE 5

Use the power reduction identities to write $\cos^2 x \sin^2 x$ in terms of cosine to the power one.

►► Solution

$\cos^2 x \sin^2 x = \left(\dfrac{1 + \cos 2x}{2}\right)\left(\dfrac{1 - \cos 2x}{2}\right)$ | Substitute the reduction identity for cosine and the reduction identity for sine.

$\cos^2 x \sin^2 x = \dfrac{1}{4}(1 + \cos 2x)(1 - \cos 2x)$ | Multiply the denominators and factor out the one fourth.

$\cos^2 x \sin^2 x = \dfrac{1}{4}\left(1 - \cos 2x + \cos 2x - \cos^2 2x\right)$ | F.O.I.L. and simplify.

$\cos^2 x \sin^2 x = \dfrac{1}{4}\left(1 - \cos^2 2x\right)$

Notice that the angle is *2x* and when we apply the reduction identity for cosine the angle is two times *2x* which is *4x*.

$\cos^2 x \sin^2 x = \dfrac{1}{4}\left(1 - \left(\dfrac{1 + \cos 4x}{2}\right)\right)$

$\cos^2 x \sin^2 x = \dfrac{1}{4}\left(1 - \left(\dfrac{1}{2} + \dfrac{1}{2}\cos 4x\right)\right)$ | Simplify.

$\cos^2 x \sin^2 x = \dfrac{1}{4}\left(1 - \dfrac{1}{2} - \dfrac{1}{2}\cos 4x\right)$

$\cos^2 x \sin^2 x = \dfrac{1}{4}\left(\dfrac{1}{2} - \dfrac{1}{2}\cos 4x\right)$ | Factor out the one-half and simplify.

$\cos^2 x \sin^2 x = \dfrac{1}{4} \cdot \dfrac{1}{2}(1 - \cos 4x) \longrightarrow \cos^2 x \sin^2 x = \dfrac{1}{8}(1 - \cos 4x)$

► EXAMPLE 6

Use the power reduction identities to write $\cos^4 x$ in terms of cosine to the power one.

►► Solution

$$\cos^2 x \cos^2 x = \left(\frac{1 + \cos 2x}{2}\right)\left(\frac{1 + \cos 2x}{2}\right)$$

Rewrite $\cos^4 x$ as a product and substitute the reduction identity for cosine.

$$\cos^2 x \cos^2 x = \frac{1}{4}\left(1 + \cos 2x\right)\left(1 + \cos 2x\right)$$

Multiply the denominators and factor out the one fourth.

$$\cos^2 x \cos^2 x = \frac{1}{4}\left(1 + \cos 2x + \cos 2x + \cos^2 2x\right)$$

F.O.I.L. and simplify.

$$\cos^2 x \cos^2 x = \frac{1}{4}\left(1 + 2\cos 2x + \cos^2 2x\right)$$

Apply the reduction identity for cosine.

$$\cos^2 x \cos^2 x = \frac{1}{4}\left(1 + 2\cos 2x + \left(\frac{1 + \cos 4x}{2}\right)\right)$$

$$\cos^2 x \cos^2 x = \frac{1}{4}\left(1 + 2\cos 2x + \left(\frac{1}{2} + \frac{1}{2}\cos 4x\right)\right)$$

Remove the parentheses and simplify.

$$\cos^2 x \cos^2 x = \frac{1}{4}\left(1 + 2\cos 2x + \frac{1}{2} + \frac{1}{2}\cos 4x\right)$$

$$\cos^2 x \cos^2 x = \frac{1}{4}\left(\frac{3}{2} + 2\cos 2x + \frac{1}{2}\cos 4x\right)$$

$$\cos^2 x \cos^2 x = \frac{1}{4}\cdot\frac{1}{2}\left(3 + 4\cos 2x + \cos 4x\right)$$

Factor out the one-half and simplify.

$$\cos^2 x \cos^2 x = \frac{1}{8}\left(3 + 4\cos 2x + \cos 4x\right)$$

► **EXAMPLE 7**

Use the power reduction identities to write $6\cos^4 x \sin^2 x$ **in terms of cosine to the power one.**

►► **Solution**

$$6\cos^2 x \cos^2 x \sin^2 x = 6\left(\frac{1+\cos 2x}{2}\right)\left(\frac{1+\cos 2x}{2}\right)\left(\frac{1-\cos 2x}{2}\right)$$

> Multiply the denominators and factor out one eighth.

$$6\cos^2 x \cos^2 x \sin^2 x = 6\cdot\frac{1}{8}(1+\cos 2x)(1+\cos 2x)(1-\cos 2x)$$

> F.O.I.L. the last two parentheses and simplify.

$$6\cos^2 x \cos^2 x \sin^2 x = \frac{3}{4}(1+\cos 2x)(1-\cos 2x+\cos 2x-\cos^2 2x)$$

$$6\cos^2 x \cos^2 x \sin^2 x = \frac{3}{4}(1+\cos 2x)(1-\cos^2 2x)$$

$$6\cos^2 x \cos^2 x \sin^2 x = \frac{3}{4}(1+\cos 2x)\left(1-\left(\frac{1+\cos 4x}{2}\right)\right)$$

> Apply the reduction identity for cosine.

$$6\cos^2 x \cos^2 x \sin^2 x = \frac{3}{4}(1+\cos 2x)\left(1-\left(\frac{1}{2}+\frac{1}{2}\cos 4x\right)\right)$$

> Remove the parentheses and simplify.

$$6\cos^2 x \cos^2 x \sin^2 x = \frac{3}{4}(1+\cos 2x)\left(1-\frac{1}{2}-\frac{1}{2}\cos 4x\right)$$

$$6\cos^2 x \cos^2 x \sin^2 x = \frac{3}{4}(1+\cos 2x)\left(\frac{1}{2}-\frac{1}{2}\cos 4x\right)$$

> Factor out the one-half and simplify.

$$6\cos^2 x \cos^2 x \sin^2 x = \frac{3}{4}\cdot\frac{1}{2}(1+\cos 2x)(1-\cos 4x)$$

> F.O.I.L

$$6\cos^2 x \cos^2 x \sin^2 x = \frac{3}{8}(1-\cos 4x+\cos 2x-\cos 2x\cos 4x)$$

HOMEWORK

Objective 1

Write each product as a sum using the product-to-sum identities.

See Example 1.

1. $6\sin(4\theta)\cos(2\theta)$

2. $\cos(4\beta)\sin(6\beta)$

3. $\sin(x+y)\sin(x-y)$

4. $3\cos 2\alpha \cos 4\alpha$

5. $2\sin(-\theta)\cos(4\theta)$

6. $\cos\left(\dfrac{3x}{4}\right)\sin\left(\dfrac{x}{4}\right)$

Objective 1

Find the exact value using product-to-sum identities.

See Example 2.

7. $10\cos(-15°)\sin 75°$

8. $\cos 75° \cos 75°$

9. $\sin 105° \sin 15°$

10. $4\cos\dfrac{5\pi}{12}\cos\dfrac{\pi}{12}$

11. $\sin\dfrac{7\pi}{8}\cos\dfrac{\pi}{8}$

12. $3\sin\dfrac{7\pi}{12}\sin\left(-\dfrac{\pi}{12}\right)$

Objective 2

Write each sum as a product using the sum-to-product identities.

See Example 3.

13. $\sin(4\theta)+\sin(2\theta)$

14. $\sin(x+y)-\sin(x-y)$

15. $\cos 6x + \cos 4x$

16. $\cos\left(\dfrac{7\theta}{6}\right)-\cos\left(\dfrac{\theta}{6}\right)$

17. $\sin\left(\dfrac{5\theta}{9}\right)+\sin\left(\dfrac{4\theta}{9}\right)$

18. $\cos \beta - \cos 4\beta$

Objective 2

Find the exact value using sum-to-product identities.

See Example 4.

19. $\sin 105° + \sin 15°$

20. $\cos 285° - \cos 195°$

21. $\cos \dfrac{17\pi}{12} + \cos \dfrac{13\pi}{12}$

22. $\sin \dfrac{3\pi}{4} + \sin \dfrac{\pi}{4}$

23. $\sin\left(\dfrac{11\pi}{12}\right) - \sin\left(\dfrac{7\pi}{12}\right)$

24. $\cos 75° - \cos 15°$

Objective 3

Use the power reduction identities to write the following expressions in terms of cosine to the power one.

See Examples 5-7.

25. $\sin^4 \theta$

26. $\cos^2 \theta \sin^4 \theta$

27. $4\cos^4 x$

28. $\cos^4 x \sin^4 x$

29. $6\cos^6 x$

30. $4\sin^6 x$

ANSWERS

1. $3[\sin(6\theta) + \sin(2\theta)]$

2. $\frac{1}{2}[\sin(10\beta) + \sin(2\beta)]$

3. $\frac{1}{2}[\cos(2y) - \cos(2x)]$

4. $\frac{3}{2}[\cos(6\alpha) + \cos(2\alpha)]$

5. $\sin(3\theta) - \sin(5\theta)$

6. $\frac{1}{2}\left[\sin(x) - \sin\left(\frac{x}{2}\right)\right]$

7. $\frac{5}{2}(\sqrt{3} + 2)$

8. $\frac{1}{4}(-\sqrt{3} + 2)$

9. $\frac{1}{4}$

10. 1

11. $\frac{\sqrt{2}}{4}$

12. $-\frac{3}{4}$

13. $2\sin(3\theta)\cos(\theta)$

14. $2\cos x \sin y$

15. $2\cos(5x)\cos x$

16. $-2\sin\left(\frac{2\theta}{3}\right)\sin\left(\frac{\theta}{2}\right)$

17. $2\sin\left(\frac{\theta}{2}\right)\sin\left(\frac{\theta}{18}\right)$

18. $2\sin\left(\frac{5\beta}{2}\right)\sin\left(\frac{3\beta}{2}\right)$

19. $\frac{\sqrt{6}}{2}$

20. $\frac{\sqrt{6}}{2}$

21. $-\frac{\sqrt{6}}{2}$

22. $\sqrt{2}$

23. $-\frac{\sqrt{2}}{2}$

24. $-\frac{\sqrt{2}}{2}$

25. $\frac{1}{8}[3 - 4\cos(2x) + \cos(4x)]$

26. $\frac{1}{16}[1 - \cos(2x) - \cos(4x) + \cos(2x)\cos(4x)]$

27. $\frac{1}{2}[3 + 4\cos(2x) + \cos(4x)]$

28. $\frac{1}{128}[3 - 4\cos(4x) + \cos(8x)]$

29. $\frac{3}{8}[5 + 7\cos(2x) + 3\cos(4x)]$

30. $\frac{1}{4}[5 - 7\cos(2x) + 3\cos(4x)]$

Section 16:   Trigonometric Equations

Learning Outcomes:

- The student will solve trigonometric equations, right triangles, and oblique triangles.
- The student will correctly memorize and apply trigonometric formulas, definitions, identities, and properties.

Objectives: At the conclusion of this lesson you should be able to:

1.   Solve trigonometric equations for roots in a specific interval.
2.   Solve trigonometric equations for roots in $(-\infty, \infty)$.

In this section we are going to use everything we have learned so far in this course along with elements of basic equation solving to solve trigonometric equations. Consider the trigonometric equation $2\cos x + \sqrt{3} = 0$ on the interval $[0, 2\pi)$. To solve this equation we will isolate $\cos x$ and find all values for x in the interval $[0, 2\pi)$ that makes the equation true. This is an algebraic approach to solving the equation.

Begin by solving for $\cos x$.

$$2\cos x + \sqrt{3} = 0$$
$$2\cos x = -\sqrt{3}$$
$$\frac{2\cos x}{2} = -\frac{\sqrt{3}}{2}$$
$$\cos x = -\frac{\sqrt{3}}{2}$$

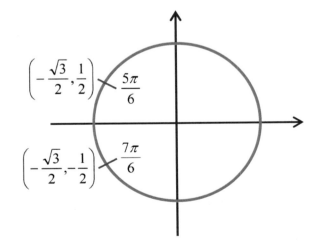

Figure 16.1

The solutions are the angles on the unit circle in the interval $[0, 2\pi)$ that have a cosine of $-\dfrac{\sqrt{3}}{2}$. Looking at **Figure 16.1** we find the solutions to be $x = \dfrac{5\pi}{6}$ and

$$x = \frac{7\pi}{6}.$$

Visualizing the Solution

If we take $2\cos x + \sqrt{3} = 0$ and move $\sqrt{3}$ to the other side of the equation we have $2\cos x = -\sqrt{3}$. Consider the graphs of $y = 2\cos x$ (**blue graph**) and $y = -\sqrt{3}$ (**red graph**) over the interval $[0, 2\pi)$ as shown in **Figure 16.2**. Notice that the two graphs intersect at $x = \dfrac{5\pi}{6}$ and $x = \dfrac{7\pi}{6}$. Therefore the solutions are $x = \dfrac{5\pi}{6}$ and $x = \dfrac{7\pi}{6}$.

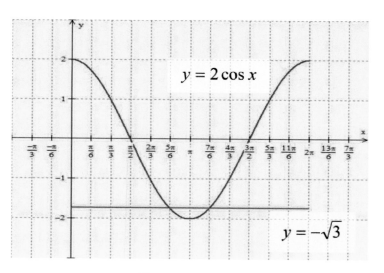

Figure 16.2

Let us consider the same trigonometric equation $2\cos x + \sqrt{3} = 0$ on the interval $(-\infty, \infty)$. The interval $(-\infty, \infty)$ means all real solutions. Since cosine is periodic, there is infinite number of solutions. We found the solutions in the interval $[0, 2\pi)$ to be $x = \dfrac{5\pi}{6}$ and $x = \dfrac{7\pi}{6}$. These solutions, plus any multiple of 2π, are the solutions on the interval $(-\infty, \infty)$. We use k to represent any integer and we have: $x = \dfrac{5\pi}{6} + 2\pi k$ and $x = \dfrac{7\pi}{6} + 2\pi k$. Ten of the infinite solutions are:

$$-\frac{19\pi}{6}, -\frac{17\pi}{6}, \quad -\frac{7\pi}{6}, -\frac{5\pi}{6}, \quad \frac{5\pi}{6}, \frac{7\pi}{6}, \quad \frac{17\pi}{6}, \frac{19\pi}{6}, \quad \frac{29\pi}{6}, \frac{31\pi}{6}$$

$$\underbrace{\qquad\qquad}_{k=-2} \quad \underbrace{\qquad\qquad}_{k=-1} \quad \underbrace{\qquad\qquad}_{k=0} \quad \underbrace{\qquad\qquad}_{k=1} \quad \underbrace{\qquad\qquad}_{k=2}$$

Trigonometric Equations Involving a Single Trigonometric Function

In the previous worked problem the trigonometric equation contained only one trigonometric function. The following examples will involve a single trigonometric function.

► EXAMPLE 1

Solve $2\sin x - \sqrt{2} = 0$ **on the interval: a.)** $[0, 2\pi)$**, b.)** $[0°, 360°)$**, c.)** $(-\infty, \infty)$.

►► Solution

$$2\sin x - \sqrt{2} = 0 \qquad \text{Solve for } \sin x.$$
$$2\sin x = \sqrt{2}$$
$$\frac{2\sin x}{2} = \frac{\sqrt{2}}{2}$$
$$\sin x = \frac{\sqrt{2}}{2} \qquad \text{Where on the unit circle}$$
$$\text{does sine equal } \frac{\sqrt{2}}{2}?$$

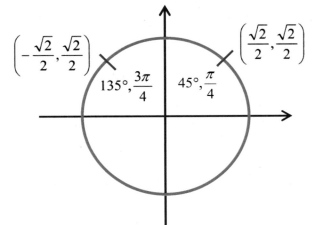

$\left(-\dfrac{\sqrt{2}}{2}, \dfrac{\sqrt{2}}{2}\right)$ $\left(\dfrac{\sqrt{2}}{2}, \dfrac{\sqrt{2}}{2}\right)$

$135°, \dfrac{3\pi}{4}$ $45°, \dfrac{\pi}{4}$

a.)   On the interval $[0, 2\pi)$, $\sin x = \dfrac{\sqrt{2}}{2}$ at $x = \dfrac{\pi}{4}$ and $x = \dfrac{3\pi}{4}$.

b.)   On the interval $[0°, 360°)$, $\sin x = \dfrac{\sqrt{2}}{2}$ at $x = 45°$ and $x = 135°$.

c.)   On the interval $(-\infty, \infty)$, $\sin x = \dfrac{\sqrt{2}}{2}$ at $x = \dfrac{\pi}{4} + 2\pi k$ and $x = \dfrac{3\pi}{4} + 2\pi k$.

▶EXAMPLE 2

Solve $4\sin^2 x - 3 = 0$ **on the interval: a.)** $[0,2\pi)$**, b.)** $[0°,360°)$**, c.)** $(-\infty,\infty)$**.**

▶▶**Solution**

$4\sin^2 x - 3 = 0$ Solve for $\sin x$.

$4\sin^2 x = 3$

$\dfrac{4\sin x}{4} = \dfrac{3}{4}$

$\sin^2 x = \dfrac{3}{4}$

$\sqrt{\sin^2 x} = \pm\sqrt{\dfrac{3}{4}}$

$\sin x = \pm\dfrac{\sqrt{3}}{2}$ Where on the unit circle does sine equal $-\dfrac{\sqrt{3}}{2}$ and $\dfrac{\sqrt{3}}{2}$?

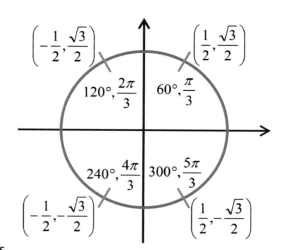

a.)   On the interval $[0,2\pi)$, $\sin x = \dfrac{\sqrt{3}}{2}$ at $x = \dfrac{\pi}{3}$ and $x = \dfrac{2\pi}{3}$,

$\sin x = -\dfrac{\sqrt{3}}{2}$ at $x = \dfrac{4\pi}{3}$ and $x = \dfrac{5\pi}{3}$.

b.)   On the interval $[0°,360°)$, $\sin x = \dfrac{\sqrt{3}}{2}$ at $x = 60°$ and $x = 120°$,

$\sin x = -\dfrac{\sqrt{3}}{2}$ at $x = 240°$ and $x = 300°$.

c.)   On the interval $(-\infty,\infty)$, $\sin x = \pm\dfrac{\sqrt{3}}{2}$ at $x = \dfrac{\pi}{3} + \pi k$ and $x = \dfrac{2\pi}{3} + \pi k$.

We add π instead of 2π because the solutions $x = \dfrac{\pi}{3}$ and $x = \dfrac{4\pi}{3}$ have an

equidistance of π and the solutions $x = \dfrac{2\pi}{3}$ and $x = \dfrac{5\pi}{3}$ have an equidistance of π.

► EXAMPLE 3

Solve $\sec^2 x - 4 = 0$ **on the interval: a.)** $[0, 2\pi)$**, b.)** $[0°, 360°)$**, c.)** $(-\infty, \infty)$**.**

►►**Solution**

$\sec^2 x - 4 = 0$ Solve for $\sec x$.

$\sec^2 x = 4$

$\sqrt{\sec^2 x} = \pm\sqrt{4}$

$\sec x = \pm 2$

$\cos x = \pm\dfrac{1}{2}$ Where on the unit circle does cosine equal $-\dfrac{1}{2}$ and $\dfrac{1}{2}$?

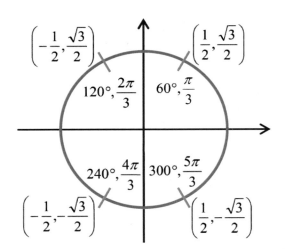

a.) On the interval $[0, 2\pi)$, $\cos x = \dfrac{1}{2}$ at $x = \dfrac{\pi}{3}$ and $x = \dfrac{5\pi}{3}$,

$\cos x = -\dfrac{1}{2}$ at $x = \dfrac{2\pi}{3}$ and $x = \dfrac{4\pi}{3}$.

b.) On the interval $[0°, 360°)$, $\cos x = \dfrac{1}{2}$ at $x = 60°$ and $x = 300°$,

$\cos x = -\dfrac{1}{2}$ at $x = 120°$ and $x = 240°$.

c.) On the interval $(-\infty, \infty)$, $\cos x = \pm\dfrac{1}{2}$ at $x = \dfrac{\pi}{3} + \pi k$ and $x = \dfrac{2\pi}{3} + \pi k$.

We add π instead of 2π because the solutions $x = \dfrac{\pi}{3}$ and $x = \dfrac{4\pi}{3}$ have an

equidistance of π and the solutions $x = \dfrac{2\pi}{3}$ and $x = \dfrac{5\pi}{3}$ have an equidistance of π.

Trigonometric Equations That Require Factoring

The previous examples involved a single trigonometric function. When a trigonometric equation contains more than one trigonometric function, factoring may be required to solve the equation.

▶ EXAMPLE 4

Solve $2\cos^2 x = \cos x$ **on the interval: a.)** $[0, 2\pi)$, **b.)** $[0°, 360°)$, **c.)** $(-\infty, \infty)$.

▶▶ Solution

$$2\cos^2 x = \cos x \qquad \textit{Get everything in descending order equal to zero.}$$

$$2\cos^2 x - \cos x = 0 \qquad \textit{Factor out the greatest common factor, cosine.}$$

$$\cos x(2\cos x - 1) = 0 \qquad \textit{Set each factor equal to zero and solve.}$$

$$\cos x = 0 \qquad\qquad 2\cos x - 1 = 0$$

Where on the unit circle does cosine equal 0 ?

$$2\cos x = 1$$

$$x = \frac{\pi}{2} \text{ and } x = \frac{3\pi}{2}$$

$$\frac{2\cos x}{2} = \frac{1}{2}$$

$$\cos x = \frac{1}{2}$$

Where on the unit circle does cosine equal $\frac{1}{2}$?

$$x = \frac{\pi}{3} \text{ and } x = \frac{5\pi}{3}$$

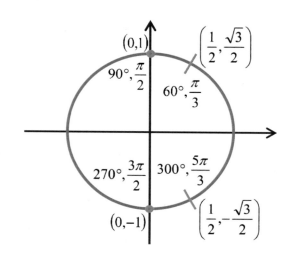

a.) $\quad x = \dfrac{\pi}{3}, \ x = \dfrac{\pi}{2}, \ x = \dfrac{3\pi}{2} \text{ and } x = \dfrac{5\pi}{3}$

b.) $\quad x = 60°, \ x = 90°, \ x = 270° \text{ and } x = 300°$

c.) $\quad x = \dfrac{\pi}{3} + 2\pi k, \ x = \dfrac{2\pi}{3} + 2\pi k \text{ and } x = \dfrac{\pi}{2} + \pi k$

► **EXAMPLE 5**

Solve $2\sin^2 x = \sin x + 1$ **on the interval: a.)** $[0,2\pi)$**, b.)** $[0°,360°)$**, c.)** $(-\infty,\infty)$**.**

►► **Solution**

$$2\sin^2 x = \sin x + 1 \qquad \textit{Get everything in descending order equal to zero.}$$

$$2\sin^2 x - \sin x - 1 = 0 \qquad \textit{Factor the trinomial.}$$

$$(2\sin x + 1)(\sin x - 1) = 0 \qquad \textit{Set each factor equal to zero and solve.}$$

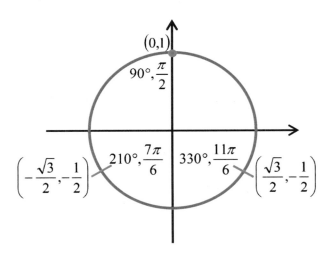

$2\sin x + 1 = 0$	$\sin x - 1 = 0$
$2\sin x = -1$	$\sin x = 1$

$\dfrac{2\sin x}{2} = -\dfrac{1}{2}$ Where on the unit circle does sine equal 1?

$\sin x = -\dfrac{1}{2}$

Where on the unit circle does sine equal $-\dfrac{1}{2}$?

$x = \dfrac{7\pi}{6}$ and $x = \dfrac{11\pi}{6}$

$x = \dfrac{\pi}{2}$

a.) $x = \dfrac{\pi}{2}, x = \dfrac{7\pi}{6}$ and $x = \dfrac{11\pi}{6}$

b.) $x = 90°$, $x = 210°$ and $x = 330°$

c.) $x = \dfrac{\pi}{2} + 2\pi k$, $x = \dfrac{7\pi}{6} + 2\pi k$ and $x = \dfrac{11\pi}{6} + 2\pi k$

► EXAMPLE 6

Solve $\cos^2 x = 2 - \cos x$ **on the interval: a.)** $[0, 2\pi)$, **b.)** $[0°, 360°)$, **c.)** $(-\infty, \infty)$.

►► Solution

$$\cos^2 x = 2 - \cos x \qquad \textit{Get everything in descending order equal to zero.}$$

$$\cos^2 x + \cos x - 2 = 0 \qquad \textit{Factor the trinomial.}$$

$$(\cos x + 2)(\cos x - 1) = 0 \qquad \textit{Set each factor equal to zero and solve.}$$

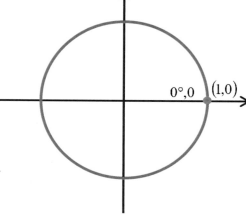

$\cos x + 2 = 0$ 　　　　　　　$\cos x - 1 = 0$

$\cos x = -2$ 　　　　　　　　$\cos x = 1$

Where on the unit circle does 　Where on the unit circle
cosine equal -2? Cosine will 　does cosine equal 1?
not equal -2 because -2 is 　　　　$x = 0$
outside the range of cosine.

a.) 　$x = 0$ 　　　　**b.)** 　$x = 0°$ 　　　**c.)** 　$x = 2\pi k$

► EXAMPLE 7

Solve $2\cos^2 x - 3 = 5\cos x$ **on the interval: a.)** $[0, 2\pi)$, **b.)** $[0°, 360°)$, **c.)** $(-\infty, \infty)$.

►► Solution

$$2\cos^2 x - 3 = 5\cos x \qquad \textit{Get everything in descending order equal to zero.}$$

$$2\cos^2 x - 5\cos x - 3 = 0 \qquad \textit{Factor the trinomial.}$$

$$(2\cos x + 1)(\cos x - 3) = 0 \qquad \textit{Set each factor equal to zero and solve.}$$

$2\cos x + 1 = 0$ 　　　　$\cos x - 3 = 0$

$2\cos x = -1$ 　　　　　　$\cos x = 3$

$\dfrac{2\cos x}{2} = -\dfrac{1}{2}$ 　　　Where on the unit circle
　　　　　　　　　　does cosine equal 3?
$\cos x = -\dfrac{1}{2}$ 　　　　Cosine will not equal 3
　　　　　　　　　　because 3 is outside the
Where on the unit circle 　range of cosine.

does cosine equal $-\dfrac{1}{2}$?

$x = \dfrac{2\pi}{3}$ and $x = \dfrac{4\pi}{3}$

a.) $x = \dfrac{2\pi}{3}$ and $x = \dfrac{4\pi}{3}$ **b.)** $x = 120°$ and $x = 240°$ **c.)** $x = \dfrac{2\pi}{3} + 2\pi k$ and $x = \dfrac{4\pi}{3} + 2\pi k$

Solving Trigonometric Equations Using Fundamental identities

We solve trigonometric equations involving two different trigonometric functions using basic algebra and trigonometric identities.   When given an equation with different trigonometric functions, rewrite the equation in a form involving only one trigonometric function.

▶ EXAMPLE 8

Solve $\sin 2x = 2\cos x$ **on the interval: a.)** $[0,2\pi)$**, b.)** $[0°,360°)$**, c.)** $(-\infty,\infty)$**.**

▶▶ **Solution**

$$\sin 2x = 2\cos x \qquad \textit{Get everything in descending order equal to zero.}$$
$$\sin 2x - 2\cos x = 0 \qquad \textit{Substitute the double angle identity for sine.}$$
$$2\sin x \cos x - 2\cos x = 0 \qquad \textit{Factor out the common factor, 2cosx.}$$
$$2\cos x(\sin x - 1) = 0 \qquad \textit{Set each factor equal to zero and solve.}$$

$2\cos x = 0$ $\sin x - 1 = 0$

$\dfrac{2\cos x}{2} = \dfrac{0}{2}$ $\sin x = 1$

$\cos x = 0$ Where on the unit circle does sine equal 1?

Where on the unit circle does cosine equal 0? $x = \dfrac{\pi}{2}$

$x = \dfrac{\pi}{2}$ and $x = \dfrac{3\pi}{2}$

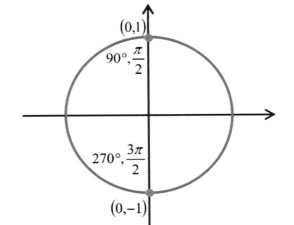

a.) $x = \dfrac{\pi}{2}$ and $x = \dfrac{3\pi}{2}$

b.) $x = 90°$ and $x = 270°$

c.) $x = \dfrac{\pi}{2} + \pi k$

►EXAMPLE 9

Solve $3\sin^2 x - \cos x = -1$ **on the interval: a.)** $[0,2\pi)$**, b.)** $[0°,360°)$**, c.)** $(-\infty,\infty)$**.**

►►**Solution**

$$3\sin^2 x - \cos x = -1$$

Get everything in descending order equal to zero.

$$3\sin^2 x - \cos x + 1 = 0$$

$$3(1 - \cos^2 x) - \cos x + 1 = 0$$

We cannot substitute an identity for cosine but we can substitute for sine squared using a Pythagorean identity.

$$3 - 3\cos^2 x - \cos x + 1 = 0$$

$$-3\cos^2 x - \cos x + 4 = 0$$

Factor out the negative.

$$-(3\cos^2 x + \cos x - 4) = 0$$

Factor.

$$-(3\cos x + 4)(\cos x - 1) = 0$$

Set each factor equal to zero and solve.

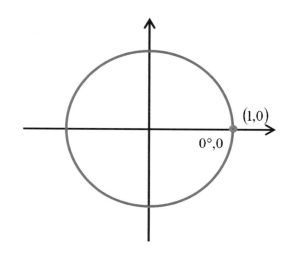

(1,0)

0°,0

$3\cos x + 4 = 0$ $\cos x - 1 = 0$

$3\cos x = -4$ $\cos x = 1$

$\dfrac{3\cos x}{3} = -\dfrac{4}{3}$ Where on the unit circle does cosine equal 1?

$\cos x = -\dfrac{4}{3}$ $x = 0$

Where on the unit circle does cosine equal $-\dfrac{4}{3}$?

Cosine will not equal $-\dfrac{4}{3}$ because $-\dfrac{4}{3}$ is outside the range of cosine.

a.) $x = 0$ **b.)** $x = 0°$ **c.)** $x = 2\pi k$

► EXAMPLE 10

Solve $4\sin x\cos x - 2\sqrt{3}\sin x - 2\cos x + \sqrt{3} = 0$ **on the interval:**

    a.) $[0,2\pi)$,   **b.)** $[0°,360°)$,   **c.)** $(-\infty,\infty)$.

►►Solution

$4\sin x\cos x - 2\sqrt{3}\sin x - 2\cos x + \sqrt{3} = 0$ *Factor by grouping.*

$$2\sin x\left(2\cos x - \sqrt{3}\right) - 1\left(2\cos x - \sqrt{3}\right) = 0$$

$$\left(2\cos x - \sqrt{3}\right)\left(2\sin x - 1\right) = 0$$ *Set each factor equal to zero and solve.*

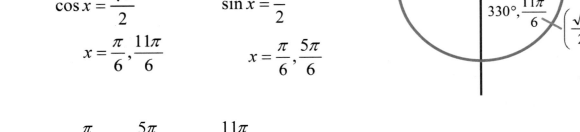

$$2\cos x - \sqrt{3} = 0$$
$$2\cos x = \sqrt{3}$$
$$\frac{2\cos x}{2} = \frac{\sqrt{3}}{2}$$
$$\cos x = \frac{\sqrt{3}}{2}$$
$$x = \frac{\pi}{6}, \frac{11\pi}{6}$$

$$2\sin x - 1 = 0$$
$$2\sin x = 1$$
$$\frac{2\sin x}{2} = \frac{1}{2}$$
$$\sin x = \frac{1}{2}$$
$$x = \frac{\pi}{6}, \frac{5\pi}{6}$$

a.) $x = \dfrac{\pi}{6}$, $x = \dfrac{5\pi}{6}$ and $x = \dfrac{11\pi}{6}$

b.) $x = 30°$, $x = 150°$ and $x = 330°$

c.) $x = \dfrac{\pi}{6} + 2\pi k$ and $x = \dfrac{5\pi}{6} + \pi k$

Solve Trigonometric Equations Involving Multiple Angles

▶EXAMPLE 11

Solve $\tan 3x - \sqrt{3} = 0$ **on the interval: a.)** $[0, 2\pi)$, **b.)** $(-\infty, \infty)$.

▶▶**Solution**

$$\tan 3x - \sqrt{3} = 0 \qquad \textit{Solve for tangent.}$$
$$\tan 3x = \sqrt{3}$$

Where on the unit circle
does tangent equal $\sqrt{3}$?

$$3x = \frac{\pi}{3} \qquad\qquad 3x = \frac{4\pi}{3}$$

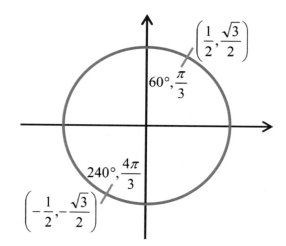

On the interval $[0, 2\pi)$ we know that $3x = \frac{\pi}{3}$ and $3x = \frac{4\pi}{3}$ are the only solutions. In general we

have $3x = \frac{\pi}{3} + 2\pi k$ and $3x = \frac{4\pi}{3} + 2\pi k$ and solving for x gives us $x = \frac{\pi}{9} + \frac{2\pi}{3}k$ and

$$x = \frac{4\pi}{9} + \frac{2\pi}{3}k.$$

> Pick positive values for k since we are
> looking for solutions in the interval $[0, 2\pi)$.

a.) Using $x = \frac{\pi}{9} + \frac{2\pi}{3}k$ and $x = \frac{4\pi}{9} + \frac{2\pi}{3}k$,

the solutions in $[0, 2\pi)$ are

$$x = \frac{\pi}{9}, \frac{4\pi}{9}, \frac{7\pi}{9}, \frac{10\pi}{9}, \frac{13\pi}{9}, \frac{16\pi}{9}.$$

$$k = 0, \ x = \frac{\pi}{9}, \frac{4\pi}{9}$$

$$k = 1, \ x = \frac{7\pi}{9}, \frac{10\pi}{9}$$

$$k = 2, \ x = \frac{13\pi}{9}, \frac{16\pi}{9}$$

Not in the interval $[0, 2\pi)$ → $k = 3, \ x = \frac{19\pi}{9}, \frac{22\pi}{9}$

b.) $x = \frac{\pi}{9} + \frac{2\pi}{3}k$ and $x = \frac{4\pi}{9} + \frac{2\pi}{3}k$

► EXAMPLE 12

Solve $2\sin 3x + \sqrt{3} = 0$ on the interval: a.) $[0, 2\pi)$, b.) $(-\infty, \infty)$.

►► Solution

$$2\sin 3x + \sqrt{3} = 0$$ *Solve for sine.*

$$2\sin 3x = -\sqrt{3}$$

$$\frac{2\sin 3x}{2} = -\frac{\sqrt{3}}{2}$$

$$\sin 3x = -\frac{\sqrt{3}}{2}$$

Where on the unit circle

does sine equal $-\dfrac{\sqrt{3}}{2}$?

$$3x = \frac{4\pi}{3} \qquad 3x = \frac{5\pi}{3}$$

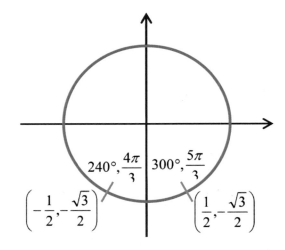

On the interval $[0, 2\pi)$ we know that $3x = \dfrac{4\pi}{3}$ and $3x = \dfrac{5\pi}{3}$ are the only solutions. In general

we have $3x = \dfrac{4\pi}{3} + 2\pi k$ and $3x = \dfrac{5\pi}{3} + 2\pi k$ and solving for x gives us $x = \dfrac{4\pi}{9} + \dfrac{2\pi}{3}k$ and

$$x = \frac{5\pi}{9} + \frac{2\pi}{3}k .$$

a.)   Using $x = \dfrac{4\pi}{9} + \dfrac{2\pi}{3}k$ and $x = \dfrac{5\pi}{9} + \dfrac{2\pi}{3}k$,

the solutions in $[0, 2\pi)$ are

$$x = \frac{4\pi}{9}, \frac{5\pi}{9}, \frac{10\pi}{9}, \frac{11\pi}{9}, \frac{16\pi}{9}, \frac{17\pi}{9} .$$

> Pick positive values for k since we are
> looking for solutions in the interval $[0, 2\pi)$.

$$k = 0, \ x = \frac{4\pi}{9}, \frac{5\pi}{9}$$

$$k = 1, \ x = \frac{10\pi}{9}, \frac{11\pi}{9}$$

$$k = 2, \ x = \frac{16\pi}{9}, \frac{17\pi}{9}$$

Not in the interval $[0, 2\pi)$ → $k = 3, \ x = \dfrac{22\pi}{9}, \dfrac{23\pi}{9}$

b.) $x = \dfrac{4\pi}{9} + \dfrac{2\pi}{3}k$ and $x = \dfrac{5\pi}{9} + \dfrac{2\pi}{3}k$

►EXAMPLE 13

Solve $8\cos\left(\dfrac{1}{2}x\right)+4\sqrt{3}=0$ **on the interval: a.)** $[0,2\pi)$, **b.)** $(-\infty,\infty)$.

►►**Solution**

$$8\cos\left(\frac{1}{2}x\right)+4\sqrt{3}=0 \quad \textit{Solve for cosine.}$$

$$8\cos\left(\frac{1}{2}x\right)=-4\sqrt{3}$$

$$\frac{8\cos\left(\dfrac{1}{2}x\right)}{8}=-\frac{4\sqrt{3}}{8}$$

$$\cos\left(\frac{1}{2}x\right)=-\frac{\sqrt{3}}{2}$$

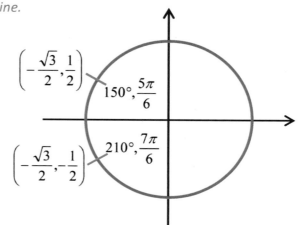

Where on the unit circle does cosine equal $-\dfrac{\sqrt{3}}{2}$?

$$\frac{1}{2}x=\frac{5\pi}{6}\qquad\qquad\frac{1}{2}x=\frac{7\pi}{6}$$

On the interval $[0,2\pi)$ we know that $\dfrac{1}{2}x=\dfrac{5\pi}{6}$ and $\dfrac{1}{2}x=\dfrac{7\pi}{6}$ are the only solutions. In general

we have $\dfrac{1}{2}x=\dfrac{5\pi}{6}+2\pi k$ and $\dfrac{1}{2}x=\dfrac{7\pi}{6}+2\pi k$ and solving for x gives us $x=\dfrac{5\pi}{3}+4\pi k$ and

$x=\dfrac{7\pi}{3}+4\pi k$.

a.) Using $x=\dfrac{5\pi}{3}+4\pi k$ and $x=\dfrac{7\pi}{3}+4\pi k$,

$\qquad$ the solution in $[0,2\pi)$ is $x=\dfrac{5\pi}{3}$.

$$k=0,\ x=\frac{5\pi}{3},\frac{7\pi}{3}$$

$$\boxed{\text{Not in the interval }[0,2\pi)}$$

b.) $x=\dfrac{5\pi}{3}+4\pi k$ and $x=\dfrac{7\pi}{3}+4\pi k$

Solve Trigonometric Equations Using the Quadratic Formula

► **EXAMPLE 14**

Solve $\sin^2 x - 2\sin x - 1 = 0$ **on the interval** $[0, 2\pi)$. **Round to the nearest ten-thousandth.**

►► **Solution**

$$\sin^2 x - 2\sin x - 1 = 0$$

The equation is in quadratic form. Since the equation cannot be factored we need to use the Quadratic Formula.

$$a = 1, \quad b = -2, \quad c = -1$$

$$\sin x = \frac{-(-2) \pm \sqrt{(-2)^2 - 4(1)(-1)}}{2(1)}$$

Substitute into the Quadratic Formula the values for a, b, and c.

$$\sin x = \frac{2 \pm \sqrt{4 + 4}}{2}$$ *Simplify.*

$$\sin x = \frac{2 \pm \sqrt{8}}{2}$$ *Get the radical in reduced radical form.*

$$\sin x = \frac{2 \pm \sqrt{4 \cdot 2}}{2}$$

$$\sin x = \frac{2 \pm 2\sqrt{2}}{2}$$ *Reduce the fration.*

$$\sin x = \frac{2\left(1 \pm \sqrt{2}\right)}{2}$$ *Factor out the common factor of 2 and simplify.*

$$\sin x = 1 \pm \sqrt{2}$$

$\sin x = 1 + \sqrt{2}$ 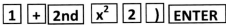 1 + 2nd x² 2) ENTER

$\sin x = 2.4142136$ Since the range of the sine function is $[-1, 1]$, there is no x such that $\sin x = 2.4142136$.

$\sin x = 1 - \sqrt{2}$ 1 - 2nd x² 2) ENTER

$\sin x = -0.4142136$ *Radian Mode* 2nd SIN - 0 . 4 1 4 2 1 3 6) ENTER

$x = -0.4271$

The reference angle for $x = -0.4271$ is $x_r = 0.4271$. Sine is negative in quadrants three and four. The solutions are $x = 3.5687$ and $x = 5.8561$.

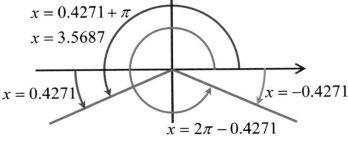

▶ EXAMPLE 15

Find all real solutions for $\cos^2(2x) + 4\cos(2x) - 3 = 0$. Round to the nearest ten-thousandth.

▶▶ Solution

$$\cos^2(2x) + 4\cos(2x) - 3 = 0$$ *The equation is in quadratic form. Since the equation*

$$a = 1, \quad b = 4, \quad c = -3$$ *cannot be factored we need to use the Quadratic Formula.*

$$\cos(2x) = \frac{-(4) \pm \sqrt{(4)^2 - 4(1)(-3)}}{2(1)}$$ *Substitute into the Quadratic Formula the values for a, b, and c.*

$$\cos(2x) = \frac{-4 \pm \sqrt{16 + 12}}{2}$$ *Simplify.*

$$\cos(2x) = \frac{-4 \pm \sqrt{28}}{2}$$ *Get the radical in reduced radical form.*

$$\cos(2x) = \frac{-4 \pm \sqrt{4 \cdot 7}}{2}$$

$$\cos(2x) = \frac{-4 \pm 2\sqrt{7}}{2}$$ *Reduce the fraction.*

$$\cos(2x) = \frac{2(-2 \pm \sqrt{7})}{2}$$ *Factor out the common factor of 2 and simplify.*

$$\cos(2x) = -2 \pm \sqrt{7}$$

$$\cos(2x) = -2 - \sqrt{7} \quad \boxed{-}\,\boxed{2}\,\boxed{-}\,\boxed{\text{2nd}}\,\boxed{x^2}\,\boxed{7}\,\boxed{)}\ \boxed{\text{ENTER}}$$

$$\cos(2x) = -4.64575131 \qquad \text{Since the range of the cosine function is } [-1,1], \text{ there is no } x \text{ such that}$$
$$\cos(2x) = -4.64575131.$$

$$\cos(2x) = -2 + \sqrt{7} \quad \boxed{-}\,\boxed{2}\,\boxed{+}\,\boxed{\text{2nd}}\,\boxed{x^2}\,\boxed{7}\,\boxed{)}\ \boxed{\text{ENTER}}$$

$$\cos(2x) = 0.64575131 \ \textit{Radian Mode}\ \boxed{\text{2nd}}\,\boxed{\text{COS}}\,\boxed{0}\,\boxed{.}\,\boxed{6}\,\boxed{4}\,\boxed{5}\,\boxed{7}\,\boxed{5}\,\boxed{1}\,\boxed{3}\,\boxed{1}\,\boxed{)}\ \boxed{\text{ENTER}}$$

$$2x = 0.8688$$

Cosine is positive in quadrants one and four.
We have $2x = 0.8688$ and $2x = 5.4144$.

$$2x = 0.8688 + 2\pi k \qquad 2x = 5.4144 + 2\pi k$$

$$\frac{2x}{2} = \frac{0.8688}{2} + \frac{2\pi k}{2} \qquad \frac{2x}{2} = \frac{5.4144}{2} + \frac{2\pi k}{2}$$

$$x = 0.4344 + \pi k \qquad x = 2.7072 + \pi k$$

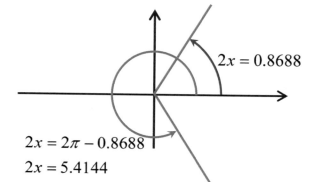

$2x = 0.8688$

$2x = 2\pi - 0.8688$

$2x = 5.4144$

Solve Trigonometric Equations by Squaring Both Sides of the Equation

Sometimes we must square each side of an equation to obtain a quadratic equation. Squaring both sides of the equation can introduce extraneous solutions and therefore each solution should be checked in the original equation for validity.

►**EXAMPLE 16**

Solve $\cos x = \sin x - 1$ **on the interval** $[0, 2\pi)$.

►►**Solution**

$$\cos x = \sin x - 1 \qquad \text{\textit{Square both sides of the equation.}}$$

$$(\cos x)^2 = (\sin x - 1)^2$$

$$\cos^2 x = (\sin x - 1)(\sin x - 1) \qquad \text{\textit{F.O.I.L.}}$$

$$\cos^2 x = \sin^2 x - \sin x - \sin x + 1 \quad \text{\textit{Simplify.}}$$

$$\cos^2 x = \sin^2 x - 2\sin x + 1 \qquad \text{\textit{Rewrite cosine in terms of sine.}}$$

$$1 - \sin^2 x = \sin^2 x - 2\sin x + 1 \qquad \text{\textit{Get the equation in standard form.}}$$

$$0 = 2\sin^2 x - 2\sin x \qquad \text{\textit{Factor out the common factor, 2sinx.}}$$

$$0 = 2\sin x(\sin x - 1) \qquad \text{\textit{Set each factor equal to zero and solve.}}$$

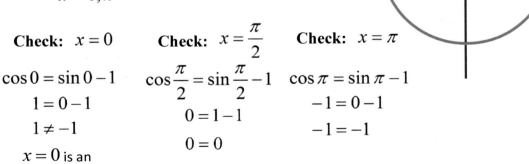

$$2\sin x = 0 \qquad\qquad \sin x - 1 = 0$$

$$\frac{2\sin x}{2} = \frac{0}{2} \qquad\qquad \sin x = 1$$

$$\sin x = 0 \qquad\qquad x = \frac{\pi}{2}$$

$$x = 0, \pi$$

Check: $x = 0$ **Check:** $x = \dfrac{\pi}{2}$ **Check:** $x = \pi$

$$\cos 0 = \sin 0 - 1 \qquad \cos\frac{\pi}{2} = \sin\frac{\pi}{2} - 1 \qquad \cos \pi = \sin \pi - 1$$

$$1 = 0 - 1 \qquad\qquad 0 = 1 - 1 \qquad\qquad -1 = 0 - 1$$

$$1 \ne -1 \qquad\qquad 0 = 0 \qquad\qquad -1 = -1$$

$x = 0$ is an extraneous solution.

Solutions: $x = \dfrac{\pi}{2}$ and $x = \pi$

Guidelines for Solving Trigonometric Equations

1. If the equation involves only one term containing a simple trigonometric function without a multiple angle, then isolate the trigonometric function and apply knowledge of the unit circle and the range of inverse functions to determine the solution.
2. If the equation is written in terms of a single trigonometric function, but contains more than one term with that function, then combine like terms and factor if necessary. Set each factor equal to 0, then solve each resulting equation using step 1 above.
3. A quadratic equation that cannot be factored may be solved using the quadratic formula. See Example 14.
4. If the equation contains more than one trigonometric function, then it may be possible to factor by grouping and solve as in step 2 above. If not, then refer to the trigonometric identities in order to write the equation in terms of a single trigonometric function. If the trigonometric functions are not raised to a power greater than 1 and there does not appear to be an identity that applies, try squaring both sides of the equation.
5. For an equation involving multiple angles, solve using the same technique from step 1, isolating the multiple angle rather than just x. Add $2\pi k$ (k = 0, 1, 2, 3, . . .) to the solution before dividing by the coefficient on the angle. Stop this process when a value beyond 2π is reached.
6. Once an equation has been solved, take note of the interval for the solutions and adjust answers to fit the given interval. Check your answers to verify that the original equation is satisfied.

HOMEWORK

Objective 1

Solve on the interval $[0, 2\pi)$.

See Examples 1-14, 16.

1. $2\sin x - 1 = 0$
2. $-4\cos x = 2\sqrt{3}$
3. $2\sin^2 x + \sin x - 1 = 0$
4. $2\sin(2x) - \sqrt{3} = 0$
5. $3\csc^2 x - 6 = 0$
6. $2\cos(2x) + 1 = 0$
7. $\cos 2x + \cos x + 1 = 0$
8. $\tan^2 x \cos x - \cos x + 1 = \tan^2 x$

Objective 1

Solve on the interval $[0°, 360°)$.

See Examples 1-10.

9. $4\sec x + 8 = 0$
10. $\sqrt{3}\tan x + 1 = 0$
11. $3\cot^2 x - 1 = 0$
12. $\cos^2 x - \sin x = 1$
13. $\sin 2x - \cos x = 0$
14. $4\sin^2 x - 1 = 0$
15. $2\cos^2 x - \cos x - 1 = 0$
16. $8\sin^2 x \cos x - 2\cos x - 4\sin^2 x + 1 = 0$

Objective 2

Solve on the interval $(-\infty,\infty)$.

See Examples 1-13.

17. $2\sin x + \sqrt{3} = 0$ 18. $3\cot x - \sqrt{3} = 0$ 19. $8\tan x + 7\sqrt{3} = -\sqrt{3}$

20. $2\sqrt{3}\cos x = -\sqrt{3}$ 21. $4\sin^2 x - 2 = 0$ 22. $\tan^2 x - 1 = 0$

23. $2\cos^2 x + 5\cos x - 3 = 0$ 24. $\sin^2 x + 2\sin x - 3 = 0$

25. $6\tan^2 x - 2\sqrt{3}\tan x = 0$ 26. $4\sin x \cos x - 2\sqrt{3}\sin x - 2\cos x + \sqrt{3} = 0$

27. $2\cos(3x) - \sqrt{3} = 0$ 28. $4\sqrt{3}\sin^2 x \sec x - \sqrt{3}\sec x + 2 = 8\sin^2 x$

29. $2\tan\left(\dfrac{x}{2}\right) - 2 = 0$ 30. $\cos(2x)\sin x + \sin x = 0$

Objective 1

Use the quadratic formula to solve on the interval $[0,2\pi)$. Round to the nearest ten-thousandth.

See Example 14.

31. $5\sin^2 x - 8\sin x = 3$ 32. $\cos^2 x + 6\cos x + 4 = 0$

33. $2\tan^2 x = 3\tan x + 7$ 34. $3\sin^2 x - 2 = 3\sin x$

Objective 2

Use the quadratic formula to solve on the interval $(-\infty,\infty)$. Round to the nearest ten-thousandth.

See Example 15.

35. $2\cos^2 x + 5\cos x - 3 = 0$ 36. $\sin^2 x + 2\sin x - 3 = 0$

37. $12\sin^2 x - \sin x - 1 = 0$ 38. $15\cos^2 x + \cos x = 6$

39. $\sec^2 x + 2\tan x = 2$ 40. $3\tan^2 x + 2\sec x = 5$

Objective 1

Solve on the interval $[0,2\pi)$.

See Example 14.

41. $\sin x + \cos x = \dfrac{\sqrt{6}}{2}$ 42. $\sin x + \cos x = \sqrt{2}$

43. $\tan x - \sec x = -1$ 44. $\cot x - \csc x = \sqrt{3}$

ANSWERS

1. $x = \frac{\pi}{6}, \frac{5\pi}{6}$

2. $x = \frac{5\pi}{6}, \frac{7\pi}{6}$

3. $x = \frac{\pi}{6}, \frac{5\pi}{6}, \frac{3\pi}{2}$

4. $x = \frac{\pi}{6} + \pi k$ and $x = \frac{\pi}{3} + \pi k$ for $k = 0,1$ Or $x = \frac{\pi}{6}, \frac{\pi}{3}, \frac{7\pi}{6}, \frac{4\pi}{3}$

5. $x = \frac{\pi}{4}, \frac{3\pi}{4}, \frac{5\pi}{4}, \frac{7\pi}{4}$

6. $x = \frac{\pi}{3} + \pi k$ and $x = \frac{2\pi}{3} + \pi k$ for $k = 0,1$ Or $x = \frac{\pi}{3}, \frac{2\pi}{3}, \frac{4\pi}{3}, \frac{5\pi}{3}$

7. $x = \frac{\pi}{2}, \frac{2\pi}{3}, \frac{4\pi}{3}, \frac{3\pi}{2}$

8. $x = 0, \frac{\pi}{4}, \frac{3\pi}{4}, \frac{5\pi}{4}, \frac{7\pi}{4}$

9. $x=120°, 240°$

10. $x=150°, 330°$

11. $x=60°,120°, 240°, 300°$

12. $x=0°,180°,270°$

13. $x=30°, 90°, 150°, 270°$

14. $x=30°, 150°, 210°, 330°$

15. $x=0°, 120°, 240°$

16. $x=30°, 60°, 150°, 210°, 300°, 330°$

17. $x = \frac{4\pi}{3} + 2\pi k$ and $x = \frac{5\pi}{3} + 2\pi k$

18. $x = \frac{\pi}{3} + \pi k$

19. $x = \frac{2\pi}{3} + \pi k$

20. $x = \frac{2\pi}{3} + 2\pi k$ and $x = \frac{4\pi}{3} + 2\pi k$

21. $x = \frac{\pi}{4} + \frac{\pi}{2} k$

22. $x = \frac{\pi}{4} + \frac{\pi}{2} k$

23. $x = \frac{\pi}{3} + 2\pi k$ and $x = \frac{5\pi}{3} + 2\pi k$

24. $x = \frac{\pi}{2} + 2\pi k$

25. $x = \frac{\pi}{6} + \pi k$ and $x = \pi k$

26. $x = \frac{\pi}{6} + 2\pi k$ and $x = \frac{5\pi}{6} + \pi k$

27. $x = \frac{\pi}{18} + \frac{2\pi}{3} k$ and $x = \frac{11\pi}{18} + \frac{2\pi}{3} k$

28. $x = \frac{\pi}{6} + \pi k$ and $x = \frac{5\pi}{6} + \pi k$

29. $x = \frac{\pi}{2} + 2\pi k$

30. $x = \pi k$ and $x = \frac{\pi}{2} + \pi k$

31. $x = 3.4606$ and $x = 5.9642$

32. $x = 2.4402$ and $x = 3.8430$

33. $x = 1.2238 , x = 2.2395, x = 4.3654$ and $x = 5.3811$

34. $x = 3.6167$ and $x = 5.8081$

35. $x = 1.0472 + 2\pi k$ and $x = 5.2360 + 2\pi k$

36. $x = 1.5708 + 2\pi k$

37. $x = 0.3398 + 2\pi k, x = 2.8018 + 2\pi k, x = 3.3943 + 2\pi k$ and $x = 6.0305 + 2\pi k$

38. $x = 0.9273 + 2\pi k, x = 2.3005 + 2\pi k, x = 3.9827 + 2\pi k$ and $x = 5.3559 + 2\pi k$

39. $x = 0.3927 + \pi k$ and $x = 1.9635 + \pi k$

40. $x = 0.7227 + 2\pi k, x = 2.0944 + 2\pi k, x = 4.1888 + 2\pi k$ and $x = 5.5605 + 2\pi k$

41. $\frac{\pi}{12}, \frac{5\pi}{12}$

42. $\frac{\pi}{4}$

43. 0

44. $\frac{4\pi}{3}$

Section 17: Oblique Triangles: Law of Sines and Law of Cosines

Learning Outcomes:

- The student will solve trigonometric equations, right triangles, and oblique triangles.
- The student will correctly memorize and apply trigonometric formulas, definitions, identities, and properties.

Objectives: At the conclusion of this lesson you should be able to:

1. Use the law of sines to solve Angle-Side-Angle (ASA) triangles.
2. Use the law of sines to solve Angle-Angle-Side (AAS) triangles.
3. Use the law of sines to solve Side-Side-Angle (SSA) triangles.
4. Use the law of cosines to solve Side-Angle-Side (SAS) triangles.
5. Use the law of cosines to solve Side-Side-Side (SSS) triangles.

A triangle that does not have a 90° angle is called an oblique triangle. To solve an oblique triangle we need to know the measure of at least one side and any two other parts of the triangle. We cannot determine the lengths of the sides of an oblique triangle when three angles are given because there are infinitely many similar triangles that are possible. We can classify the five possible cases as follows:

1. Two angles and the included side (ASA)
2. Two angles and not the included side (AAS)
3. Two sides and not the included angle (SSA)
4. Two sides and the included angle (SAS)
5. Three sides (SSS)

The first three cases can be solved using the law of sines and the last two using the law of cosines. When solving oblique triangles it is useful to draw a triangle and label what is given. Recall the properties of a triangle form Section 3.

Properties of a Triangle

I. The sum of the angles of a triangle is 180°:
$$A + B + C = 180°$$

II. The combined length of any two sides exceeds that of the third side: $a + b > c$, $a + c > b$, and $b + c > a$

III. The largest angle is opposite the largest side: $B > A$, then $b > a$.

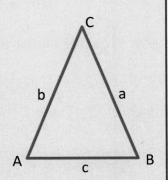

Law of Sines

Consider the oblique triangle as shown in **Figure 17.1**. If we drop the perpendicular from angle **B** to the base of the triangle we have two right triangles with a common side of **h**. For angle **A** the sine of angle **A** is $\sin A = \dfrac{h}{c}$ and for angle **C** the sine of angle **C** is $\sin C = \dfrac{h}{a}$. If we solve

for h in both we have $c\sin A = h$ and $a\sin C = h$. Since h is the same measure in both, we can set the two equal to each other: $c\sin A = a\sin C$. If we multiply both

sides by $\dfrac{1}{ca}$ we have $\dfrac{\sin A}{a} = \dfrac{\sin C}{c}$.

If we had drawn the perpendicular from vertex C to side c, we would have $\dfrac{\sin A}{a} = \dfrac{\sin B}{b}$. Using the transitive

property of equality, we conclude that

$$\dfrac{\sin A}{a} = \dfrac{\sin B}{b} = \dfrac{\sin C}{c}.$$

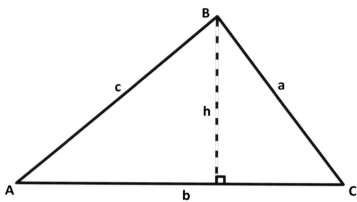

Figure 17.1

Law of Sines

If ABC is a triangle with sides a, b, and c, then

$$\dfrac{\sin A}{a} = \dfrac{\sin B}{b} = \dfrac{\sin C}{c}.$$

CASE #1: Angle-Side-Angle (ASA)

Known: Two angles and the included side

Solve: Use the Law of Sines.

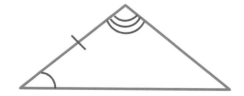

► **EXAMPLE 1**

Solve the triangle: $B = 47°, C = 67°, a = 12.5\,m$. **Round your answers to the nearest tenth.**

►►**Solution**

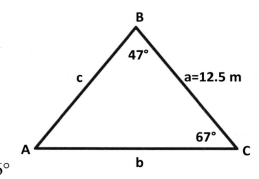

Step 1: Draw a triangle and label what we know. We are given two angles and the included side. This is Case #1 and we will use the Law of Sines to solve the triangle.

Step 2: We can find the third angle, A, by adding the two given angles and subtracting the sum from 180°. $A = 180° - (47° + 67°) = 66°$

Step 3: Now that we have the measure of angle A we can find the measure of one of the two remaining sides using the Law of Sines. We will find the measure of side b first:

$$\frac{\sin 66°}{12.5} = \frac{\sin 47°}{b}$$ *Substitute into the Law of Sines formula.*

$$b\sin 66° = 12.5\sin 47°$$ *Cross multiply.*

$$\frac{b\sin 66°}{\sin 66°} = \frac{12.5\sin 47°}{\sin 66°}$$ *Solve for b.*

$$b = \frac{12.5\sin 47°}{\sin 66°}$$ $\boxed{1}\boxed{2}\boxed{.}\boxed{5}\boxed{\text{SIN}}\boxed{4}\boxed{7}\boxed{)}\boxed{÷}\boxed{\text{SIN}}\boxed{6}\boxed{6}\boxed{)}\boxed{\text{ENTER}}$

$$b = 10.0$$

Step 4: We need to find the remaining side c using the Law of Sines.

$$\frac{\sin 66°}{12.5} = \frac{\sin 67°}{c}$$ *Substitute into the Law of Sines formula.*

$$c\sin 66° = 12.5\sin 67°$$ *Cross multiply.*

$$\frac{c\sin 66°}{\sin 66°} = \frac{12.5\sin 67°}{\sin 66°}$$ *Solve for c.*

$$c = \frac{12.5\sin 67°}{\sin 66°}$$ $\boxed{1}\boxed{2}\boxed{.}\boxed{5}\boxed{\text{SIN}}\boxed{6}\boxed{7}\boxed{)}\boxed{÷}\boxed{\text{SIN}}\boxed{6}\boxed{6}\boxed{)}\boxed{\text{ENTER}}$

$$c = 12.6$$

Step 5: Solution: $A = 66°$ $B = 47°$ $C = 67°$

$$a = 12.5\,m \quad b = 10.0\,m \quad c = 12.6\,m$$

CASE #2: Angle-Angle-Side (AAS)

Known: Two angles and not the included side

Solve: Use the Law of Sines.

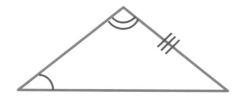

▶EXAMPLE 2

Solve the triangle: $A = 43°, C = 85°, a = 19.5\,m$. **Round your answers to the nearest tenth.**

▶▶Solution

Step 1: Draw a triangle and label what we know. We are given two angles and not the included side. This is Case #2 and we will use the Law of Sines to solve the triangle.

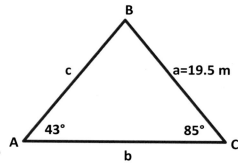

Step 2: We can find the third angle, B, by adding the two given angles and subtracting the sum from 180°. $A = 180° - (43° + 85°) = 52°$

Step 3: Now that we have the measure of angle B we can find the measure of one of the two remaining sides using the Law of Sines. We will find the measure of side b first:

$$\frac{\sin 43°}{19.5} = \frac{\sin 52°}{b}$$ *Substitute into the Law of Sines formula.*

$$b\sin 43° = 19.5\sin 52°$$ *Cross multiply.*

$$\frac{b\sin 43°}{\sin 43°} = \frac{19.5\sin 52°}{\sin 43°}$$ *Solve for b.*

$$b = \frac{19.5\sin 52°}{\sin 43°}$$ [1][9][.][5][SIN][5][2][)][÷][SIN][4][3][)][ENTER]

$$b = 22.5$$

Step 4: We need to find the remaining side c using the Law of Sines.

$$\frac{\sin 43°}{19.5} = \frac{\sin 85°}{c}$$ *Substitute into the Law of Sines formula.*

$$c\sin 43° = 19.5\sin 85°$$ *Cross multiply.*

$$\frac{c\sin 43°}{\sin 43°} = \frac{19.5\sin 85°}{\sin 43°}$$ *Solve for c.*

$$c = \frac{19.5\sin 85°}{\sin 43°}$$ [1][9][.][5][SIN][8][5][)][÷][SIN][4][3][)][ENTER]

$$c = 28.5$$

Step 5: Solution: $A = 43°$ $B = 52°$ $C = 85°$

$a = 19.5\,m$ $b = 22.5\,m$ $c = 28.5\,m$

Ambiguous Case

In case 1 and case 2 we were given two angles and one side. If two sides and one opposite angle are given we have three possibilities: (1) no triangle exists, (2) one triangle exists, or (3) two distinct triangles exist.

CASE #3: Side-Side-Angle (SSA) (Ambiguous Case)

Known: Two sides and not the included angle

Solve: Use the Law of Sines.

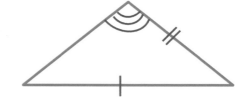

▶EXAMPLE 3

Solve the triangle: $B = 67°$, $a = 36.5\,m$, $b = 12.9\,m$. **Round your answers to the nearest tenth.**

▶▶Solution

Step 1:   Draw a triangle and label what we know.
We are given two sides and not the included
angle. This is Case 3 and we will use the
Law of Sines to solve the triangle. We are
dealing with the ambiguous case.

Step 2:   We have enough information to find the
measure of angle A.

$$\frac{\sin 67°}{12.9} = \frac{\sin A}{36.5}$$ *Substitute into the Law of Sines formula.*

$12.9 \sin A = 36.5 \sin 67°$ *Cross multiply.*

$$\frac{12.9 \sin A}{12.9} = \frac{36.5 \sin 67°}{12.9}$$ *Solve for the sine of A.*

$$\sin A = \frac{36.5 \sin 67°}{12.9}$$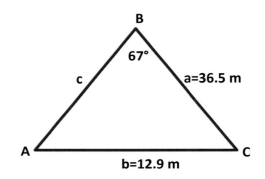

$$A = ERROR$$

$$\sin A = \frac{36.5 \sin 67°}{12.9}$$

$$\sin A = 2.604529$$

2.604529 is not within the range $[-1,1]$. There is no triangle.

►EXAMPLE 4

Solve the triangle: $B = 48°$, $a = 33$ *ft*, $b = 25$ *ft* . **Round your answers to the nearest tenth.**

►►Solution

Step 1:   Draw a triangle and label what we know. We are given two sides and not the included side. This is Case 3 and we will use the Law of Sines to solve the triangle. We are dealing with the ambiguous case.

Step 2:   We have enough information to find the measure of angle A.

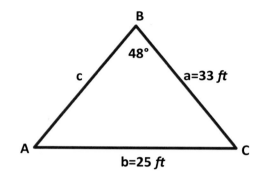

$$\frac{\sin 48°}{25} = \frac{\sin A}{33}$$ *Substitute into the Law of Sines formula.*

$$25\sin A = 33\sin 48°$$ *Cross multiply.*

$$\frac{25\sin A}{25} = \frac{33\sin 48°}{25}$$ *Solve for the sine of A.*

$$\sin A = \frac{33\sin 48°}{25}$$ $\boxed{2^{\text{nd}}}\ \boxed{\text{SIN}}\ \boxed{3}\boxed{3}\ \boxed{\text{SIN}}\ \boxed{4}\boxed{8}\boxed{)}\ \boxed{÷}\boxed{2}\boxed{5}\ \boxed{)}\boxed{\text{ENTER}}$

$$A = 78.8°$$

Since sine is positive in Quadrants I and II, we need to find the angle in Quadrant II. The angle in Quadrant II is 101.2°.   Adding the angle in Quadrant II to the given angle of 48 °gives us 48°+101.2°=149.2° which is less than 180°. Since the sum is less than 180°, we have two triangles. We will label the first triangle with subscripts of 1 and the second triangle with subscripts of 2: $A_1 = 78.8°$ and $A_2 = 101.2°$.

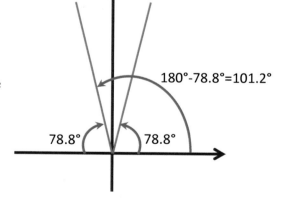

Step 3:   We can find the remaining angle C for both triangles :

$$C_1 = 180° - (48° + 78.8°) = 53.2° \text{ and } C_2 = 180° - (48° + 101.2°) = 30.8°$$

Step 4:   We need to find the remaining side c for both triangles using the Law of Sines.

$$\frac{\sin 48°}{25} = \frac{\sin 53.2°}{c_1}$$ *Substitute into the Law of Sines formula.*

$$c_1 \sin 48° = 25 \sin 53.2°$$ *Cross multiply.*

$$\frac{c_1 \sin 48°}{\sin 48°} = \frac{25 \sin 53.2°}{\sin 48°}$$ *Solve for c_1.*

$$c_1 = \frac{25 \sin 53.2°}{\sin 48°}$$ ⎡2⎤⎡5⎤⎡SIN⎤⎡5⎤⎡3⎤⎡.⎤⎡2⎤⎡)⎤⎡÷⎤⎡SIN⎤⎡4⎤⎡8⎤⎡)⎤⎡ENTER⎤

$$c_1 = 26.9$$

$$\frac{\sin 48°}{25} = \frac{\sin 30.8°}{c_2}$$ *Substitute into the Law of Sines formula.*

$$c_2 \sin 48° = 25 \sin 30.8°$$ *Cross multiply.*

$$\frac{c_2 \sin 48°}{\sin 48°} = \frac{25 \sin 30.8°}{\sin 48°}$$ *Solve for c_2.*

$$c_2 = \frac{25 \sin 30.8°}{\sin 48°}$$ ⎡2⎤⎡5⎤⎡SIN⎤⎡3⎤⎡0⎤⎡.⎤⎡8⎤⎡)⎤⎡÷⎤⎡SIN⎤⎡4⎤⎡8⎤⎡)⎤⎡ENTER⎤

$$c_2 = 17.2$$

Step 5:   Solution:

$$A_1 = 78.8° \qquad B_1 = 48° \qquad C_1 = 53.2°$$
$$a_1 = 33 \text{ ft} \qquad b_1 = 25 \text{ ft} \qquad c_1 = 26.9 \text{ ft}$$

$$A_2 = 101.2° \qquad B_2 = 48° \qquad C_2 = 30.8°$$
$$a_2 = 33 \text{ ft} \qquad b_2 = 25 \text{ ft} \qquad c_2 = 17.2 \text{ ft}$$

►**EXAMPLE 5**

Solve the triangle: $C = 59°$, $a = 465\,ft$, $c = 538\,ft$. **Round your answers to the nearest tenth.**

►►**Solution**

Step 1:   Draw a triangle and label what we know. We are given two sides and not the included side. This is Case 3 and we will use the Law of Sines to solve the triangle. We are dealing with the ambiguous case.

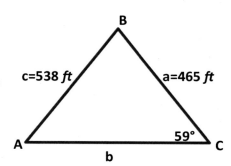

Step 2:   We have enough information to find the measure of angle A.

$$\frac{\sin 59°}{538} = \frac{\sin A}{465}$$ *Substitute into the Law of Sines formula.*

$$538\sin A = 465\sin 59°$$ *Cross multiply.*

$$\frac{538\sin A}{538} = \frac{465\sin 59°}{538}$$ *Solve for the sine of A.*

$$\sin A = \frac{465\sin 59°}{538}$$ 2nd SIN 4 6 5 SIN 5 9) ÷ 5 3 8) ENTER

$$A = 47.8°$$

Since sine is positive in Quadrants I and II, we need to find the angle in Quadrant II. The angle in Quadrant II is 132.2°. Adding the angle in Quadrant II to the given angle of 59° gives us 59°+132.2°=191.2° which is greater than 180°. Since the sum is greater than 180°, we have only one triangle.

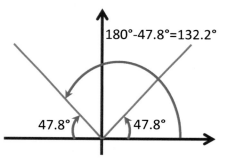

Step 3:   We can find the remaining angle B: $B = 180° - (59° + 47.8°) = 73.2°$

Step 4:   We need to find the remaining side b using the Law of Sines.

$$\frac{\sin 59°}{538} = \frac{\sin 73.2°}{b}$$ *Substitute into the Law of Sines formula.*

$$b\sin 59° = 538\sin 73.2°$$ *Cross multiply.*

$$\frac{b\sin 59°}{\sin 59°} = \frac{538\sin 73.2°}{\sin 59°}$$ *Solve for c_1.*

$$b = \frac{538\sin 73.2°}{\sin 59°}$$ 5 3 8 SIN 7 3 . 2) ÷ SIN 5 9) ENTER

$$b = 600.9$$

Step 5:   Solution: $A = 47.8°$ $B = 73.2°$ $C = 59°$

$\qquad\qquad\qquad\qquad$ $a = 465\,ft$ $b = 600.9\,ft$ $c = 538\,ft$

Law of Cosines

We have looked at the first three cases ASA, AAS, and SSA for solving triangles using the law of sines. The last two cases SAS and SSS cannot be handled with the law of sines but require the law of cosines.

Consider the oblique triangle shown in **Figure 17.2**. Notice that vertex B has coordinates (x, y) and vertex **C** has the coordinates $(b, 0)$. Because **a** is the distance from vertex **B** to vertex **C**, we can use the distance formula to find the measure of **a**.

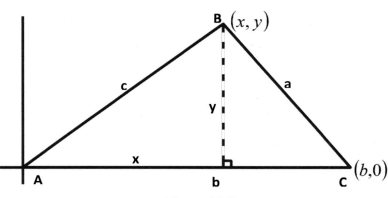

Figure 17.2

$a = \sqrt{(x - b)^2 + (y - 0)^2}$

$a^2 = (x - b)^2 + y^2$ *Square both sides.*

$\cos A = \dfrac{x}{c} \qquad \sin A = \dfrac{y}{c}$

$c \cos A = x \qquad c \sin A = y$

$a^2 = (c \cos A - b)^2 + (c \sin A)^2$ *Substitute for x and y.*

$a^2 = c^2 \cos^2 A - bc \cos A - bc \cos A + b^2 + c^2 \sin^2 A$ *Expand.*

$a^2 = c^2 \cos^2 A + c^2 \sin^2 A - 2bc \cos A + b^2$ *Rewrite and combine like terms.*

$a^2 = c^2(\cos^2 A + \sin^2 A) - 2bc \cos A + b^2$ *Factor out c^2.*

$a^2 = c^2(1) - 2bc \cos A + b^2$ *Substitute using $\cos^2 A + \sin^2 A = 1$ and rewrite.*

$a^2 = b^2 + c^2 - 2bc \cos A$ *One of the Law of Cosines formulas.*

Law of Cosines

If ABC is a triangle with sides a, b, and c, then

$$a^2 = b^2 + c^2 - 2bc \cos A$$

$$b^2 = a^2 + c^2 - 2ac \cos B$$

$$c^2 = a^2 + b^2 - 2ab \cos C$$

We can derive the alternate form of the law of cosines by solving for $\cos A$ in the law of cosine formula $a^2 = b^2 + c^2 - 2bc \cos A$.

$$a^2 = b^2 + c^2 - 2bc \cos A$$

$$2bc \cos A + a^2 = b^2 + c^2 \qquad \textit{Move 2bccosA to the other side of the equation.}$$

$$2bc \cos A = b^2 + c^2 - a^2 \qquad \textit{Move } a^2 \textit{ to the other side of the equation.}$$

$$\frac{2bc \cos A}{2bc} = \frac{b^2 + c^2 - a^2}{2bc} \qquad \textit{Solve for cosA.}$$

$$\cos A = \frac{b^2 + c^2 - a^2}{2bc} \qquad \textit{Alternate form of the law of cosines.}$$

Alternate form of the Law of Cosines

If ABC is a triangle with sides a, b, and c, then

$$\cos A = \frac{b^2 + c^2 - a^2}{2bc}$$

$$\cos B = \frac{a^2 + c^2 - b^2}{2ac}$$

$$\cos C = \frac{a^2 + b^2 - c^2}{2ab}$$

CASE #4: Side –Angle-Side (SAS)

Known: Two sides and the included angle

Solve: Use the Law of Cosines.

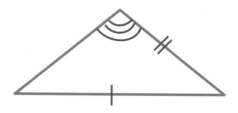

▶ EXAMPLE 6

Solve the triangle: $B = 85°, a = 15\,m, c = 12\,m$. **Round your answers to the nearest tenth.**

▶▶ Solution

Step 1: Draw a triangle and label what we know. We are given two angles and the included side. This is Case #4 and we will use the law of cosines in solving the triangle.

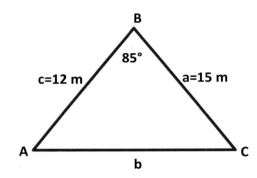

Step 2: We will use the law of cosines and find the third side.

$b^2 = a^2 + c^2 - 2ac\cos B$

$b^2 = (15)^2 + (12)^2 - 2(15)(12)\cos 85°$ *Substitute into the Law of Cosines formula.*

$\sqrt{b^2} = \sqrt{(15)^2 + (12)^2 - 2(15)(12)\cos 85}$ *Take the square root of both sides.*

$b = \sqrt{(15)^2 + (12)^2 - 2(15)(12)\cos 85°}$

$b = 18.4$

| 2ⁿᵈ | x² | 1 | 5 | x² | + | 1 | 2 | x² | - | 2 | × | 1 | 5 |
| × | 1 | 2 | COS | 8 | 5 |) |) | ENTER |

Step 3: We will use the law of sines to find the smallest remaining angle. Finding the smallest remaining angle avoids an ambiguous case.

$\dfrac{\sin 85°}{18.4} = \dfrac{\sin C}{12}$ *Substitute into the Law of Sines formula.*

$18.4\sin C = 12\sin 85°$ *Cross multiply.*

$\dfrac{18.4\sin C}{18.4} = \dfrac{12\sin 85°}{18.4}$ *Solve for sinC.*

$\sin C = \dfrac{12\sin 85°}{18.4}$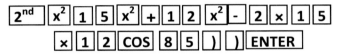

$C = 40.5°$

Step 4: We can find the third angle, A, by adding the two given angles and subtracting the sum from 180°. $A = 180° - (85° + 40.5°) = 54.5°$

Step 5: Solution: $A = 54.5°$ $B = 85°$ $C = 40.5°$

                     $a = 15\,m$ $b = 18.4\,m$ $c = 12\,m$

► EXAMPLE 7

Solve the triangle: $a = 12\,m, b = 15\,m, c = 8\,m$. **Round your answers to the nearest tenth.**

►►Solution

Step 1: Draw a triangle and label what we know. We are given three sides. This is Case #5 and we will use the alternate form of the law of cosines in solving the triangle.

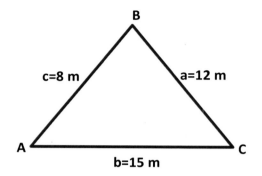

Step 2: We will use the alternate form of the law of cosines and find the largest angle.

$$\cos B = \frac{a^2 + c^2 - b^2}{2ac}$$

$$\cos B = \frac{(12)^2 + (8)^2 - (15)^2}{2(12)(8)}$$ *Substitute into the Alternate Form of the Law of Cosines.*

$B = 95.1°$

| 2ⁿᵈ | COS | (| (| 1 | 2 | x² | + | 8 | x² | - | 1 | 5 | x² |) | ÷ |

| (| 2 | × | 1 | 2 | × | 8 |) |) | ENTER |

Step 3: We will use the law of sines to find one of the remaining angles.

$$\frac{\sin 95.1°}{15} = \frac{\sin A}{12}$$ *Substitute into the Law of Sines formula.*

$15 \sin A = 12 \sin 95.1°$ *Cross multiply.*

$$\frac{15 \sin A}{15} = \frac{12 \sin 95.1°}{15}$$ *Solve for sinA.*

$$\sin A = \frac{12 \sin 95.1°}{15}$$

| 2ⁿᵈ | SIN | 1 | 2 | SIN | 9 | 5 | . | 1 |) | ÷ | 1 | 5 |) | ENTER |

$A = 52.8°$

Step 4: We can find the third angle, C, by adding the two given angles and subtracting the sum from 180°. $A = 180° - (52.8° + 95.1°) = 32.1°$

Step 5: Solution: $A = 52.8°$ $B = 95.1°$ $C = 32.1°$

$a = 12\,m$ $b = 15\,m$ $c = 8\,m$

▶ **EXAMPLE 8**

Given $A = 68.4°$, $B = 54.3°$, $a = 12\,m$ find c. Round your answer to the nearest tenth.

▶▶ **Solution**

Step 1:   Draw a triangle and label what we know. We are given two angles and not the included side. This is Case 2 and we will use the Law of Sines to aid in solving for c.

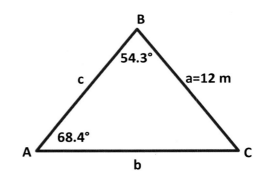

Step 2:   We can find angle C by adding the two given angles and subtracting the sum from 180°.

$$A = 180° - (68.4° + 54.3°) = 57.3°$$

Step 3:   Now that we have the measure of angle B we can find the measure of one of the two remaining sides using the Law of Sines. We will find the measure of side b first:

$$\frac{\sin 68.4°}{12} = \frac{\sin 57.3°}{c} \qquad \text{\textit{Substitute into the Law of Sines formula.}}$$

$$c \sin 68.4° = 12 \sin 57.3° \qquad \text{\textit{Cross multiply.}}$$

$$\frac{c \sin 68.4°}{\sin 68.4°} = \frac{12 \sin 57.3°}{\sin 68.4°} \qquad \text{\textit{Solve for c.}}$$

$$c = \frac{12 \sin 57.3°}{\sin 68.4°}$$

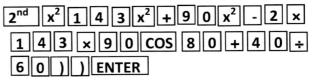

$$c = 10.9$$

▶ **EXAMPLE 9**

Given $A = 80°40'$, $b = 143\,m$, $c = 90\,m$, find a. Round your answer to the nearest tenth.

▶▶ **Solution**

Step 1:   Draw a triangle and label what we know. We are given two angles and the included side. This is Case #4 and we will use the law of cosines in solving the triangle.

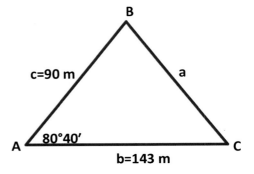

Step 2:   We will use the law of cosines and find the third side a.

$$a^2 = b^2 + c^2 - 2bc \cos A$$

$$a^2 = (143)^2 + (90)^2 - 2(143)(90)\cos 80°40' \qquad \text{\textit{Substitute into the Law of Cosines formula.}}$$

$$\sqrt{a^2} = \sqrt{(143)^2 + (90)^2 - 2(143)(90)\cos 80°40'} \qquad \text{\textit{Take the square root of both sides.}}$$

$$a = \sqrt{(143)^2 + (90)^2 - 2(143)(90)\cos 80°40'}$$

$$a = 156.1\,m$$

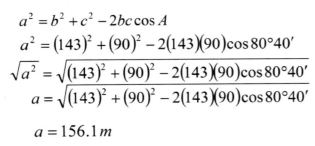

►**EXAMPLE 10**

Given $B = 48.2°$, $a = 890m$, $b = 697m$, **find** A. **Round your answer to the nearest tenth.**

►►**Solution**

Step 1: Draw a triangle and label what we know. We are given two sides and not the included angle. This is Case 3(ambiguous case) and we will use the Law of Sines to aid in solving for A.

Step 2: We have enough information to find the measure of angle A.

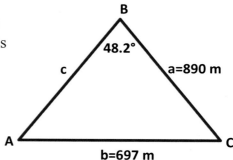

$$\frac{\sin 48.2°}{697} = \frac{\sin A}{890}$$ *Substitute into the Law of Sines formula.*

$$697 \sin A = 890 \sin 48.2°$$ *Cross multiply.*

$$\frac{697 \sin A}{697} = \frac{890 \sin 48.2°}{697}$$ *Solve for the sine of A.*

$$\sin A = \frac{890 \sin 48.2°}{697}$$ $\boxed{2^{nd}}$ $\boxed{\text{SIN}}$ $\boxed{8}\boxed{9}\boxed{0}$ $\boxed{\text{SIN}}$ $\boxed{4}\boxed{8}\boxed{.}\boxed{2}\boxed{)}$ $\boxed{÷}\boxed{6}\boxed{9}\boxed{7}\boxed{)}$ $\boxed{\text{ENTER}}$

$$A = 72.2°$$

Since sine is positive in Quadrants I and II, we need to find the angle in Quadrant II. The angle in Quadrant II is 107.8°. Adding the angle in Quadrant II to the given angle of 48.2°gives us 48.2°+107.8°=156° which is less than 180°. Since the sum is less than 180°, we have two triangles and two values for angle A: $A_1 = 72.2°$ and $A_1 = 107.8°$.

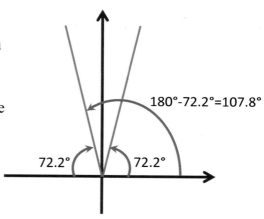

Five Cases for Solving Oblique Triangles

Oblique Triangle		Procedure for Solving	
Case 1	(ASA) Two angles and the included side are known.	Step 1	Find the remaining angle using the angle sum formula $A + B + C = 180°$.
		Step 2	Find the remaining sides using the law of sines.
Case 2	(SAA) Two angles and not the included side are known.	Step 1	Find the remaining angle using the angle sum formula $A + B + C = 180°$.
		Step 2	Find the remaining sides using the law of sines.
Case 3	(SSA) Two sides and not the included angle are known. This is the ambiguous case. 0, 1 or 2 Triangles	Step 1	Find an angle using the law of sines.
		Step 2	Determine if 0, 1 or 2 triangles.
		Step 3	Find the remaining angle using the angle sum formula $A + B + C = 180°$.
		Step 4	Find the remaining side using the law of sines.
Case 4	(SAS) Two sides and the included angle are known.	Step 1	Find the third side using the law of cosines.
		Step 2	Find the smaller of the two remaining angles using the law of sines.
		Step 3	Find the remaining angle using the angle sum formula $A + B + C = 180°$.
Case 5	(SSS) All three sides are known.	Step 1	Find the largest angle using the alternate form of the law of cosines.
		Step 2	Find either remaining angle using the law of sines.
		Step 3	Find the remaining angle using the angle sum formula $A + B + C = 180°$.

HOMEWORK

Objectives 1-5

Solve the triangle using the Law of Sines and/or the Law of Cosines. Round your answers to the nearest tenth.

See Examples 1-7.

1. $A = 30°, B = 45°, a = 20\,cm$
2. $A = 24.3°, C = 54.6°, c = 2.7\,m$
3. $C = 145°, b = 4\,in, c = 14\,in$
4. $A = 60°, a = 9\,m, c = 10\,m$
5. $a = 11\,in, b = 14\,in, c = 20\,in$
6. $C = 11°, a = 40\,in, c = 30\,in$
7. $B = 76°, a = 62\,m, c = 95\,m$
8. $a = 78\,cm, b = 52\,cm, c = 52\,cm$
9. $A = 76°, B = 41°, b = 54\,mi$
10. $A = 100°, a = 15.1\,m, c = 21.2\,m$
11. $A = 52°, a = 32\,m, b = 36\,m$
12. $A = 135°, a = 36.4\,cm, c = 44.7\,cm$
13. $a = 3.2\,in, b = 4.8\,in, c = 6.4\,in$
14. $A = 135°, b = 24\,cm, c = 30\,cm$
15. $C = 29°, a = 538\,mi, b = 465\,mi$
16. $B = 110°, C = 32°, a = 39\,cm$
17. $A = 40°, a = 220\,ft, c = 200\,ft$
18. $B = 27°, C = 98°, b = 7.2\,yd$
19. $A = 13°, C = 22°, b = 126.2\,in$
20. $B = 67°, a = 490\,cm, b = 385\,cm$
21. $C = 59°, b = 67\,m, c = 58\,m$
22. $A = 80°, b = 50\,m, c = 26.6\,m$
23. $A = 30°, a = 816\,m, b = 1115\,m$
24. $a = 12.5\,yd, b = 8\,yd, c = 15\,yd$
25. $B = 67°, a = 36.5\,cm, b = 12.9\,cm$
26. $B = 45°, a = 32.8\,yd, b = 24.9\,yd$

Objectives 1-5

Find the indicated part of each triangle. Round your answers to the nearest tenth.

See Examples 8-10.

27. $A = 68.41°, B = 54.23°, a = 12.75m$, *find c.*
28. $A = 31.3°, B = 71.2°, c = 184m$, *find a.*
29. $A = 23°, C = 131°, b = 10\,in$, *find c.*
30. $C = 39.7°, a = 9.8cm, c = 23.5cm$, *find A.*
31. $A = 43.6°, a = 15\,yd, c = 28\,yd$, *find C.*

32. $B = 29°, b = 15\,m, c = 20\,m,$ *find* $C.$

33. $B = 45.4°, a = 71.3m, b = 61.2m,$ *find* $A.$

34. $a = 8m, b = 5m, c = 12m,$ *find the measure of the largest angle.*

35. $a = 3m, b = 5m, c = 6m,$ *find the measure of the largest angle.*

36. $A = 131.7°, b = 41.3m, c = 76.8m,$ *find* $a.$

37. $A = 80°40', b = 143m, c = 89.6m,$ *find* $a.$

ANSWERS

1.
$A = 30°$ $B = 45°$ $C = 105°$
$a = 20\,cm$ $b = 28.3\,cm$ $c = 38.6\,cm$

2.
$A = 24.3°$ $B = 101.1°$ $C = 54.6°$
$a = 1.4\,m$ $b = 3.3\,m$ $c = 2.7\,m$

3.
$A = 25.6°$ $B = 9.4°$ $C = 145°$
$a = 10.5\,in$ $b = 4\,in$ $c = 14\,in$

4.
$A_1 = 60°$ $B_1 = 45.8°$ $C_1 = 74.2°$
$a_1 = 9\,m$ $b_1 = 7.5\,m$ $c_1 = 10\,m$

$A_2 = 60°$ $B_2 = 14.2°$ $C_2 = 105.8°$
$a_2 = 9\,m$ $b_2 = 2.5\,m$ $c_2 = 10\,m$

5.
$A = 32°$ $B = 42.4°$ $C = 105.6°$
$a = 11\,in$ $b = 14\,in$ $c = 20\,in$

6.
$A = 14.7°$ $B = 154.3°$ $C = 11°$
$a = 40\,in$ $b = 68.2\,in$ $c = 30\,in$

7.
$A = 36.9°$ $B = 76°$ $C = 67.1°$
$a = 62\,m$ $b = 100.1\,m$ $c = 95\,m$

8.
$A = 97.2°$ $B = 41.4°$ $C = 41.4°$
$a = 78\,cm$ $b = 52\,cm$ $c = 52\,cm$

9.

$A = 76°$ $B = 41°$ $C = 63°$

$a = 79.9\,mi$ $b = 54\,mi$ $c = 73.3\,mi$

10.   No Triangle

11.

$A_1 = 52°$ $B_1 = 62.4°$ $C_1 = 65.6°$

$a_1 = 32m$ $b_1 = 36\,m$ $c_1 = 37\,m$

$A_2 = 52°$ $B_2 = 117.6°$ $C_2 = 10.4°$

$a_2 = 32m$ $b_2 = 36\,m$ $c_2 = 7.3\,m$

12.   No Triangle

13.

$A = 28.9°$ $B = 46.6°$ $C = 104.5°$

$a = 3.2\,in$ $b = 4.8\,in$ $c = 6.4\,in$

14.

$A = 135°$ $B = 19.9°$ $C = 25.1°$

$a = 49.9\,cm$ $b = 24\,cm$ $c = 30\,cm$

15.

$A = 91.2°$ $B = 59.8°$ $C = 29°$

$a = 538\,mi$ $b = 465\,mi$ $c = 260.9\,mi$

16.

$A = 38°$ $B = 110°$ $C = 32°$

$a = 39\,cm$ $b = 59.5\,cm$ $c = 33.6\,cm$

17.

$A = 40°$ $B = 104.2°$ $C = 35.8°$

$a = 220\,ft$ $b = 331.8\,ft$ $c = 200\,ft$

18.

$A = 55°$ $B = 27°$ $C = 98°$

$a = 13.0\,yd$ $b = 7.2\,yd$ $c = 15.7\,yd$

19.

$A = 13°$ $B = 145°$ $C = 22°$

$a = 49.5\,in$ $b = 126.2\,in$ $c = 82.4\,in$

20.   No Triangle

21.
$A_1 = 39°$ $B_1 = 82°$ $C_1 = 59°$
$a_1 = 42.6m$ $b_1 = 67\,m$ $c_1 = 58\,m$

$A_2 = 23°$ $B_2 = 98°$ $C_2 = 59°$
$a_2 = 26.4m$ $b_2 = 67\,m$ $c_2 = 58\,m$

22.
$A = 80°$ $B = 70°$ $C = 30°$
$a = 52.4m$ $b = 50\,m$ $c = 26.6\,m$

23.
$A = 30°$ $B = 43.1°$ $C = 106.9°$
$a = 816m$ $b = 1115\,m$ $c = 1561.5\,m$

24.
$A = 56.4°$ $B = 32.2°$ $C = 91.4°$
$a = 12.5\,yd$ $b = 8\,yd$ $c = 15\,yd$

25.   No Triangle

26.
$A = 68.7°$ $B = 45°$ $C = 66.3°$
$a = 32.8\,yd$ $b = 24.9\,yd$ $c = 32.2\,yd$

27. $c = 11.5\,m$ 28. $a = 97.9\,m$
29. $c = 17.2\,in$ 30. $A = 15.4°$
31.   No Triangle   32. $C = 40.3°, 139.7°$
33. $A = 56.1°, 123.9°$ 34. $C = 133.4°$
35. $C = 93.8°$ 36. $a = 108.7\,m$
37. $a = 156\,m$

Section 18: Applications and Area of Oblique Triangles

Learning Outcomes:

- The student will solve trigonometric equations, right triangles, and oblique triangles.
- The student will correctly memorize and apply trigonometric formulas, definitions, identities, and properties.

Objectives: At the conclusion of this lesson you should be able to:

1. Use the law of sines to solve applications.
2. Use the law of cosines to solve applications.
3. Find the area of an oblique triangle.

In the last section we solved oblique triangles using the law of sines and the law of cosines. In this section we are going to solve applications involving the law of sines and the law of cosines and find the area of oblique triangles. We are going to use the same steps for solving an application that we used in section 5.

Steps for Solving an Application

Step 1 Identify all given quantities. Make a sketch and label the quantities given.

Step 2 Label the quantity that you are trying to find with a variable.

Step 3 Use the information from the sketch to write an equation relating what is given to the variable.

Step 4 Solve the equation.

Step 5 Check your solution to make sure it makes sense.

► EXAMPLE 1

A tree grew so that it was leaning 5° from the vertical. At a point 42 meters from the tree, the angle of elevation to the top of the tree is 24°. Find the height of the tree to the nearest tenth of a meter.

►►Solution

Step 1:   We know the distance from the tree is 42 m and the angle of elevation to the top of the tree is 24°. Since the tree grew leaning 5° from the vertical we know the angle formed with the ground and tree is 95°.

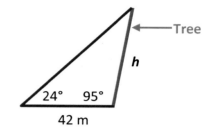

Step 2: We want to find the height of the tree labeled *h* in the drawing.

Step 3: We have ASA (angle-side-angle) and we will use the law-of-sines to find *h*. First we will find the angle opposite the 42m: 180°-(24°+95°)=61°. Now we can apply the law of sines.

Step 4: $\dfrac{\sin 61°}{42} = \dfrac{\sin 24°}{h}$ *Substitute into the Law of Sines formula.*

$h\sin 61° = 42\sin 24°$ *Cross multiply.*

$\dfrac{h\sin 61°}{\sin 61°} = \dfrac{42\sin 24°}{\sin 61°}$ *Solve for h.*

$h = \dfrac{42\sin 24°}{\sin 61°}$

$h = 19.5$

4 2 SIN 2 4) ÷ SIN 6 1) ENTER

Step 5: The height of the tree is 19.5 m.

► EXAMPLE 2

Points B and C are on opposite sides of a pond. The distance from point A to B is 1.3 miles and from point A to C is 1.2 miles. What is the distance from point B to C to the nearest tenth of a mile if $\angle BAC = 55°$.

►►Solution

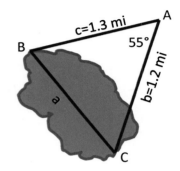

Step 1:   We know the distance from A to B is 1.3 miles and the distance from A to C is 1.2 miles. We also know $\angle BAC = 55°$.

Step 2: We want to find the distance from B to C which is labeled *a* in the sketch.

Step 3: We have SAS (side-angle-side) and we will use the law-of-cosines to find *a*.

Step 4: $a^2 = b^2 + c^2 - 2bc \cos A$ *Substitute into the Law of Cosines.*

$a^2 = (1.2)^2 + (1.3)^2 - 2(1.2)(1.3)\cos 55°$ *Take the square root of both sides.*

$\sqrt{a^2} = \sqrt{(1.2)^2 + (1.3)^2 - 2(1.2)(1.3)\cos 55°}$

2nd | x² | 1 | . | 2 | x² | + | 1 | . | 3 | x² | -
2 | × | 1 | . | 2 | × | 1 | . | 3 | COS | 5 | 5
) |) | ENTER

$a = 1.2$

Step 5: The distance from B to C is 1.2 miles.

▶ EXAMPLE 3

A pilot flew 100 miles on a bearing of 18° from Fairbanks Airport toward Fort Yukon. The pilot then flew due east on a bearing of 90° to drop food and medical supplies to a snowbound group of travelers. After the drop, the return to Fairbanks had a bearing of 225°. What was the pilot's maximum distance from Fairbanks to the nearest tenth of a mile?

▶▶ Solution

Step 1: Recall bearing from Section 5 and sketch a graph. Using alternate interior angles (see Section 5) we can get $\angle A = 27°$, $\angle B = 108°$ and $\angle C = 45°$. We also know the distance from A to B is 100 miles.

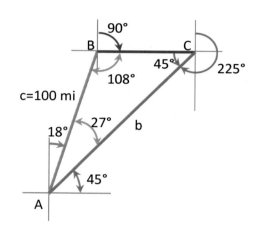

Step 2: We want to find the distance from A to C which is labeled **b** in the sketch.

Step 3: We have ASA (angle-side-angle) and also AAS (angle-angle-side) we will use the law- of-sines to find **b**.

Step 4: $$\frac{\sin C}{c} = \frac{\sin B}{b}$$

$$\frac{\sin 45°}{100} = \frac{\sin 108°}{b}$$ *Substitute into the Law of Sines formula.*

$$b \sin 45° = 100 \sin 108°$$ *Cross multiply.*

$$\frac{b \sin 45°}{\sin 45°} = \frac{100 \sin 108°}{\sin 45°}$$ *Solve for b.*

$$b = \frac{100 \sin 108°}{\sin 45°}$$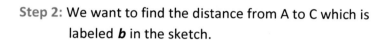

$$b = 134.5$$

Step 5: The pilot's maximum distance is 134.5 miles.

▶EXAMPLE 4

The bearing from the Cat Paw fire tower to the Pine Comb fire tower is N 65° E and the two towers are 18 miles apart. A fire is spotted N 80°E from the Cat Paw fire tower and S 70° E from the Pine Comb fire tower. How far is the fire from the Cat Paw fire tower?

▶▶Solution

Step 1:   Sketch the graph and label what we know. We can find all three of the angles of the triangle using bearing, alternate interior angles, properties of triangles, etc.

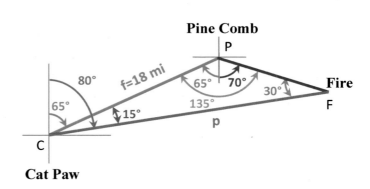

Step 2: We want to find the distance from Cat Paw, C, to the fire, F labeled **p** in the drawing.

Step 3: We have ASA (angle-side-angle) and also AAS (angle-angle-side) we will use the law-of-sines to find **p**.

Step 4: $$\frac{\sin F}{f} = \frac{\sin P}{p}$$

$$\frac{\sin 30°}{18} = \frac{\sin 135°}{p}$$ *Substitute into the Law of Sines formula.*

$$p \sin 30° = 18 \sin 135°$$ *Cross multiply.*

$$\frac{p \sin 30°}{\sin 30°} = \frac{18 \sin 135°}{\sin 30°}$$ *Solve for p.*

$$p = \frac{18 \sin 135°}{\sin 30°}$$ 1 8 SIN 1 3 5) ÷ SIN 3 0) ENTER

$$p = 25.5$$

Step 5: The distance from Cat Paw to the fire is 25.5 miles.

►EXAMPLE 5

A ship travels 60 miles due east, then adjusts its course northward and travels 80 miles. The ship is now 139 miles from its point of departure. What bearing to the nearest tenth of a degree did the ship take when it turned northward?

►►Solution

Step 1:   Sketch the graph and label what we know. We know the measure of all three sides.

Step 2:   We want to find the bearing the ship took went it turned northward, so we need to find the measure of angle B.

Step 3:   We have SSS (side-side-side) and we will use the alternate form of the law-of-cosines to find angle *B*.

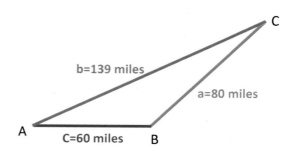

Step 4:

$$\cos B = \frac{a^2 + c^2 - b^2}{2ac}$$ *Substitute into the alternate form of the Law of Cosines.*

$$\cos B = \frac{(80)^2 + (60)^2 - (139)^2}{2(80)(60)}$$ 2nd COS ((8 0 x^2 + 6 0 x^2 - 1 3 9 x^2)

÷ ((2 × 8 0 × 6 0))) ENTER

$$B = 166.2°$$

Step 5:   We found angle B to equal 166.2° and subtracting 90° gives us a bearing of 76.2°.

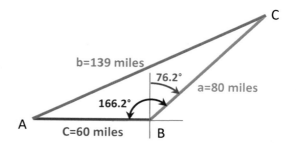

Area of Oblique Triangles

Given Two Sides and the Included Angle (SAS)

Recall the area formula for a triangle, $A = \dfrac{1}{2}bh$ where b represents the base of the triangle and h represents the height which is the length of the altitude of the triangle. Consider the triangle in

Figure 18.1 where h represents the height. If we are given sides a and c and the included angle B, we can derive a formula for finding the area. The sine of angle B, $\sin B = \dfrac{h}{a}$ can be solved for h giving us $a\sin B = h$. Substituting $a\sin B$ for h and c for the base into the area formula give us $A = \dfrac{1}{2}(c)(a\sin B)$. Simplifying gives us an area

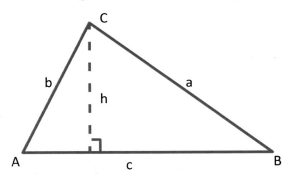

Figure 18.1

formula to use when we are given two sides and the included angle: $A = \dfrac{1}{2}ac\sin B$.

Area of an Oblique Triangle

Given Two Sides and the Included Angle (SAS)

If ABC is a triangle with sides a, b, and c, then

$$A = \frac{1}{2}ab\sin C$$

$$A = \frac{1}{2}ac\sin B$$

$$A = \frac{1}{2}bc\sin A$$

► EXAMPLE 6

A surveillance camera is set up to monitor activity in the parking lot of a college. If the camera has a 36° field of vision, how many square feet of the parking lot can it tape using the dimensions given in the sketch? Round to the nearest tenth.

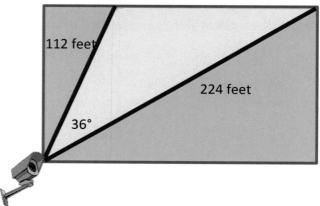

►►Solution

We are given two sides and the included angle which is side-angle-side (SAS). We will use the formula $A = \dfrac{1}{2}ab\sin C$ to find the area.

$A = \dfrac{1}{2}ab\sin C$ *Substitute our values into the formula.*

$A = \dfrac{1}{2}(112\ ft)(224\ ft)\sin 36°$ [1] [÷] [2] [×] [1][1][2] [×] [2][2][4] [SIN] [3][6] [)] [ENTER]

$A = 7373.2$

The camera can tape 7373.2 ft² of the parking lot.

Given Two Angles and a Side (AAS) (ASA)

Using the area formulas we have already developed for SAS, we can derive formulas for when we are given two angles and a side. Solving

for b in $A = \dfrac{1}{2}bc\sin A$ gives us $\dfrac{2A}{c\sin A} = b$

and solving for a in $A = \dfrac{1}{2}ac\sin B$ gives us

$\dfrac{2A}{c\sin B} = a$. See Figure 18.2.

$A = \dfrac{1}{2}bc\sin A \qquad A = \dfrac{1}{2}ac\sin B$

$2A = bc\sin A \qquad 2A = ac\sin B$

$\dfrac{2A}{c\sin A} = b \qquad \dfrac{2A}{c\sin B} = a$

Figure 18.2

We can substitute these values for *a* and *b* in the third formula and have an area formula for AAS and ASA.

$$A = \frac{1}{2}ab\sin C$$

$$A = \frac{1}{2} \cdot \frac{2A}{c\sin B} \cdot \frac{2A}{c\sin A} \cdot \sin C \qquad \textit{Substitute for a and b.}$$

$$(2)(c\sin B)(c\sin A)(A) = \left(\frac{1}{2} \cdot \frac{2A}{c\sin B} \cdot \frac{2A}{c\sin A} \cdot \sin C\right)(2)(c\sin B)(c\sin A) \qquad \textit{Multiply both sides by 2(csinB)(csinA).}$$

$$2Ac^2\sin B\sin A = 4A^2\sin C \qquad \textit{Simplify.}$$

$$\frac{2Ac^2\sin B\sin A}{4A\sin c} = \frac{4A^2\sin C}{4A\sin C} \qquad \textit{Divide both sides by 4AsinC.}$$

$$\frac{c^2\sin B\sin A}{2\sin C} = A \qquad \textit{An area formula for AAS and ASA.}$$

Area of an Oblique Triangle

Given Two Angles and a Side (AAS)(ASA)

If ABC is a triangle with sides a, b, and c, then

$$A = \frac{c^2\sin A\sin B}{2\sin C}$$

$$A = \frac{a^2\sin B\sin C}{2\sin A}$$

$$A = \frac{b^2\sin A\sin C}{2\sin B}$$

► EXAMPLE 7

A surveyor is surveying a field for a farmer. Find the area of the field. Round to the nearest tenth.

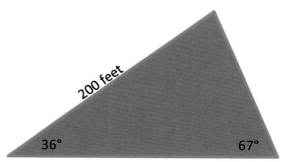

►► Solution

We are given two angles and not the included side which is angle-angle-side (AAS). First we

will find the third angle: $180°-(36°+67°)=77°$.   We will use the formula $A = \dfrac{c^2 \sin A \sin B}{2 \sin C}$ to

find the area.

$$A = \frac{c^2 \sin A \sin B}{2 \sin C}$$ *Substitute our values into the formula.*

$$A = \frac{(200)^2 \sin 36° \sin 77°}{2 \sin 67°}$$

2 0 0 x^2 SIN 3 6) SIN 7 7)
÷ (2 SIN 6 7)) ENTER

$$A = 12{,}443.6$$

The area of the field is 12,443.6 ft^2.

Given All Three Sides (SSS)

To derive the area formula for an oblique triangle when given three sides, we will begin with the

formula $A = \dfrac{1}{2} ab \sin C$.

$$A = \frac{1}{2} ab \sin C$$

$$A = \frac{1}{2} ab \sqrt{1 - \cos^2 C}$$ *We can substitute $\sqrt{1 - \cos^2 C}$ for sinC. Solve for sinC in the Pythagorean Identity: $\sin^2 C + \cos^2 C = 1$.*

$$A = \frac{1}{2} ab \sqrt{1 - \left(\frac{a^2 + b^2 - c^2}{2ab} \right)^2}$$ *We can substitute the alternate form of the Law of Cosines for cosC.*

$$(A)^2 = \left(\frac{1}{2}ab\sqrt{1-\left(\frac{a^2+b^2-c^2}{2ab}\right)^2}\right)^2 \qquad \textit{Square both sides of the equation.}$$

$$A^2 = \frac{1}{4}a^2b^2\left[1-\left(\frac{a^2+b^2-c^2}{2ab}\right)^2\right]$$

$$A^2 = \frac{1}{4}a^2b^2\left[1-\left(\frac{a^2+b^2-c^2}{2ab}\right)\right]\left[1+\left(\frac{a^2+b^2-c^2}{2ab}\right)\right] \qquad \begin{array}{l}\textit{Factor using a difference between}\\ \textit{two perfect squares.}\end{array}$$

$$A^2 = \frac{1}{4}a^2b^2\left[\frac{2ab-a^2-b^2+c^2}{2ab}\right]\left[\frac{2ab+a^2+b^2-c^2}{2ab}\right] \qquad \begin{array}{l}\textit{Get a common denominator}\\ \textit{and simplify.}\end{array}$$

$$A^2 = \frac{1}{16}\left(2ab-a^2-b^2+c^2\right)\left(2ab+a^2+b^2-c^2\right) \qquad \textit{Reduce and factor out the } \frac{1}{4}.$$

$$A^2 = \frac{1}{16}\left(c^2-\left(a^2-2ab+b^2\right)\right)\left(\left(a^2+2ab+b^2\right)-c^2\right) \qquad \textit{Regroup.}$$

$$A^2 = \frac{1}{16}\left(c^2-(a-b)^2\right)\left((a+b)^2-c^2\right) \qquad \textit{Factor the perfect square trinomials.}$$

$$A^2 = \frac{1}{16}\left(c-a+b\right)\left(c+a-b\right)\left(a+b-c\right)\left(a+b+c\right) \qquad \begin{array}{l}\textit{Factor using a difference between}\\ \textit{two perfect squares.}\end{array}$$

The perimeter formula for a triangle is $P = a+b+c$ and the semi-perimeter is $S = \dfrac{P}{2} = \dfrac{a+b+c}{2}$.

We can manipulate the perimeter formula to match what we have in the parentheses in our equation.

$P = a+b+c$	*Perimeter formula.*
$P-2a = a+b+c-2a$	*Subtract 2a from both sides.*
$P-2a = c-a+b$	*Simplify.*
$P = a+b+c$	*Perimeter formula.*
$P-2b = a+b+c-2b$	*Subtract 2b from both sides.*
$P-2b = c+a-b$	*Simplify.*
$P = a+b+c$	*Perimeter formula.*
$P-2c = a+b+c-2c$	*Subtract 2c from both sides.*
$P-2c = a+b-c$	*Simplify.*

$$A^2 = \frac{1}{16}(c-a+b)(c+a-b)(a+b-c)(a+b+c)$$

$$A^2 = \frac{1}{16}(P-2a)(P-2b)(P-2c)(p) \quad \textit{Substitute.}$$

$$A^2 = \frac{p}{2}\left(\frac{P-2a}{2}\right)\left(\frac{P-2b}{2}\right)\left(\frac{P-2c}{2}\right) \quad \textit{Rewrite. Recall } \frac{1}{16} = \left(\frac{1}{2}\right)^4.$$

$$A^2 = \frac{P}{2}\left(\frac{P}{2}-\frac{2a}{2}\right)\left(\frac{P}{2}-\frac{2b}{2}\right)\left(\frac{P}{2}-\frac{2c}{2}\right) \quad \textit{Rewrite the fractions and simplify.}$$

$$A^2 = \frac{P}{2}\left(\frac{P}{2}-a\right)\left(\frac{P}{2}-b\right)\left(\frac{P}{2}-c\right) \quad \textit{We know that the semi-perimeter is } S = \frac{P}{2}. \textit{ Substitute S for } \frac{P}{2}.$$

$$A^2 = S(S-a)(S-b)(S-c)$$

$$\sqrt{A^2} = \sqrt{S(S-a)(S-b)(S-c)} \quad \textit{Take the square root of both sides.}$$

$$A = \sqrt{S(S-a)(S-b)(S-c)}$$

$A = \sqrt{S(S-a)(S-b)(S-c)}$ is the formula for finding the area of an oblique triangle when given three sides. This is called Heron's Formula, named after Heron of Alexandria, who is believed to have discovered it in about A.D. 75.

Area of an Oblique Triangle

Given Three Sides (SSS)

If ABC is a triangle with sides a, b, and c, the area of the triangle is

$$A = \sqrt{S(S-a)(S-b)(S-c)} \text{ where } S = \frac{a+b+c}{2}.$$

► EXAMPLE 8

Find the area of the triangle. Round to the nearest tenth.

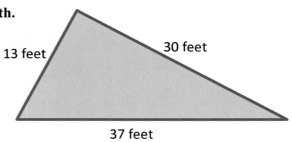

13 feet

30 feet

37 feet

►►Solution

We are given three sides which is side-side-side (SSS).   We will use Heron's Formula

$A = \sqrt{S(S-a)(S-b)(S-c)}$ to find the area. First we will find $S = \dfrac{a+b+c}{2}$.

$S = \dfrac{a+b+c}{2}$ *Substitute our values into the formula.*

$S = \dfrac{13+30+37}{2}$ *Simplify.*

$S = \dfrac{80}{2}$

$S = 40$

$A = \sqrt{S(S-a)(S-b)(S-c)}$ *Substitute our values into the formula.*

$A = \sqrt{40(40-13)(40-30)(40-37)}$ *Simplify.*

$A = \sqrt{40(27)(10)(3)}$

$A = \sqrt{32400}$

$A = 180$

The area is 180 ft^2.

Area of an Oblique Triangle

Given Two Sides and the Included Angle (SAS)

If ABC is a triangle with sides a, b, and c, then

$$A = \frac{1}{2}ab\sin C$$

$$A = \frac{1}{2}ac\sin B$$

$$A = \frac{1}{2}bc\sin A$$

Given Two Angles and a Side (AAS)(ASA)

If ABC is a triangle with sides a, b, and c, then

$$A = \frac{c^2 \sin A \sin B}{2\sin C}$$

$$A = \frac{a^2 \sin B \sin C}{2\sin A}$$

$$A = \frac{b^2 \sin A \sin C}{2\sin B}$$

Given Three Sides (SSS)

If ABC is a triangle with sides a, b, and c, the area of the triangle is

$$A = \sqrt{S(S-a)(S-b)(S-c)} \text{ where } S = \frac{a+b+c}{2}.$$

$$\cos^{-1}\left(\frac{a^2 - b^2 - c^2}{2bc}\right)$$

HOMEWORK

Objectives 1& 2

Solve. Round your answers to the nearest tenth.

See Examples 1-5.

1. A helicopter left the station on a bearing of 210°. After flying for 15 miles the helicopter headed due east along a river bank. After some time flying along the river bank, the helicopter headed back to the station on a bearing of 310°. How far to the nearest tenth of a mile did the helicopter travel along the river bank?

2. A surveyor locating the corners of a triangular piece of property started at one corner and walked 500 feet in the direction N36°W to reach the second corner. When the surveyor reached the second corner she turned and walked S21°W to get to the third corner of the property. To return to the original corner, the surveyor walked in the direction N82°E. How far did the surveyor walk from the third corner back to the corner where she started?

3. A triangular parcel of land has 115 meters of frontage, and the other boundaries have lengths of 76 meters and 92 meters. What angle is formed by the two boundaries?

4. Two ships leave a port at 10 a.m. One ship travels at a bearing of S67°W at 18 miles per hour and the other travels at a bearing of N53°W at 15 miles per hour. Approximate how far apart the ships are at 4 p.m.

5. Airports *A* and *B* are 450 km apart, on an east-west line. A pilot flies in a northeast direction from *A* to another Airport *C*. From *C* he flies 380 km on a bearing of 129° to Airport *B*. How far is Airport *A* from Airport *C*?

6. Points *A* and *B* are on opposite sides of a lake. From a third point, *C*, the angle between the lines of sight to *A* and *B* is 46.3°. If *AC* is 350 m long and *BC* is 286 m long, find the length of *AB*.

7. On a baseball diamond with 90 foot sides, the distance from home plate to the pitcher's mound is 60.5 feet. How far is third base from the pitcher's mound?

8. A plane flies 500 kilometers with a bearing of 316° from Naples to Elgin. The plane then flies on a bearing of 240° from Elgin to Canton. If Canton and Naples are on an east-west line, find the distance the plane flew from Elgin to Canton.

9. A bridge is to be built across a small lake from a picnic area to a small dock. The bearing from the picnic area to the small dock is S41°W. From a tree 100 meters from the picnic area, the angle of rotation from the picnic area to the small dock is 46°. The distance from the tree to the small dock is 120 meters. How long will the bridge be from the picnic area to the small dock?

10.   A triangular parcel of land has sides of lengths 815 feet, 618 feet, and 475 feet. Find the measure of the largest angle.

11.   A fisherman on a boat spots two lighthouses on shore. He estimates he is 8 miles from one lighthouse and 6 miles from the other lighthouse. The angle between the line of sight to the lighthouses is 120°. What is the distance between the two lighthouses?

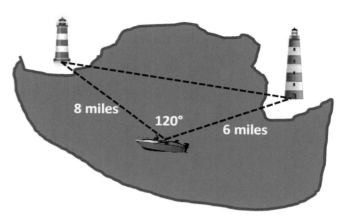

12.

On a map, Albany, NY is 165 millimeters due east of Minneapolis, MN and 368 millimeters from Phoenix, AZ. Minneapolis is 216 millimeters from Phoenix. What is the bearing of Albany from Phoenix?

13.   A satellite measures its distance from Portland and from Green Bay using radio waves. If the angle of rotation from Portland to Green Bay is 99°, how far apart are the two cities?

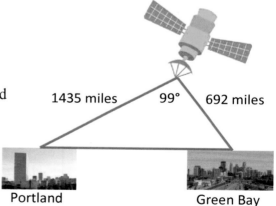

14.   A ship is sailing due north. At a certain point the bearing to a lighthouse 12.5 kilometers away is N38°E. A couple of hours later the captain notices that the bearing of the lighthouse has become S44°E. How far did the ship travel between the two observations of the lighthouse?

15.   The bearing of a lighthouse from a ship traveling due south is N37°E. Find the distance from the ship and lighthouse after the ship sails for 3 miles and the new bearing is N25°E.

16.   Two ranger stations are on an east-west line 110 miles apart. A forest fire is located on a bearing of N52°E from the western station and a bearing of N28°W from the eastern station. How far is the fire from the eastern station?

17.   Two ships leave a harbor together, traveling on courses that have an angle of 140° between them. If each ship travels 613 miles, how far apart are they?

18.   If a triangular truss has sides of 16 feet, 13 feet, and 20 feet, what is the measure of the largest angle?

19.

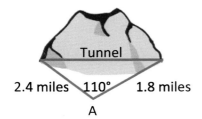

2.4 miles   110°   1.8 miles

A

To measure the distance through a mountain for a new tunnel, a point A is chosen that can be reached from each end of the tunnel. See the figure. Find the length of the tunnel.

20.   A surveyor wants to find the distance across a canyon to construct a bridge. The surveyor marks a point from a ledge 1200 meters from one end of the proposed bridge and 1800 from the other end. He finds the angle of rotation from one end to the other end to be 66°. How long will the bridge need to be?

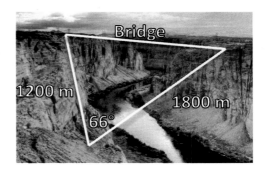

Objective 3

Solve. Round your answers to the nearest tenth.

See Examples 6-8.

21.   Find the area of a triangular plot of land with sides of 789 feet, 1289 feet, and 1526 feet.

22.   The distance from a gazebo to a swing set is 110 feet and the distance from the gazebo to a sand box is 210 feet. Find the area of the triangle if the angle of rotation from the swing set to the sand box is 43°.

23.   Find the area of a triangular field having angles of 72°and 23° and the included side of length 62 meters.

24.   Find the area of a flower bed with sides of length 12 meters, 6 meters, and 8 meters.

25. A surveillance camera is set up to monitor activity in the parking lot of a mall. If the camera has a 37° field of vision, how many square feet of the parking lot can it tape using the dimensions given in the sketch? Round to the nearest tenth.

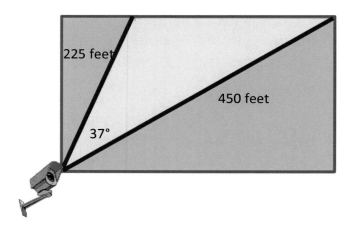

Objective 3

Find the area of each triangle with the given parts. Round your answers to the nearest tenth.

See Examples 6-8.

26. $B = 45°, a = 32.8\,yd, c = 24.9\,yd$

27. $A = 68.41°, B = 54.23°, a = 12.75m$

28. $A = 31.3°, B = 71.2°, c = 184m$

29. $A = 23°, C = 131°, b = 10\,in$

30. $C = 39.7°, a = 9.8cm, b = 23.5cm$

31. $A = 43.6°, b = 15\,yd, c = 28\,yd$

32. $A = 29°, b = 15\,m, c = 20\,m$

33. $B = 45.4°, a = 71.3m, c = 61.2m$

34. $a = 8m, b = 5m, c = 12m$

35. $a = 3m, b = 5m, c = 6m$

ANSWERS

1.	23.0 miles	2.	479.4 feet	3.	85.8°
4.	100.2 miles	5.	284.8 km	6.	256.9 m
7.	63.7 feet	8.	719.3 km	9.	87.9 m
10.	95.5°	11.	12.2 miles	12.	N72.8°E
13.	1687.8 miles	14.	17.8 km	15.	8.7 miles
16.	68.8 miles	17.	1152.1 miles	18.	86.6°
19.	3.5 miles	20.	1709.6 m	21.	508,388.1 ft^2
22.	7877.1 ft^2	23.	717.0 m^2	24.	21.3 m^2
25.	30,466.9 ft^2	26.	288.8 yd^2	27.	59.7 m^2
28.	8527.4 m^2	29.	33.6 in^2	30.	73.6 cm^2
31.	144.8 yd^2	32.	72.7 m^2	33.	1553.5 m^2
34.	14.5 m^2	35.	7.5 m^2		

Section 19:   Vectors

Learning Outcomes:

- The student will perform applications on vectors and complex numbers.
- The student will solve trigonometric equations, right triangles, and oblique triangles.
- The student will correctly memorize and apply trigonometric formulas, definitions, identities, and properties.

Objectives: At the conclusion of this lesson you should be able to:

1. Write vectors in component forms, linear combinations of unit vectors, and trigonometric forms.
2. Perform basic vector operations.
3. Find the magnitudes and direction angles of vectors.
4. Find unit vectors.
5. Find the horizontal and vertical components of a vector.
6. Use vectors to model and solve applications.

The statement "the driving distance between Cornelia, Georgia and Demorest, Georgia is 4.17 miles" tells us only the distance between the two cities. The statement "a car travels northwest for 4.17 miles from Cornelia, Georgia to Demorest, Georgia" tells us both the distance and the direction between the two cities. A quantity which indicates magnitude (size) but no direction is called a scalar quantity. Examples of scalar quantities are distance, length, area, volume, temperature and time. A quantity which indicates both magnitude and direction is called a vector. Examples of vectors are force, velocity and acceleration. Vector quantities can be graphically represented by directed line segments called vectors. The directed line segment $\overrightarrow{AB}$ is represented by an arrow that indicates the direction and contains an initial point A and a terminal point B as shown in **Figure 19.1**. The magnitude is represented by $\|\overrightarrow{AB}\|$. Vectors are denoted by boldface lowercase letters such as **u**, **v**, and **w** and their magnitudes are denoted by $\|\boldsymbol{u}\|, \|\boldsymbol{v}\|$ and $\|\boldsymbol{w}\|$. Two vectors that have the same magnitude and direction are equivalent. In **Figure 19.2** the vectors are equivalent.

Figure 19.1

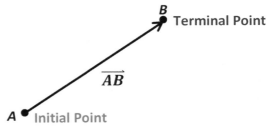

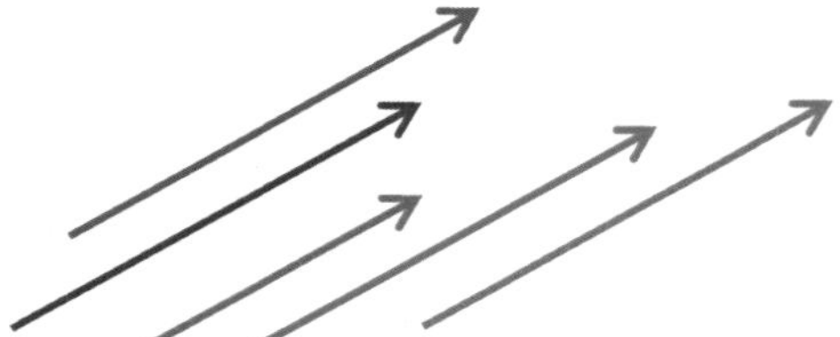

Figure 19.2

Component Form of a Vector

A vector **v** with initial point at the origin $(0,0)$ and terminal point (a,b) is in standard position. The component form of a vector **v** in standard position is written $v =< a,b >$ where a is the horizontal component and b is the vertical component. Consider the two vectors in **Figure 19.3.** Vector **v** is in standard position and is written in component form. Vector **u** is not in standard position but we can write vector **u** in component form.

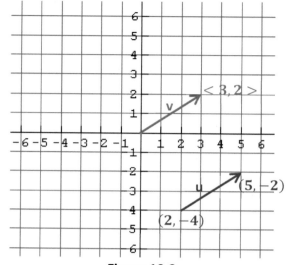

Initial Point of a vector: (x_1, y_1)

Terminal point of a vector: (x_2, y_2)

Component Form: $< \underbrace{x_2 - x_1}_{a}, \underbrace{y_2 - y_1}_{b} >$

$$u =< 5 - 2, -2 - (-4) > = < 3, 2 >$$

Figure 19.3

Vector **u** in component form is $u =< 3,2 >$. Notice vector **u** and vector **v** or equivalent vectors.

►EXAMPLE 1

Find the component form of vector u with initial point at (-4,-4) and terminal point at (2,3). Graph the component form of vector u with the initial point at the origin.

►►Solution

Label the ordered pairs: initial point: (-4,-4)   terminal point: (2,3)
                              x_1 y_1 x_2 y_2

Substitute into the formula for the Component Form: $< x_2 - x_1, y_2 - y_1 >$

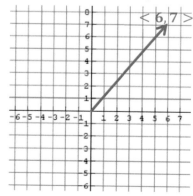

$$< 2 - (-4), 3 - (-4) >$$
$$< 2 + 4, 3 + 4 >$$
$$< 6, 7 >$$

Term = Init + comp

_+r_

toh = Comp + rhid

► EXAMPLE 2

Find the terminal point of vector v with initial point at (5,-2) and component form v = <6,-3>.

►►Solution

Label the initial point: initial point: $\underset{x_1\ y_1}{(5,\ -2)}$ terminal point: $(x_2,\ y_2)$

Component Form

$< \underset{\substack{\uparrow \\ x_2 - x_1}}{6}, \underset{\substack{\nwarrow \\ y_2 - y_1}}{-3} >$

$x_2 - x_1 = 6$

$x_2 - 5 = 6$ *Substitute 5 for x_1.*

$x_2 = 11$ *Solve for x_1.*

$y_2 - y_1 = -3$

$y_2 - (-2) = -3$ *Substitute -2 for y_1.*

$y_2 + 2 = -3$

$y_2 = -5$ *Solve for y_1.*

The terminal point is (11, -5).

Magnitude and Direction of a Vector

The length of a vector represents the magnitude of the vector and can be found using the distance formula. Given vector **v** with initial point (x_1, y_1) and terminal point (x_1, y_1), the magnitude of vector **v** is $\|v\| = \sqrt{(x_2 - x_1)^2 + (y_2 - y_1)^2}$. Given the same vector **v** in component form

$v = <a, b>$, as shown in **Figure 19.4**, the magnitude of vector **v** is $\|v\| = \sqrt{a^2 + b^2}$.

The direction of vector **v** is given by the angle θ (measured counterclockwise) from the x-axis to **v**. The angle θ is called the direction angle of vector **v** and the reference angle θ_r is found by

using $\theta_r = \tan^{-1}\left|\dfrac{b}{a}\right|$. The horizontal component,

a, of vector **v** can be found by using $a = \|v\|\cos\theta$ and the vertical component, **b**, of

vector v can be found by using $b = \|v\|\sin\theta$.

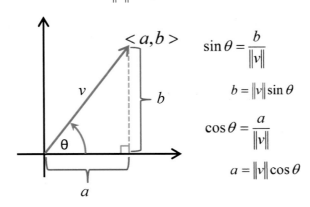

$\sin\theta = \dfrac{b}{\|v\|}$

$b = \|v\|\sin\theta$

$\cos\theta = \dfrac{a}{\|v\|}$

$a = \|v\|\cos\theta$

Figure 19.4

Vector Components in Trigonometric Form

For a vector $v = <a, b>$ and angle θ, we have horizontal component: $a = \|v\| \cos \theta$ and

vertical component: $b = \|v\| \sin \theta$, where $\theta_r = \tan^{-1} \left| \dfrac{b}{a} \right|$ and $\|v\| = \sqrt{a^2 + b^2}$.

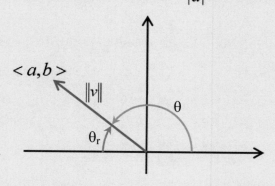

►EXAMPLE 3

Find the horizontal and vertical components of vector u with the given magnitude and directional angle: $\|u\| = 105$ **and** $\theta = 102°$. **Round to the nearest tenth.**

►►Solution

Horizontal Component	Vertical Component
$a = \|u\| \cos \theta$	$b = \|u\| \sin \theta$
$a = 105 \cos 102°$	$b = 105 \sin 102°$
$a = -21.8$	$b = 102.7$

$$u = <-21.8, 102.7>$$

►EXAMPLE 4

Find the horizontal and vertical components of vector u with magnitude $\|u\| = 25$, **angle** $\theta = 32°$ **formed with the nearest x-axis, and the terminal point of vector u is in Quadrant III. Round to the nearest tenth.**

►►Solution

Vector u is in Quadrant III. Sketch and label what we know.
The reference angle is $32°$ and the angle we need is $212°$.

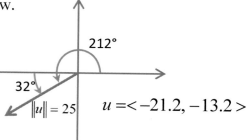

Horizontal Component	Vertical Component
$a = \|u\| \cos \theta$	$b = \|u\| \sin \theta$
$a = 25 \cos 212°$	$b = 25 \sin 212°$
$a = -21.2$	$b = -13.2$

$$u = <-21.2, -13.2>$$

► EXAMPLE 5

Find the magnitude and direction angle of vector u, $u = <8, -6>$. Round to the nearest tenth.

►► Solution

Vector u is in Quadrant IV. Sketch and label what we know.

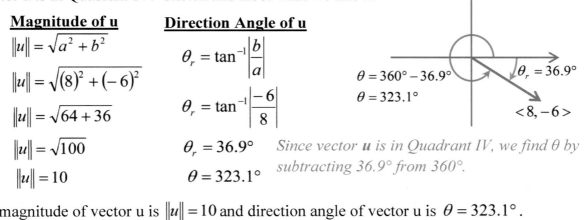

Magnitude of u	Direction Angle of u	
$\|u\| = \sqrt{a^2 + b^2}$	$\theta_r = \tan^{-1}\left\|\dfrac{b}{a}\right\|$	
$\|u\| = \sqrt{(8)^2 + (-6)^2}$		$\theta = 360° - 36.9°$
$\|u\| = \sqrt{64 + 36}$	$\theta_r = \tan^{-1}\left\|\dfrac{-6}{8}\right\|$	$\theta = 323.1°$
$\|u\| = \sqrt{100}$	$\theta_r = 36.9°$	*Since vector **u** is in Quadrant IV, we find θ by*
$\|u\| = 10$	$\theta = 323.1°$	*subtracting 36.9° from 360°.*

The magnitude of vector u is $\|u\| = 10$ and direction angle of vector u is $\theta = 323.1°$.

The Resultant of a Vector

We can perform a variety of mathematical operations with and upon vectors. One of these operations is the addition of vectors. Two vectors can be added together to determine the result which is called the resultant. The resultant is the vector sum of two or more vectors. Consider vector **v** as shown in **Figure19.5**. The initial point we will call the tail and the terminal point we will call the head. The sum of two vectors **u** and **v**, **u** + **v**, is found by first moving **v** without changing its magnitude or direction, so that the tail of **v** is placed at the head of **u**. Draw the resultant from the tail of **u** to the head of **v** as shown in **Figure 19.6**. The sum of the two vectors **u** and **v** equals the resultant vector **w**.

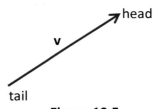

Figure 19.5

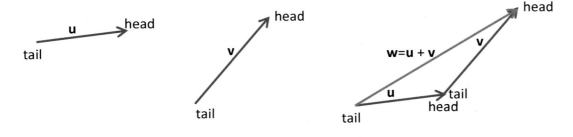

Figure 19.6

Vector Addition

If $u = <u_1, u_2>$ and $v = <v_1, v_2>$, then $u + v = <u_1 + v_1, u_2 + v_2>$.

Vector Subtraction

If $u = <u_1, u_2>$ and $v = <v_1, v_2>$, then $u - v = <u_1 - v_1, u_2 - v_2>$.

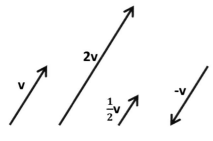

Figure 19.7

The product of a scalar k and a vector **u**, k**u**, is represented by drawing the vector whose length is k times the length of **u**. If k is a positive number, **u** and k**u** have the same direction and if k is a negative number, **u** and k**u** have opposite directions. In **Figure 19.7** we have some representations of scalar multiples of vector **v**. Notice that $-$**v** is going in the opposite direction of vector **v**.

Scalar Multiplication

For real numbers k and a vector $u = <u_1, u_2>$, the scalar product of k and **u** is $ku = <ku_1, ku_2>$. The vector k**u** is a scalar multiple of the vector u.

Properties of Vector Addition and Scalar Multiplication

For all vectors **u**, **v**, and **w** and for all scalars m and n:

1. $u + v = v + u$

2. $u + (v + w) = (u + v) + w$

3. $u + 0 = 0 + u = u$

4. $1u = u, \quad 0u = 0$

5. $u + (-u) = -u + u = 0$

6. $m(nu) = (mn)u$

7. $(m + n)u = mu + nu$

8. $m(u + v) = mu + mv$

► EXAMPLE 6

Perform the following calculations, where $u = <3, -2>$ **and** $v = <-4, 6>$.

a.) $2u$ b.) $u + 3v$ c.) $2u + 3v$ d.) $4u - v$

►►Solution

a.) $2u = < 2(3), 2(-2) >$ b.) $u + 3v = < 3 + 3(-4), -2 + 3(6) >$

   $2u = < 6, -4 >$ $u + 3v = < 3 - 12, -2 + 18 >$

                                                                     $u + 3v = < -9, 16 >$

c.) $2u + 3v = < 2(3) + 3(-4), 2(-2) + 3(6) >$

   $2u + 3v = < 6 - 12, -4 + 18 >$

   $2u + 3v = < -6, 14 >$

d.) $4u - v = < 4(3) - (-4), 4(-2) - 6 >$

   $4u - v = < 12 + 4, -8 - 6 >$

   $4u - v = < 16, -14 >$

A zero vector is a vector that has magnitude of 0 and a unit vector is a vector that has magnitude of 1. The vector $u = <\frac{1}{2}, \frac{\sqrt{3}}{2}>$ is a unit vector because it has magnitude of 1:

$$\|u\| = \sqrt{\left(\frac{1}{2}\right)^2 + \left(\frac{\sqrt{3}}{2}\right)^2}$$

$$\|u\| = \sqrt{\frac{1}{4} + \frac{3}{4}}$$

$$\|u\| = \sqrt{1}$$

$$\|u\| = 1$$

Two very important unit vectors that are parallel to the x-axis and y-axis are defined as $i = < 1, 0 >$ and $j = < 0, 1 >$. Any vector can be expressed as a linear combination of unit vectors i and j. For example, let $u = < u_1, u_2 >$, then

$$u = < u_1, u_2 > = < u_1, 0 > + < 0, u_2 >$$

$$u = < u_1, u_2 > = u_1 < 1, 0 > + u_2 < 0, 1 >$$

$$u = < u_1, u_2 > = u_1 i + u_2 j$$

We can write $u = < -3, 4 >$ in linear combination form $u = -3i + 4j$ and we can write the vector $v = -2i - 4j$ in component form $v = < -2, -4 >$.

► EXAMPLE 7

Perform the following calculations, where $u = 3i - 2j$ and $v = 4i + j$.

a.) $3u$ b.) $3u + v$ c.) $2u + 3v$ d.) $u - 2v$

►► Solution

a.) $3u = 3(3)i + 3(-2)j$ b.) $3u + v = [3(3) + 4]i + [3(-2) + 1]j$

   $3u = 9i - 6j$ $3u + v = (9 + 4)i + (-6 + 1)j$

                                         $3u + v = 13i - 5j$

c.) $2u + 3v = [2(3) + 3(4)]i + [2(-2) + 3(1)]j$

   $2u + 3v = (6 + 12)i + (-4 + 3)j$

   $2u + 3v = 18i - j$

d.) $u - 2v = [3 - 2(4)]i + [-2 - 2(1)]i$

   $u - 2v = (3 - 8)i + (-2 - 2)j$

   $u - 2v = -5i - 4j$

Sometimes it is useful to write a vector **u** in terms of its magnitude and direction angle rather than component form or linear combination form. The form $u = \|u\|(cos\theta i + sin\theta j)$ is called the trigonometric form of a vector. We arrive at this form by using the linear combination form and the formulas for the horizontal and vertical components:

   $u = ai + bj$ *Linear Combination Form of a vector.*

   $u = (\|u\|cos\theta)i + (\|u\|sin\theta)j$ *Substitute the formula for the horizontal component for* **a** *and the formula for the vertical component for* **b**.

   $u = \|u\|(cos\theta i + sin\theta j)$ *Factor out the magnitude of* **u**.

► EXAMPLE 8

Write vector $u = -4i - 3j$ in trigonometric form. Round to the nearest tenth.

►► Solution

To write vector **u** in trigonometric form we need to know both its magnitude and direction angle.

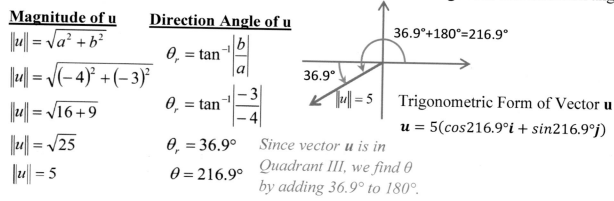

Magnitude of u	**Direction Angle of u**
$\|u\| = \sqrt{a^2 + b^2}$	$\theta_r = \tan^{-1}\left\|\dfrac{b}{a}\right\|$
$\|u\| = \sqrt{(-4)^2 + (-3)^2}$	
$\|u\| = \sqrt{16 + 9}$	$\theta_r = \tan^{-1}\left\|\dfrac{-3}{-4}\right\|$
$\|u\| = \sqrt{25}$	$\theta_r = 36.9°$
$\|u\| = 5$	$\theta = 216.9°$

36.9°+180°=216.9°

36.9°

$\|u\| = 5$

Since vector **u** *is in Quadrant III, we find* θ *by adding 36.9° to 180°.*

Trigonometric Form of Vector **u**

$u = 5(cos216.9°i + sin216.9°j)$

Every nonzero vector has a corresponding unit vector, which has the same direction as that vector but a magnitude of 1. We denote the unit vector for vector u with a circumflex in the Latin script which is chevron-shaped, $\hat{u}$. To find the unit vector of a vector, divide the vector by its magnitude: $\hat{u} = \dfrac{u}{\|u\|}$ or $\hat{u} = \dfrac{a}{\sqrt{a^2+b^2}}i + \dfrac{b}{\sqrt{a^2+b^2}}j$.

▶ **EXAMPLE 9**

Find the unit vector for $v = <-4, 6>$. Verify that a unit vector was found.

▶▶ **Solution**

The unit vector formula $\hat{v} = \dfrac{v}{\|v\|}$ can be written as $\hat{v} = \dfrac{1}{\|v\|}v$.

Using the rewritten formula we have:

$\hat{v} = \dfrac{1}{\sqrt{(-4)^2 + (6)^2}} <-4, 6>$ *Substitute into the magnitude formula and simplify.*

$\hat{v} = \dfrac{1}{\sqrt{52}} <-4, 6>$ *Rewrite the $\sqrt{52}$ in reduced radical form.*

$\hat{v} = \dfrac{1}{\sqrt{4 \cdot 13}} <-4, 6>$ *The largest perfect square factor of 52 is 4.*

$\hat{v} = \dfrac{1}{2\sqrt{13}} <-4, 6>$ *Multiply the horizontal and vertical components by the scalar $\dfrac{1}{2\sqrt{13}}$.*

$\hat{v} = <\dfrac{-4}{2\sqrt{13}}, \dfrac{6}{2\sqrt{13}}>$ *Simplify.*

$\hat{v} = <-\dfrac{2}{\sqrt{13}}, \dfrac{3}{\sqrt{13}}>$ *The unit vector for vector v.*

Verify that a unit vector was found.

$\|\hat{v}\| = \sqrt{\left(-\dfrac{2}{\sqrt{13}}\right)^2 + \left(\dfrac{3}{\sqrt{13}}\right)^2}$ *Find the magnitude of the unit vector.*

$\|\hat{v}\| = \sqrt{\dfrac{4}{13} + \dfrac{9}{13}}$

$\|\hat{v}\| = \sqrt{\dfrac{13}{13}}$

$\|\hat{v}\| = \sqrt{1} = 1$ *Since the magnitude is one it is a unit vector.*

There are infinite applications of vectors with many of them in the applied sciences. We are going to look at a couple of applications in this section and more in Section 20.

▶EXAMPLE 10

An airplane travels at airspeed of 300 miles per hour on a bearing of 100°. A wind is blowing 52 miles per hour from 220°. Find the groundspeed of the airplane and its course over the ground. Round to the nearest tenth.

▶▶Solution

Step 1:   We draw a sketch and label what we know. The wind speed is represented by the **blue** vector **u** in the drawing and the airplane vector is represented by the **red** vector **v** in the drawing.

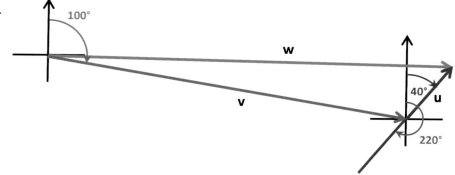

Step 2: We want to find the resultant, the **green** vector **w**. Once we find vector **w** we can get its direction angle and its magnitude. The direction angle is the course over the ground and the magnitude is the groundspeed.

Step 3: We are going to convert the bearings to standard angles as shown in **purple** below.

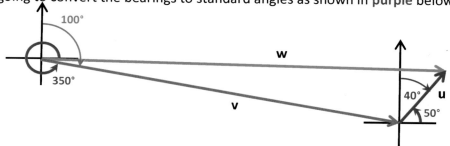

We can write vectors **u** and **v** in trigonometric form and simplify to get the linear combination forms.

$$u = 52(cos50°i + sin50°j) \qquad v = 300(cos350°i + sin350°j)$$
$$u = 52cos50°i + 52sin50°j \qquad v = 300cos350°i + 300sin350°j$$
$$u = 53.4i + 39.8j \qquad\qquad v = 295.4i - 52.1j$$

We can find the resultant vector **w**.

$$w = u + v = (53.4 + 295.4)i + \big(39.8 + (-52.1)\big)j$$
$$w = u + v = 328.8i - 12.3j$$

Step 4: We need to find the direction angle which is the course over the ground and the magnitude which is the groundspeed.

$$\|w\| = \sqrt{(328.8)^2 + (-12.3)^2} \qquad \theta_r = tan^{-1}\left|\frac{-12.3}{328.8}\right|$$
$$\|w\| = 329.0 \qquad\qquad \theta_r = 2.1°$$

Since vector **w** is in Quadrant IV the bearing will be $\theta_r + 90°$.

$$\theta = 90° + 2.1° = 92.1°$$

Step 5: The course over the ground is 92.1°and the groundspeed of the plane is 329.0 miles per hour.

▶ EXAMPLE 11

Two tugboats are pulling a large ship into dry dock. The first is pulling with a force of 1450 N (Newton) and the second with a force of 2150 N. Determine the angle θ for the second tugboat that will keep the ship moving straight forward and into the dock.

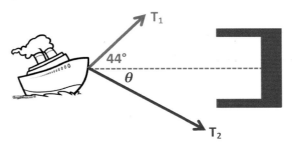

▶▶ Solution

Step 1: Using the given sketch we can label what we know. We will drop the perpendiculars from each tugboat to form right triangles.

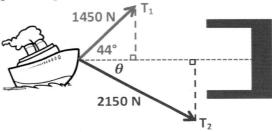

Step 2: We want to find the angle of the second tugboat so that the ship will move straight forward into the dock.

Step 3: We need to find the vertical components of each tugboat.

$$T_1 = 1450sin44° \qquad\qquad T_2 = 2150sin\theta$$

Step 4: For the ship to move straight being pulled by the two tugboats, the two tugboats will have to have the same vertical component. We will set the vertical components from Step 3 equal to each other and solve for θ.

$$1450sin44° = 2150sin\theta$$

$$\frac{1450sin44°}{2150} = \frac{2150sin\theta}{2150}$$

$$\frac{1450sin44°}{2150} = sin\theta \quad \boxed{2^{nd}}\ \boxed{SIN}\ \boxed{1}\boxed{4}\boxed{5}\boxed{0}\ \boxed{SIN}\ \boxed{4}\boxed{4}\boxed{)}\ \boxed{\div}\boxed{2}\boxed{1}\boxed{5}\boxed{0}\ \boxed{)}\boxed{ENTER}$$

$$27.9° = \theta$$

Step 5: The second tugboat will have an angle of 27.9°.

►EXAMPLE 12

A ball is thrown with an initial velocity of 80 feet per second, at an angle of 25° with the horizontal. Find the vertical and horizontal components of the velocity. Round to the nearest tenth.

►►Solution

Step 1:   Sketch and label what we know. We know the direction angle is 25° and the magnitude is 80 feet.

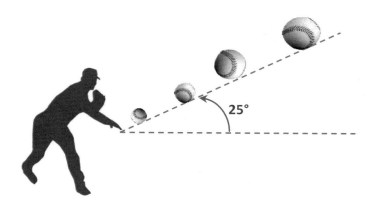

Step 2: We want to find the vertical and horizontal components.

Step 3: We can substitute into the formula for the horizontal component and the formula for the vertical component.

Horizontal Component

$$a = \|u\| \cos \theta$$

$$a = 80 \cos 25°$$

$$a = 72.5$$

Vertical Component

$$b = \|u\| \sin \theta$$

$$b = 80 \sin 25°$$

$$b = 33.8$$

Step 4:   The horizontal component is 72.5 feet and the vertical component is 33.8feet.

▶EXAMPLE 13

Three forces with magnitudes of 100 pounds, 150 pounds, and 75 pounds act on an object at angles of 43°, 110°, and 140°, respectively, with the positive x-axis. Find the direction and magnitude of the resultant of these forces. Round to the nearest tenth.

▶▶Solution

Step 1: Sketch and label what we know. We will label the first vector **u**, the second vector **v** and the third vector **w**.

Step 2: We want to find the direction and magnitude of the resultant of these three vectors.

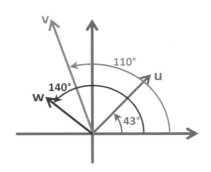

Step 3: We can write all three vectors in trigonometric form and simplify to get their linear combination forms

$u = 100(cos43°i + sin43°j)$ $v = 150(cos110°i + sin110°j)$ $w = 75(cos140°i + sin140°j)$

$u = 100cos43°i + 100sin40°j$ $v = 150cos110°i + 150sin110°j$ $w = 75cos140°i + 75sin140°j$

$u = 73.1i + 68.2j$ $v = -51.3i + 141.0j$ $w = -57.5i + 48.2j$

We can find the resultant vector **z**.

$$z = u + v + w = \left(73.1 + (-51.3) + (-57.5)\right)i + (68.2 + 141.0 + 48.2)j$$

$$z = u + v + w = -35.7i + 257.4j$$

Step 4: We need to find the direction angle and the magnitude of the resultant.

$$\|z\| = \sqrt{(-35.7)^2 + (257.4)^2}$$

$$\|z\| = 259.9$$

$$\theta_r = tan^{-1}\left|\frac{257.4}{-35.7}\right|$$

$$\theta_r = 82.1°$$

$$\theta = 180° - 82.1° = 97.9°$$

Since vector **z** is in Quadrant II we need to find θ in Quadrant II.

97.9°

82.1°   82.1°

Step 5: The direction angle is 97.9° and the magnitude is 259.5 N.

HOMEWORK

Objective 1

Find the component form of the vector u given the initial point and terminal point.

See Example 1.

1.   initial point $(-4,5)$; terminal point $(3,-2)$

2.   initial point $(2,3)$; terminal point $(6,-1)$

3.   initial point $(-2,-3)$; terminal point $(4,2)$

4.   initial point $(1,-6)$; terminal point $(2,7)$

5.   initial point $(2,-1)$; terminal point $(4,-2)$

Objective 1

Find the coordinates of the terminal point, given the initial point and vector in component form.

See Example 2.

6. $v = <-6,1>$; initial point $(5,-2)$

7. $v = <7,3>$; initial point $(-3,2)$

8. $v = <6,7>$; initial point $(-4,-4)$

9. $v = <-6,5>$; initial point $(5,-3)$

10. $v = <-3,-2>$; initial point $(1,4)$

Objective 3

Find the magnitude and direction angle for each vector u and write the vector in trigonometric form. Round to the nearest tenth.

See Examples 5 and 8.

11. $u = <-6,8>$		12. $u = <-7,6>$
13. $u = <-5,-2>$		14. $u = <4,-3>$
15. $u = <6,0>$		16. $u = 3i - 2j$
17. $u = -4i - 3j$		18. $u = 12i + 5j$
19. $u = -i + 2j$		20. $u = 20i - 21j$

Objective 2

For each pair of vectors u and v given, compute (a) through (d).

    a. $u + v$ **b.** $2u + v$ **c.** $u - v$ **d.** $2u - 3v$

See Examples 6 and 7.

21. $u = <-6,8>; v = <2,-3>$ 22. $u = <3,4>; v = <0,6>$

23. $u = <-4,2>; v = <1,4>$ 24. $u = -4i - 3j; v = 2i + 5j$

25. $u = 2i + j; v = -i - 3j$ 26. $u = i - 6j; v = -2i - 4j$

Objective 4

Find a unit vector pointing in the same direction as the vector given. Verify that a unit vector was found. Write answers in reduced radical form.

See Example 9.

27. $u = <-4,2>$ 28. $v = 2i + j$ 29. $w = 3i - 2j$

30. $p = <-20,21>$ 31. $q = <15,36>$ 32. $r = 3i - 3j$

Objective 5

For each vector, θ, represents the acute angle formed by the vector and the x-axis. (a) find the horizontal and vertical components, (b) write the vector in component form, (c) write the vector in linear combination form, (d) write the vector in trigonometric form. Round to the nearest tenth.

See Examples 3 and 4.

33.   u is in Quadrant II, $\|u\| = 16, \theta_r = 24°$

34.   v is in Quadrant III, $\|v\| = 25, \theta_r = 30°$

35.   w is in Quadrant IV, $\|w\| = 32, \theta_r = 72°$

36.   p is in Quadrant III, $\|p\| = 42, \theta_r = 64°$

Objective 6

Solve. Round to the nearest tenth.

See Examples 10-13.

37.   A vector **u** with a magnitude of 165 lb is inclined to the right and upward of 58° from the horizontal. Find the horizontal and vertical components.

38.   An airplane travels on a bearing of 115° at an airspeed of 200 km/h while a wind is blowng 60 km/h from 200°. Find the ground speed of the airplane and the direction of its course over the ground.

39.   An airplane has an airspeed of 300 mph. It is to make a flight on a bearing of 70° while there is a 25 mph wind from 340°. What will the airplane's actual bearing be?

40.   An airplane is flying at 250 mph on a bearing of 75°. There is a wind 35 mph blowing from the southwest on a bearing of 10°. What is the ground speed of the airplane and the course over the ground?

41.   An airplane's airspeed is set at 275 km/h and its bearing is N 60° E. A 40 km/h wind is blowing with a bearing of S 48° E. Determine the ground speed and true bearing of the airplane.

42.   A ball is thrown with an initial velocity of 92 feet per second, at an angle of 35° with the horizontal. Find the vertical and horizontal components of the vector representing the velocity.

43.   An arrow is shot into the air at an angle of 42° with an initial velocity of 120 ft/sec. Find the vertical and horizontal components of the vector.

44.   A kickball is kicked into the air at an angle of 19° with an initial velocity of 19 m/sec. Compute the vertical and horizontal components of the representative vector.

45.   A large van has run off the road into a ditch. Two tow trucks are attepmting to pull the van out of the ditch using winches. The cable from the first winch exerts a force of

        900 lb, while the cable from the second winch exerts a force of 700 lb at an angle of 32° with the horizontal. Find the angle for the first tow truck that will bring the van straight out of the ditch.

46.   Two tugboats are pulling an oil tanker into dock. The first tugboat is pulling with a force of 1500 N at an angle of 42° with the horizontal and the second is pulling with a force of 1750 N. Determine the angle of the second tugboat that will keep the tanker moving straight forward into the dock.

47.   A loaded barge is being towed by two tugboats. The first tugboat is pulling at a force 3611 lb at an angle of 18° with the horizontal and the second is pulling with a force of 2169 lb. Find the angle of the second tugboat that will keep the barge moving straight.

48.   Three forces with magnitudes of 90 pounds, 210 pounds, and 125 pounds act on an object at angles of 43°, 160°, and 230°, respectively, with the positive x-axis. Find the direction and magnitude of the resultant of these forces.

49.   Three forces with magnitudes of 25 pounds, 60 pounds, and 45 pounds act on an object at angles of 50°, 110°, and -62°, respectively, with the positive x-axis. Find the direction and magnitude of the resultant of these forces.

ANSWERS

1. $< 7, -7 >$

2. $< 4, -4 >$

3. $< 6, 5 >$

4. $< 1, 13 >$

5. $< 2, -1 >$

6. $(-1, -1)$

7. $(4, 5)$

8. $(2, 3)$

9. $(-1, 2)$

10. $(-2, 2)$

11. $10, 126.9°$

12. $9.2, 139.4°$

13. $5.4, 201.8°$

14. $5, 323.1°$

15. $6, 0°$

16. $3.6, 326.3°$

17. $5, 216.9°$

18. $13, 22.6°$

19. $2.2, 116.6°$

20. $29, 313.6°$

21. $< -4, 5 >, < -10, 13 >, < -8, 11 >, < -18, 25 >$

22. $< 3, 10 >, < 6, 14 >, < 3, -2 >, < 6, -10 >$

23. $< -3, 6 >, < -7, 8 >, < -5, -2 >, < -11, -8 >$

24. $-2i + 2j, -6i - j, -6i - 8j, -14i - 21j$

25. $i - 2j, 3i - j, 3i + 4j, 7i + 11j$

26. $i - 10j, -16j, 3i - 2j, 8i$

27. $< -\dfrac{2}{\sqrt{5}}, \dfrac{1}{\sqrt{5}} >$

28. $\dfrac{2}{\sqrt{5}} i + \dfrac{1}{\sqrt{5}} j$

29. $\dfrac{3}{\sqrt{13}} i - \dfrac{2}{\sqrt{13}} j$

30. $< -\dfrac{20}{29}, \dfrac{21}{29} >$

31. $< \dfrac{5}{13}, \dfrac{12}{13} >$

32. $\dfrac{1}{\sqrt{2}} i - \dfrac{1}{\sqrt{2}} j$

33. $< -14.6, 6.5 >, -14.6i + 6.5j, 16(cos156°i + sin156°j)$

34. $< -21.7, -12.5 >, -21.7i - 12.5j, 25(cos210°i + sin210°j)$

35. $< 9.9, -30.4 >, 9.9i - 30.4j, 32(cos288°i + sin288°j)$

36. $< -18.4, -37.7 >, -18.4i - 37.7j, 42(cos244°i + sin244°j)$

37. Horizontal Component: 87.4, Vertical Component: 139.9

38. Direction: 97.9°, Groundspeed: 203.7 km/h   39. 65.2°

40. Direction: 68.2°, Groundspeed: 266.7 mph

41. Direction: 58.5°, Groundspeed: 314.3 km/h

42.   Horizontal Component: 75.4, Vertical Component: 52.8

43.   Horizontal Component: 89.2, Vertical Component: 80.3

44.   Horizontal Component: 18.0, Vertical Component: 6.2

45.   24.3°   46.   35°   47.   31°

48.   215.2, 170°   49.   39.5, 65°

Section 20:   Vectors and Dot Products

Learning Outcomes:

- The student will perform applications on vectors and complex numbers.
- The student will solve trigonometric equations, right triangles, and oblique triangles.
- The student will correctly memorize and apply trigonometric formulas, definitions, identities, and properties.

Objectives: At the conclusion of this lesson you should be able to:

1.   Find a vector for equilibrium to occur.
2.   Compute the dot product.
3.   Find the angle between two vectors.
4.   Determine if vectors are orthogonal.
5.   Find the components and projections of one vector along another.
6.   Solve applications involving work.
7.   Use vectors to model and solve applications.

In Section 19 we introduced and performed operations on vectors. In this section we will not only introduce additional ideas to enable us to solve a variety of new applications but lay a strong foundation for future studies involving vectors.

Equilibrium

In the last section we talked about finding the resultant of two or more vectors. Sometimes it is necessary to find a vector that will counterbalance the resultant. This opposite vector is the equilibrant. For example, the equilibrant of vector **u** is –**u**. Two or more vectors are in equilibrium when their sum is the zero vector.

Vectors and Equilibrium

Given vectors F_1, F_2,. . .,F_n acting on a point P,

1.   The resultant vector is $F = F_1 + F_2 + . . . + F_n$.

2.   Equilibrium for these forces requires the vector –**F**, where $F + (-1F) = 0$

▶ EXAMPLE 1

The force vectors $F_1 = <-2, 7>$ and $F_2 = <5, 3>$ are acting on a common point. Find an additional force vector, F_3, so that equilibrium takes place. Graph $F_1, F_2,$ and F_3.

▶▶ Solution

Begin by graphing F_1 and F_2.
Find the resultant F_3 of F_1 and F_2.

$$F_1 + F_2 = <-2+5, 7+3>$$

$$F_1 + F_2 = <3, 10>$$

Find and graph F_3 by taking the negative of the resultant.

$$F_3 = <-3, -10>$$

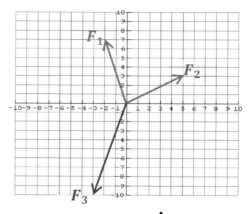

▶ EXAMPLE 2

A game of tug war between three teams $T_1, T_2,$ and T_3 has begun. Teams 1 and 2 are pulling with the magnitude and at the angles indicated in the diagram. If the three teams are at a stalemate, find the magnitude and direction of the rope held by Team 3. Round to the nearest tenth.

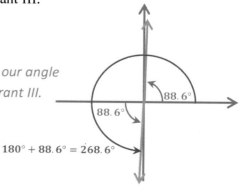

▶▶ Solution

We will begin by getting the trigonometric forms of the vectors T_1 and T_2 and then convert them to linear combination form.

$T_1 = 1500(cos34°i + sin34°j)$ $T_2 = 1850(cos130°i + sin130°j)$

$T_1 = 1500cos34°i + 1500sin34°j$ $T_2 = 1850cos130°i + 1850sin130°j$

$T_1 = 1243.6i + 838.8j$ $T_2 = -1189.2 + 1417.2j$

Since Team 2 is in Quadrant II, we will use 130°, 180°-50°=130°.

Find the resultant of T_1 and T_2.

$T_1 + T_2 = (1243.6 + (-1189.2))i + (838.8 + 1417.2)j$

$T_1 + T_2 = 54.4i + 2256.0j$

We find T_3 by taking the negative of the resultant.

$$T_3 = -54.4i - 2256.0j$$

Find the magnitude and direction angle of T_3. T_3 is in Quadrant III.

$\|T_3\| = \sqrt{(-54.4)^2 + (-2256.0)^2}$

$\|T_3\| = \sqrt{2959.4 + 5089536}$

$\|T_3\| = \sqrt{5092495.4}$

$\|T_3\| = 2256.7$

$\theta_r = tan^{-1}\left|\dfrac{-2256.0}{-54.4}\right|$

$\theta_r = 88.6°$ *We need our angle*
$\theta = 268.6°$ *for Quadrant III.*

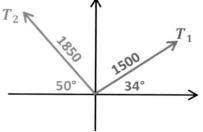

$180° + 88.6° = 268.6°$

Magnitude of T_3: $\|T_3\| = 2256.7$
Direction Angle of T_3: $\theta = 268.6°$

▶ EXAMPLE 3

Two tugboats are pulling a disabled boat into port with forces of 1400 lb and 1680 lb. The angle between these two forces is 40°. Find the direction and magnitude of the equilibrant. Round to the nearest tenth.

▶▶ Solution

We will begin by graphing the forces, with F_1 on the x-axis and F_2 at an angle of 40°.

Write the forces in trigonometric form and simplify them into linear combination form.

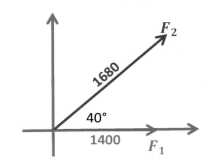

$$F_1 = 1400(cos0°i + sin0°j)$$
$$F_1 = 1400cos0°i + 1400sin0°j$$
$$F_1 = 1400i + 0j$$

$$F_2 = 1680(cos40°i + sin40°j)$$
$$F_2 = 1680cos40°i + 1680sin40°j$$
$$F_2 = 1287.0i + 1079.9j$$

Find the resultant of F_1 and F_2.

$$F_1 + F_2 = (1400 + 1287.0)i + (0 + 1079.9)j$$
$$F_1 + F_2 = 2687.0i + 1079.9j$$

The equilibrant is the negative of the resultant.

$$F_3 = -2687.0i - 1079.9j$$

We will find the magnitude and the direction angle of the equilibrant.

$$\|F_3\| = \sqrt{(-2687.0)^2 + (-1079.9)^2}$$
$$\|F_3\| = \sqrt{8386153.01}$$
$$\|F_3\| = 2895.9$$

$$\theta_r = tan^{-1}\left|\frac{-1079.9}{-2687.0}\right|$$
$$\theta_r = 21.9°$$
$$\theta = 180° + 21.9°$$
$$\theta = 201.9°$$

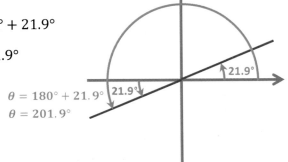

$$\theta = 180° + 21.9°$$
$$\theta = 201.9°$$

Magnitude: 2895.9 lb

Direction Angle: 201.9°

Dot Product of Two Vectors

In the previous section we noted when a vector is multiplied by a scalar, the result is a vector and when we add two vectors, the result is a vector. When we multiply two vectors, the result is not a vector, but a real number (scalar). We call the product of two vectors the dot product.

Dot Product

Let $u =< u_1, u_2 >$ and $v =< v_1, v_2 >$, then $u \cdot v = u_1 v_1 + u_2 v_2$.

►EXAMPLE 4

Find $u \cdot v$ if $u =< -2, 7 >$ and $v =< 5, 3 >$

►►Solution

Use the dot product:

$$u \cdot v = (-2)(5) + (7)(3)$$
$$u \cdot v = -10 + 21$$
$$u \cdot v = 11$$

Properties of the Dot Product

Let u, v and w be vectors in the plane and let k be a scalar.

1. $u \cdot v = v \cdot u$

2. $u \cdot (v + w) = u \cdot v + u \cdot w$

3. $0 \cdot v = 0$

4. $u \cdot u = \|u\|^2$

5. $k(u \cdot v) = ku \cdot v = u \cdot kv$

►EXAMPLE 5

Find $(u \cdot v)w$ if $=< -2, 3 >$, $v =< 2, -4 >$ and , $w =< 1, -5 >$. Is the result a vector or a scalar?

►►Solution

$$(u \cdot v)w = [(-2)(2) + (3)(-4)] < 1, -5 >$$
$$(u \cdot v)w = [-4 - 12] < 1, -5 >$$
$$(u \cdot v)w = -16 < 1, -5 >$$
$$(u \cdot v)w =< -16, 80 > \quad \text{The result is a vector.}$$

▶ EXAMPLE 6

Find $u \cdot 2v$ if $u =< 4, -3 >$ and $v =< -5, -6 >$. Is the result a vector or a scalar?

▶▶ Solution

$$u \cdot 2v = 2(u \cdot v) = 2[(4)(-5) + (-3)(-6)]$$

$$u \cdot 2v = 2(u \cdot v) = 2[-20 + 18]$$

$$u \cdot 2v = 2(u \cdot v) = 2[-2]$$

$$u \cdot 2v = 2(u \cdot v) = -4 \qquad \text{The result is a scalar.}$$

▶ EXAMPLE 7

Use the dot product to find the magnitude of vector u, $u =< -2, 3 >$.

▶▶ Solution

Because $u \cdot u = \|u\|^2$, it follows that:

$$u \cdot u = \|u\|^2$$

$$\sqrt{u \cdot u} = \sqrt{\|u\|^2}$$

$$\sqrt{u \cdot u} = \|u\|$$

We can now use the dot product to find the magnitude:

$$\|u\| = \sqrt{u \cdot u} = \sqrt{(-2)(-2) + (3)(3)}$$

$$\|u\| = \sqrt{u \cdot u} = \sqrt{4 + 9}$$

$$\|u\| = \sqrt{u \cdot u} = \sqrt{13}$$

Angle between Two Vectors

One application of the dot product is calculating the angle between two vectors. Let **u** and **v** be two vectors with the same initial point at the origin and angle θ between them as shown in **Figure 20.1**. The vectors **u**, **v** and **u − v** form a triangle. The sides of the triangle have lengths $\|u\|$, $\|v\|$ and $\|u - v\|$ and we can use the Law of Cosines to find cosθ.

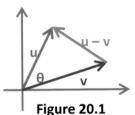

Figure 20.1

$$\|u - v\|^2 = \|u\|^2 + \|v\|^2 - 2\|u\|\|v\|\cos\theta \qquad \textit{Substitute into the Law of Cosines.}$$

$$(u - v)(u - v) = u \cdot u + v \cdot v - 2\|u\|\|v\|\cos\theta \qquad \textit{Use Property 4 and rewrite the equation in terms of dot products.}$$

$$u \cdot (u - v) - v \cdot (u - v) = u \cdot u + v \cdot v - 2\|u\|\|v\|\cos\theta$$

$$u \cdot u - u \cdot v - v \cdot u + v \cdot v = u \cdot u + v \cdot v - 2\|u\|\|v\|\cos\theta \qquad \textit{Apply Property 1 and simplify.}$$

$$u \cdot u + v \cdot v - 2u \cdot v = u \cdot u + v \cdot v - 2\|u\|\|v\|\cos\theta \qquad \textit{Isolate } 2\|u\|\|v\|\cos\theta.$$

$$-2u \cdot v = -2\|u\|\|v\|\cos\theta$$

$$\frac{-2u \cdot v}{-2\|u\|\|v\|} = \frac{-2\|u\|\|v\|\cos\theta}{-2\|u\|\|v\|} \qquad \textit{Solve for } \cos\theta.$$

$$\frac{u \cdot v}{\|u\|\|v\|} = \cos\theta \qquad \textit{Formula for the angle between two vectors.}$$

> **Angle Between Two Vectors**
>
> Let u and v be vectors in the plane.
>
> $$\cos\theta = \frac{u \cdot v}{\|u\|\|v\|}$$

►**EXAMPLE 8**

Find the angle θ between $u = 5i - 3j$ and $v = 2i + 4j$.

►►**Solution**

$$\cos\theta = \frac{u \cdot v}{\|u\|\|v\|}$$

$$\cos\theta = \frac{(5)(2)+(-3)(4)}{\sqrt{(5)^2+(-3)^2}\sqrt{(2)^2+(4)^2}} \qquad \textit{Substitute into the angle between vectors formula.}$$

$$\cos\theta = \frac{10-12}{\sqrt{25+9}\sqrt{4+16}} \qquad \textit{Simplify.}$$

$$\cos\theta = \frac{-2}{\sqrt{34}\sqrt{20}}$$

[2nd] [COS] [(-)] [2] [÷] [(] [2nd] [x²] [3] [4] [)] [2nd] [x²]

$$\theta = 94.4°$$

[2] [0] [)] [)] [ENTER]

Parallel and Orthogonal Vectors

Two vectors **u** and **v** are parallel if there is a nonzero scalar k so that $u = kv$ and the angle θ between **u** and **v** is 0° or 180°. If the angle between two nonzero vectors **u** and **v** is 90°, the vectors **u** and **v** are orthogonal (perpendicular). To determine if vectors **u** and **v** are orthogonal their dot product will equal 0, $u \cdot v = 0$.

> **Orthogonal and Parallel Vectors**
>
> Given vectors **u** and **v** and angle θ between **u** and **v**,
>
> **Orthogonal Vectors:** $u \cdot v = 0$
>
> **Parallel Vectors:** $\theta = 0°$ or $\theta = 180°$

▶ EXAMPLE 9

Determine if the following vectors u and v are orthogonal, parallel or neither.

a.) $u = <2, -3>, v = <6, 4>$ b.) $u = 2i - j, v = 4i - 2j$

▶▶ Solution

a.) $u = <2, -3>, v = <6, 4>$ Check for orthogonal first.

Orthogonal

$u \cdot v = (2)(6) + (-3)(4)$ *Find the dot product.*

$u \cdot v = 12 - 12$

$u \cdot v = 0$ *Since the dot product is 0, the vectors are orthogonal.*

The vectors are orthogonal.

b.) $u = 2i - j, v = 4i - 2j$

Orthogonal

$u \cdot v = (2)(4) + (-1)(-2)$ *Find the dot product.*

$u \cdot v = 8 + 2$

$u \cdot v = 10$ *Since the dot product does not equal 0, the vectors are not orthogonal.*

Parallel

$$cos\theta = \frac{(2)(4) + (-1)(-2)}{\sqrt{(2)^2 + (-1)^2}\sqrt{(4)^2 + (-2)^2}}$$ *Find the angle between the two*

$$cos\theta = \frac{8 + 2}{\sqrt{4 + 1}\sqrt{16 + 4}}$$

$$cos\theta = \frac{10}{\sqrt{5}\sqrt{20}}$$ $\boxed{2^{nd}}$ $\boxed{COS}$ $\boxed{1}$ $\boxed{0}$ $\boxed{\div}$ $\boxed{(}$ $\boxed{2^{nd}}$ $\boxed{x^2}$ $\boxed{5}$ $\boxed{)}$ $\boxed{2^{nd}}$ $\boxed{x^2}$ $\boxed{2}$ $\boxed{0}$ $\boxed{)}$ $\boxed{)}$ $\boxed{ENTER}$

$\theta = 0°$ *Since the angle between the two vectors is 0°, the vectors are parallel.*

The vectors are parallel.

Component of a Vector along another Vector

The component of **u** along **v**, written $comp_v u$, is defined as $\|u\|cos\theta$ where θ is the angle between vectors **u** and **v** as shown in **Figure 20.2**. The $comp_v u$ is not a vector but a scalar quantity that measures how much of **u** is in the **v** direction. Note that even when the components of a vector do not lie along the x- or y-axis, they are still orthogonal.

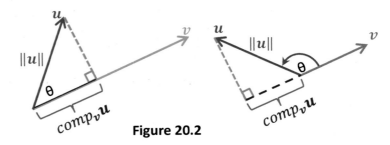

Figure 20.2

Vectors and the $comp_v u$

Given vectors **u** and **v** and angle θ where they meet.

1. $comp_v u = \|u\|\cos\theta$

2.   If $0° < \theta < 90°$, $comp_v u > 0$

3.   If $90° < \theta < 180°$, $comp_v u < 0$

4.   If $\theta = 0°$, **u** and **v** have the same direction and $comp_v u = \|u\|$

5.   If $\theta = 90°$, **u** and **v** are orthogonal and $comp_v u = 0$

6.   If $\theta = 180°$, **u** and **v** have opposite directions and $comp_v u = -\|u\|$

▶ EXAMPLE 10

Find the $comp_v u$ for the vectors u and v given. Round to the nearest tenth.

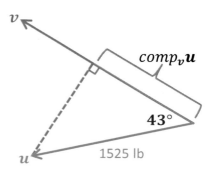

▶▶ Solution

$comp_v u = \|u\|\cos\theta$ *Substitute into the formula.*

$comp_v u = 1525\cos 43°$

$comp_v u = 1115.3$

An alternate form of the $comp_v u$ is found by using the formula for finding the angle between two vectors.

$comp_v u = \|u\|\cos\theta$ *Substitute the formula for finding the angle between two vectors.*

$comp_v u = \|u\| \left(\dfrac{u \cdot v}{\|u\|\|v\|} \right)$ *Simplify.*

$comp_v u = \dfrac{u \cdot v}{\|v\|}$

An application of the $comp_v u$ involves the force of gravity, G, acting on an object placed on an inclined plane (ramp). The steeper the inclined plane the more force it takes to move or hold an object stationary. The amount of force depends on the weight of the object, incline of the plane and friction. Since friction can be minimized, we will always assume that there is no friction in our models of inclined plane applications.

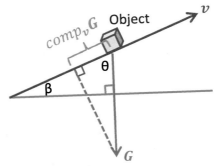

Figure 20.3

The weight of an object (the force of gravity) on an inclined plane is always modeled as the vertical vector G and its length is fixed regardless of the incline of the plane. The force required to move the object or to hold it stationary is the $comp_v G = \|G\|cos\theta$ where $\|G\|$ is the weight of the object (magnitude of G) and $\theta = 90° - \beta$ where β is the angle of incline. See **Figure 20.3**.

▶EXAMPLE 11

A 700-lb crate is sitting on a ramp that is inclined at 25°. Find the force needed to hold the crate stationary. Round to the nearest tenth.

▶▶Solution

Step 1: Sketch and label what we know. We know that $\|G\| = 700\ lb$ and the angle of incline is 25°.

Step 2: We want to find $comp_v G$.

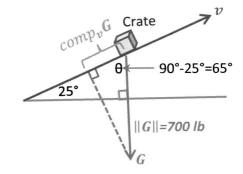

Step 3: We can find θ (see sketch) and substitute what we know into the formula to find the $comp_v G$.

$comp_v G = \|G\|cos\theta$ *Substitute into the formula.*

$comp_v G = 700cos65°$

$comp_v G = 295.8$

Step 4: The force required to keep the crate stationary is 295.8 lb.

► EXAMPLE 12

A 325-kg box is sitting on a ramp, held stationary by 225 kg of tension in a restraining rope. Find the ramp's angle of incline to the nearest tenth.

►► Solution

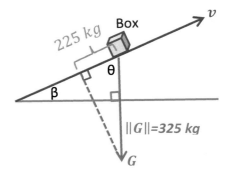

Step 1: Sketch and label what we know. We know that $\|G\| = 325\ kg$ and $comp_v G = 225\ kg$.

Step 2: We want to find β (the angle of incline).

Step 3: We can find θ by substituting into the $comp_v G$ formula and solving.

$comp_v G = \|G\|cos\theta$ *Substitute into the formula.*

$225 = 325cos\theta$

$\dfrac{225}{325} = \dfrac{325cos\theta}{325}$ *Solve for $cos\theta$.*

$\dfrac{225}{325} = cos\theta$ *Solve for θ.*

$46.2° = \theta$

Step 4: Since $\theta = 46.2°$ we can find β by subtracting θ from 90°.

$$\beta = 90° - 46.2°$$
$$\beta = 43.8°$$

Step 5: The angle of incline of the ramp is 43.8°.

Projection of a Vector along another Vector

The $proj_v u$ is a vector in the same direction of **v**, with magnitude of $comp_v u$ as shown in **Figure 20.4**. Recall that the unit vector $\widehat{v} = \dfrac{v}{\|v\|}$ has a length of one and points in the same direction of **v**, so we can define the $proj_v u$ as $proj_v u = comp_v u \cdot \dfrac{v}{\|v\|}$. If we substitute $\dfrac{u \cdot v}{\|v\|}$, the alternate form of $comp_v u$, we have $proj_v u = \dfrac{u \cdot v}{\|v\|} \cdot \dfrac{v}{\|v\|}$. Simplify and we have $proj_v u = \left(\dfrac{u \cdot v}{\|v\|^2}\right) v$.

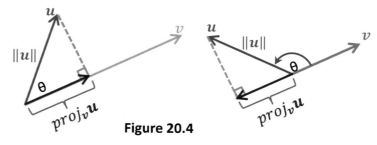

Figure 20.4

> ## *Vector Projection*
>
> Given vectors **u** and **v**, the projection of **u** along **v** is the vector
>
> $$proj_v \boldsymbol{u} = \left(\frac{\boldsymbol{u} \cdot \boldsymbol{v}}{\|\boldsymbol{v}\|^2}\right) \boldsymbol{v}$$

► EXAMPLE 13

Given vectors $\boldsymbol{u} = <2, 6>$ and $\boldsymbol{v} = <-3, 2>$ find the $proj_v \boldsymbol{u}$.

►► Solution

$$proj_v \boldsymbol{u} = \left(\frac{\boldsymbol{u} \cdot \boldsymbol{v}}{\|\boldsymbol{v}\|^2}\right) \boldsymbol{v} \qquad \textit{Substitute and simplify.}$$

$$proj_v \boldsymbol{u} = \left(\frac{(2)(-3) + (6)(2)}{\left(\sqrt{(-3)^2 + (2)^2}\right)^2}\right) <-3, 2>$$

$$proj_v \boldsymbol{u} = \left(\frac{-6 + 12}{\left(\sqrt{9 + 4}\right)^2}\right) <-3, 2>$$

$$proj_v \boldsymbol{u} = \left(\frac{6}{\left(\sqrt{13}\right)^2}\right) <-3, 2>$$

$$proj_v \boldsymbol{u} = \left(\frac{6}{13}\right) <-3, 2>$$

$$proj_v \boldsymbol{u} = <-\frac{18}{13}, \frac{12}{13}>$$

▶ EXAMPLE 14

Find the $comp_v u$ and the $proj_v u$ for vectors $u =< 2, 6 >$ and $v =< -3, 2 >$. Round to the nearest tenth.

▶▶ Solution

$$comp_v u = \frac{u \cdot v}{\|v\|} \qquad\qquad proj_v u = \left(\frac{u \cdot v}{\|v\|^2}\right) v$$

$$comp_v u = \frac{(2)(-3) + (6)(2)}{\sqrt{(-3)^2 + (2)^2}} \qquad proj_v u = \left(\frac{(2)(-3) + (6)(2)}{\left(\sqrt{(-3)^2 + (2)^2}\right)^2}\right) < -3, 2 >$$

$$comp_v u = \frac{-6 + 12}{\sqrt{9 + 4}} \qquad\qquad proj_v u = \left(\frac{-6 + 12}{\left(\sqrt{9 + 4}\right)^2}\right) < -3, 2 >$$

$$comp_v u = \frac{6}{\sqrt{13}} \qquad\qquad proj_v u = \left(\frac{6}{\left(\sqrt{13}\right)^2}\right) < -3, 2 >$$

$$comp_v u = 1.7 \qquad\qquad\qquad proj_v u = \left(\frac{6}{13}\right) < -3, 2 >$$

$$proj_v u = (.5) < -3, 2 >$$

$$proj_v u =< -1.5, 1.0 >$$

Work

In elementary physics, the work W done by a constant force F in moving an object is defined as $W = \|F\| \cdot D$ (Work equals magnitude of force times the distance the object is moved). If the force is given in Newtons and the distance in meters, the amount of work is measured in Newton-meters (N-m). If the force is given in pounds and the distance in feet, the amount of work is measured in a unit called foot-pounds (ft-lb).

▶ EXAMPLE 15

Determine the work done by a crane lifting a 2500-pound car 8 feet.

▶▶ Solution

The force is 2500-pound and the distance is 8 feet. $W = \|F\| \cdot D$

The work done in lifting the car is 20,000 ft-lb. $W = (2500\ lb)(8\ ft)$

$W = 20000\ ft - lb$

In Example 13, it is assumed that the force F is applied along the line of motion. When the force F is not applied along the line of motion but is at angle θ to the direction of the motion, the amount of work done is the component of force along vector D times the distance the object is moved, $comp_D F \cdot D$. From our previous discussion about the component of **u** along vector **v**, we can write $comp_D F$ as $\|F\|\cos\theta$. In **Figure 20.5** a truck is pulling a crate with a force of **F** applied along the line of motion at the acute angle θ to the crate and D the distance it is moved, $W = \|F\|\cos\theta \cdot D$. Notice that if the angle θ measures 0°, we have our first formula.

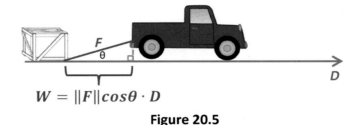

$$W = \|F\|\cos\theta \cdot D$$

Figure 20.5

Work

The work **W** done by a constant force **F** as it moves along a vector **D** is

$$W = \|F\| \cdot D$$

$$W = \|F\|\cos\theta \cdot D$$

▶ EXAMPLE 16

Determine the work done by a man mowing his lawn if he pushes his lawn mower 400 ft at a constant force of 50 lb. The handle of the lawn mower is held an angle of 43° with the level ground. Round to the nearest whole number.

▶▶ Solution

The force is 50-pound, the angle θ is 43° and the distance is 400 feet.

$$W = \|F\|\cos\theta \cdot D$$

$$W = (50\ lb)(\cos 43°)(400\ ft)$$

$$W = 14627\ ft - lb$$

The work done in mowing the lawn is 14,627 ft-lb.

HOMEWORK

Objective 1

The force vectors given are acting on a common point. Find and additional force vector so that equilibrium takes place.

See Example 1.

1. $F_1 =< -6,8 >, F_2 =< -4, -5 >$

2. $F_1 =< 4, -2 >, F_2 =< 3, -1 >$

3. $F_1 =< 6, -5 >, F_2 =< -3,7 >, F_3 =< -1,2 >$

4. $F_1 = 3i - 2j, F_2 = 4i + 2j, F_3 = -2i - 6j$

5. $F_1 = i + 3j, F_2 = -3i + 2j, F_3 = -5i - 7j$

Objective 1

Solve. Round to the nearest tenth.

See Examples 2 and 3.

6.   Three cowboys have roped a bull and are attempting to hold him steady. The first and second cowboys are pulling with the magnitude and at the angles indicated in the diagram. Find the magnitude and direction of the rope of the third cowboy in order to hold the bull steady.

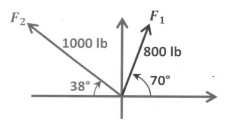

7.   Three teams are playing tug of war. Teams 1 and 2 are pulling with the magnitude and at the angles indicated in the diagram. If the three teams are at a stalemate, find the magnitude and direction of the rope held by Team 3.

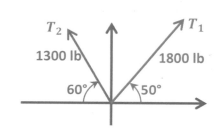

8.   Three farmers are pulling on ropes attached around the neck of a donkey. One farmer pulls with a force of 75 lbs and another pulls with a force of 100 lbs. The angle between these two forces is 35°. Find the magnitude of the third farmer to hold the donkey steady.

9.   Three children are holding to ropes attached to the tail of a kite. One child holds on with a force of 60 lbs and another child holds with a force of 80 lbs. The angle between these two forces is 65°. Find the magnitude of the third child to hold the kite steady.

Objective 2

Find the dot product of u and v.

See Example 4.

10. $u = < -6, 8 >$, $v = < 7, 6 >$

11. $u = < -5, -2 >$, $v = < 4, -3 >$

12. $u = -4i - 3j$, $v = 3i - 2j$

13. $u = -i + 2j$, $v = 12i + 5j$

Objective 2

Use the vectors $u = < -3, 2 >$ and $v = < 4, -3 >$ to find the indicated quantity and state whether the result is a vector or a scalar. Round to the nearest tenth.

See Examples 5 and 6.

14. $u \cdot u$	15. $3u \cdot v$	16. $(u \cdot v)v$
17. $\|u\| \cdot u$	18. $\|v\| \cdot u$	19. $u \cdot 2v$
20. $\|u\| + 4$	21. $\|v\| + 2$	

Objective 2

Use the dot product to find the magnitude of u. Round to the nearest tenth.

See Example 7.

22. $u = < 3, 4 >$	23. $u = < -4, 2 >$	24. $u = < 5, -12 >$
25. $u = 2i + j$	26. $u = i - 6j$	27. $u = -4i - 3j$

Objective 3

Find the angle θ between the two vectors. Round to the nearest tenth.

See Example 8.

28.　$u = <-6,8>, v = <7,6>$　　　　29.　$u = <-5,-2>, v = <4,-3>$

30.　$u = -4i - 3j, v = 3i - 2j$　　　31.　$u = -i + 2j, v = 12i + 5j$

32.　$u = -2i + 3j, v = 4i + 2j$

Objective 4

Determine if u and v are orthogonal, parallel or neither.

See Example 9.

33.　$u = <7,-2>, v = <4,14>$　　　34.　$u = <-6,-3>, v = <-2,-1>$

35.　$u = <7,-2>, v = <14,-4>$　　36.　$u = <5,-3>, v = <2,4>$

37.　$u = -2i - 6j, v = 9i - 3j$　　38.　$u = 2i - 2j, v = -i - j$

39.　$u = -4i - 3j, v = 3i - 4j$　　40.　$u = -4i - 3j, v = -8i - 6j$

Objective 5

Find the $comp_v u$ and the $proj_v u$ for vectors u and v. Round to the nearest hundredth.

See Examples 10, 13, and 14.

41.　$u = <7,-2>, v = <4,14>$　　　42.　$u = <3,5>, v = <7,1>$

43.　$u = <-5,4>, v = <-9,-11>$　　44.　$u = -2i - 6j, v = 9i - 3j$

45.　$u = 4i + 3j, v = 2i - 4j$　　　46.　$u = 6i - j, v = -3i - 2j$

Objectives5, 6, and 7

Solve. Round to the nearest tenth.

See Examples 11, 12, 15, and 16.

47.　A car with gross weight of 5300 lb is parked on a street with a grade (incline) of 8°. Find the force necessary to keep the car from rolling down the street.

48.　The amount of force required to push a block of ice up an ice covered driveway that is inclined at 30° is 145 pounds. What is the weight of the block of ice?

49. A metal ball is placed on a 12° incline. If the force required to keep the ball in place is 3.2 lb, then what is the weight of the ball?

50. If Superman exerts 1500 pounds of force to prevent an 8000-lb boulder from rolling down a hill and crushing innocent bystanders, then what is the angle of inclination of the hill?

51. Find the amount of force required for a winch to pull a 2800-lb car up a ramp that is inclined at 22°.

52. Find the amount of force required to hold a 3000-lb crate stationary on a hill with incline of 15°.

53. A force of 60 lb is required to hold a 130 lb box stationary on a ramp. What is angle of incline of the ramp?

54. A 63 kg weight is on a ramp which is inclined at 38°. What is the force needed to hold the weight stationary on the ramp?

55. A 600 lb boat is being lowered into the water down an inclined ramp which is at an angle of 22° with the horizontal. What is the force needed to hold the boat at rest?

56. A force of 39 lb is required to hold a 275 crate from sliding down an inclined ramp. At what angle is the ramp inclined?

57. Rearranging the living room furniture, a table is pushed 20 ft at a constant force of 40 lb. How much work is done?

58. A car runs out of gas 40m from a gas station pump. How much work is done if the car is pushed at a constant force of 180 N to the gas pump?

59. Find the work done by a crane lifting a 3000 lb car 6 feet.

60. Find the work done by pushing a 600 lb boulder 8 feet.

61. A wagon loaded with 250 lbs of patio bricks is pulled 200 yd over level ground by a handle which makes an angle of 41° with the horizontal. Find the work done if a force of 28 lb is exerted on the handle.

62. A mother pulls a wagon containing her 3 small children 400 ft . The handle of the wagon makes a 30° with the horizontal. How much work is done if a force of 25 lb is exerted on the hangle?

63. A wagon is pulled horizontally by exerting 40 lb force on the handle at angle of 35° with the horizontal. How much work is done in pulling the wagon 200 feet?

64. A gardner pushes a wheelbarrow 100 m at a force of 30 N. Find the amount work done if while in motion the wheelbarrow makes an angle 22° with level ground.

ANSWERS

1. $< 10, -3 >$ 2. $< -7,3 >$ 3. $< -2, -4 >$

4. $-5i + 6j$ 5. $7i + 2j$ 6. $1461.0\ lb, 290.6°$

7. $2555.5\ lb, 258.6°$ 8. $167.1\ lb$ 9. $118.6\ lb$

10. 6   11. -14   12. -6

13. -2   14. 13.0 scalar   15. -54.0 scalar

16. $< -72.0, 54.0 >$ vector   17. $< -10.8, 7.2 >$ vector

18. $< -15.0, 10.0 >$ vector   19. -36.0 scalar

20. 7.6 scalar   21. 7.0   22. 5.0

23. 4.5   24. 13.0   25. 2.2

26. 6.1   27. 5.0   28. 86.3°

29. 121.3°   30. 109.4°   31. 93.9°

32. 97.1°   33. Orthogonal   34. Parallel

35. Parallel   36. Neither   37. Orthogonal

38. Orthogonal   39. Orthogonal   40. Parallel

41. $comp_v u = 0, proj_v u = \langle 0, 0 \rangle$

42. $comp_v u = 3.68, proj_v u = \langle 3.64, 0.52 \rangle$

43. $comp_v u = 0.07, proj_v u = \langle -0.04, -0.05 \rangle$

44. $comp_v u = 0, proj_v u = 0i + 0j$

45. $comp_v u = -0.89, proj_v u = -0.40i + 0.80j$

46. $comp_v u = -4.44, proj_v u = 3.69i + 2.64j$

47. 737.6 lb   48. 290.0 lb   49. 15.4 lb

50. 10.8°   51. 1048.9 lb   52. 776.5 lb

53. 27.5°   54. 38.8 kg   55. 224.8 lb

56. 8.2°   57. 800 ft-lb   58. 7200 N-m

59. 18000.0 ft-lb   60. 4800.0 ft-lb   61. 12679.0 ft-lb

62. 8660.3 ft-lb   63. 6553.2 ft-lb   64. 2781.6 N-m

Section 21:   Trigonometric Form of Complex Numbers

Learning Outcomes:

- The student will perform applications on vectors and complex numbers.
- The student will correctly memorize and apply trigonometric formulas, definitions, identities, and properties.

Objectives: At the conclusion of this lesson you should be able to:

1.   Graph complex numbers.
2.   Find the modulus and argument of a complex number.
3.   Write a complex number in both rectangular form and trigonometric form.
4.   Multiply and divide complex numbers in trigonometric form.
5.   Use De Moivre's Theorem to find powers of complex numbers.
6.   Find nth roots of complex numbers.

Graph Representation of a Complex Number

The complex number $a + bi$ where a is the real part and b is the imaginary part can be graphed in the same manner that we graph an ordered pair (a, b) of real numbers in the Cartesian plane. To graph a complex number we will not use the Cartesian plane but instead use the complex plane. In the complex plane, the x-axis (horizontal axis) is the real axis and the y-axis (vertical axis) is the imaginary axis as shown in **Figure 21.1**. The graph of $a + bi$ is shown in **Figure 21.2**.

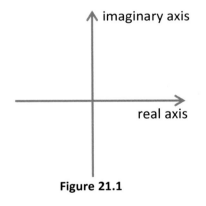

Figure 21.1

imaginary axis

bi $a + bi$

a real axis

Figure 21.2

► EXAMPLE 1

Graph $3 - 2i$ in the complex plane.

►► Solution

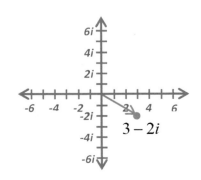

Rectangular Form of a Complex Number

The rectangular form of a complex number is $z = a + bi$. Recall that the absolute value of a real number represents the distance from the origin on the real number line to the real number. The absolute value of the complex number $z = a + bi$ is the distance from the origin to $z = a + bi$. This distance is called the modulus, r, and is found using the Pythagorean Theorem: $r = \|z\| = \|a + bi\| = \sqrt{a^2 + b^2}$.

The angle θ formed with the positive real axis to the line segment connecting the origin and the complex number is called the argument. The argument is found using $\theta = \tan^{-1}\left|\dfrac{b}{a}\right|$. See **Figure 21.3**.

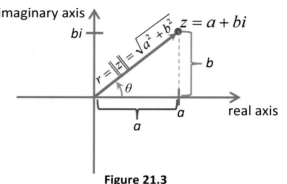

Figure 21.3

Modulus of a Complex number

The modulus, r, of the complex number $z = a + bi$ is $r = \|z\| = \sqrt{a^2 + b^2}$.

Argument of a Complex number

The argument, θ, of the complex number $z = a + bi$ is $\theta = \tan^{-1}\left|\dfrac{b}{a}\right|$.

► EXAMPLE 2

Graph $z = -4 - 3i$ as a point in the complex plane and find the modulus and the argument. Round to the nearest tenth if necessary.

►► Solution

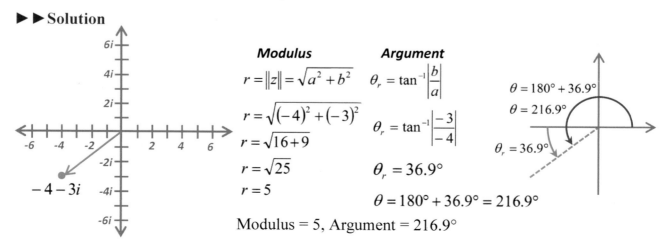

Modulus

$$r = \|z\| = \sqrt{a^2 + b^2}$$

$$r = \sqrt{(-4)^2 + (-3)^2}$$

$$r = \sqrt{16 + 9}$$

$$r = \sqrt{25}$$

$$r = 5$$

Argument

$$\theta_r = \tan^{-1}\left|\frac{b}{a}\right|$$

$$\theta_r = \tan^{-1}\left|\frac{-3}{-4}\right|$$

$$\theta_r = 36.9°$$

$$\theta = 180° + 36.9°$$
$$\theta = 216.9°$$
$$\theta_r = 36.9°$$

$$\theta = 180° + 36.9° = 216.9°$$

Modulus = 5, Argument = 216.9°

Trigonometric Form (Polar Form) of a Complex Number

It can be rather complicated to multiply and divide complex numbers as well as find powers and roots of complex numbers when they are in rectangular form, $a + bi$. When complex numbers are expressed in trigonometric form (polar form) these operations become simpler. To convert from the rectangular form of a complex number to trigonometric form of a complex number, we need to know the argument and the modulus. In **Figure 21.4** the graph of the complex number $z = a + bi$ is given. We can solve for both a and b as shown in **Figure 21.5**.

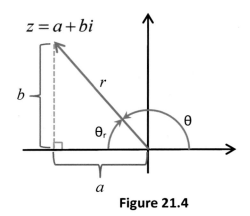

$$\sin \theta = \frac{b}{r} \qquad\qquad \cos \theta = \frac{a}{r}$$

$$b = r \sin \theta \qquad\qquad a = r \cos \theta$$

Figure 21.5

Figure 21.4

Substituting for a and b in the rectangular from, we have $z = r \cos \theta + r \sin \theta \, i$. If we factor out the r we have the trigonometric form of a complex number $z = r(\cos \theta + i \sin \theta)$. An alternate form of the trigonometric form of a complex number is $z = r cis \theta$.

Rectangular Form of a Complex Number

$$z = a + bi$$

Trigonometric (Polar) Form of a Complex Number

$$z = r(cos\theta + isin\theta)$$

$$z = rcis\theta$$

r = modulus and θ = argument

▶ EXAMPLE 3

Write $z = 4 - 2i$ in trigonometric form. Round to the nearest tenth if necessary.

▶▶ Solution

We need to find the modulus and the argument for $z = 4 - 2i$.

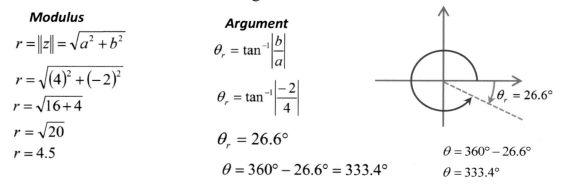

Modulus

$$r = \|z\| = \sqrt{a^2 + b^2}$$

$$r = \sqrt{(4)^2 + (-2)^2}$$

$$r = \sqrt{16 + 4}$$

$$r = \sqrt{20}$$

$$r = 4.5$$

Argument

$$\theta_r = \tan^{-1}\left|\frac{b}{a}\right|$$

$$\theta_r = \tan^{-1}\left|\frac{-2}{4}\right|$$

$$\theta_r = 26.6°$$

$$\theta = 360° - 26.6° = 333.4°$$

$\theta_r = 26.6°$

$\theta = 360° - 26.6°$

$\theta = 333.4°$

Trigonometric Form: $z = 4.5(\cos 333.4° + 333.4°i)$ or $z = 4.5 \text{cis} 333.4°$

▶ EXAMPLE 4

Write $z = 12(\cos 150° + i\sin 150°)$ in rectangular form. Write answer with exact values.

▶▶ Solution

$z = 12(\cos 150° + i\sin 150°)$ *Trigonometric Form*

$z = 12\cos 150° + i12\sin 150°$

$z = 12\left(-\dfrac{\sqrt{3}}{2}\right) + i12\left(\dfrac{1}{2}\right)$

$z = -6\sqrt{3} + 6i$ *Rectangular Form*

Products and Quotients in Trigonometric Form

In trigonometric form, multiplication and division of complex numbers is simple. Consider the complex numbers: $z_1 = r_1(\cos\theta_1 + i\sin\theta_1)$ and $z_2 = r_2(\cos\theta_2 + i\sin\theta_2)$. Multiplying the two gives us: $z_1 \cdot z_2 = r_1(\cos\theta_1 + i\sin\theta_1) \cdot r_2(\cos\theta_2 + i\sin\theta_2)$

$z_1 \cdot z_2 = r_1 \cdot r_2(\cos\theta_1 + i\sin\theta_1)(\cos\theta_2 + i\sin\theta_2)$

$z_1 \cdot z_2 = r_1 \cdot r_2(\cos\theta_1\cos\theta_2 + \cos\theta_1 i\sin\theta_2 + i\sin\theta_1\cos\theta_2 + i^2\sin\theta_1\sin\theta_2)$

$z_1 \cdot z_2 = r_1 \cdot r_2(\cos\theta_1\cos\theta_2 + \cos\theta_1 i\sin\theta_2 + i\sin\theta_1\cos\theta_2 - \sin\theta_1\sin\theta_2)$

$z_1 \cdot z_2 = r_1 \cdot r_2(\cos\theta_1\cos\theta_2 - \sin\theta_1\sin\theta_2 + \cos\theta_1 i\sin\theta_2 + i\sin\theta_1\cos\theta_2)$

$z_1 \cdot z_2 = r_1 \cdot r_2[(\cos\theta_1\cos\theta_2 - \sin\theta_1\sin\theta_2) + i(\cos\theta_1\sin\theta_2 + \sin\theta_1\cos\theta_2)]$

Using the sum identities for cosine and sine gives us $z_1 z_2 = r_1 r_2[\cos(\theta_1 + \theta_2) + i\sin(\theta_1 + \theta_2)]$.

The formula for multiplying two complex numbers is given by

$$z_1 z_2 = r_1 r_2 [\cos(\theta_1 + \theta_2) + i\sin(\theta_1 + \theta_2)]$$

From the proof, to find the multiplication of two complex numbers in trigonometric form, multiply the moduli and add the arguments.

We find the quotient of two complex numbers as follows given $z_1 = r_1(\cos\theta_1 + i\sin\theta_1)$ and $z_2 = r_2(\cos\theta_2 + i\sin\theta_2)$:

$$\frac{z_1}{z_2} = \frac{r_1(\cos\theta_1 + i\sin\theta_1)}{r_2(\cos\theta_2 + i\sin\theta_2)}$$

$$\frac{z_1}{z_2} = \frac{r_1(\cos\theta_1 + i\sin\theta_1)}{r_2(\cos\theta_2 + i\sin\theta_2)} \cdot \frac{(\cos\theta_2 - i\sin\theta_2)}{(\cos\theta_2 - i\sin\theta_2)}$$

$$\frac{z_1}{z_2} = \frac{r_1(\cos\theta_1\cos\theta_2 - i\cos\theta_1\sin\theta_2 + i\sin\theta_1\cos\theta_2 - i^2\sin\theta_1\sin\theta_2)}{r_2(\cos^2\theta_2 - i^2\sin^2\theta_2)}$$

$$\frac{z_1}{z_2} = \frac{r_1(\cos\theta_1\cos\theta_2 - i\cos\theta_1\sin\theta_2 + i\sin\theta_1\cos\theta_2 + \sin\theta_1\sin\theta_2)}{r_2(\cos^2\theta_2 + \sin^2\theta_2)}$$

$$\frac{z_1}{z_2} = \frac{r_1[(\cos\theta_1\cos\theta_2 + \sin\theta_1\sin\theta_2) + i(\sin\theta_1\cos\theta_2 - \cos\theta_1\sin\theta_2)]}{r_2(1)}$$

Using the sum identities for cosine and sine gives us: $\frac{z_1}{z_2} = \frac{r_1}{r_2}[\cos(\theta_1 - \theta_2) + i\sin(\theta_1 - \theta_2)]$.

The formula for dividing two complex numbers is given by

$$\frac{z_1}{z_2} = \frac{r_1}{r_2}[\cos(\theta_1 - \theta_2) + i\sin(\theta_1 - \theta_2)]$$

From the proof, to find the division of two complex numbers in trigonometric form, divide the moduli and subtract the arguments.

The Product and Quotient of Complex Numbers

If $z_1 = r_1(\cos\theta_1 + i\sin\theta_1)$ and $z_2 = r_2(\cos\theta_2 + i\sin\theta_2)$, then

$$z_1 z_2 = r_1 r_2 [\cos(\theta_1 + \theta_2) + i\sin(\theta_1 + \theta_2)]$$

$$z_1 z_2 = r_1 r_2 [cis(\theta_1 + \theta_2)]$$

$$\frac{z_1}{z_2} = \frac{r_1}{r_2}[\cos(\theta_1 - \theta_2) + i\sin(\theta_1 - \theta_2)]$$

$$\frac{z_1}{z_2} = \frac{r_1}{r_2}[cis(\theta_1 - \theta_2)]$$

► EXAMPLE 5

Find $z_1 \cdot z_2$ and $\dfrac{z_1}{z_2}$ given $z_1 = -6i\sqrt{2}$ and $z_2 = \dfrac{3\sqrt{2}}{2} + \dfrac{3\sqrt{6}}{2}i$. Write results in both rectangular form and trigonometric form. Give exact values (no decimals).

►► Solution

We need to get z_1 and z_2 in trigonometric form by finding their modulus and argument.

$$z_1 = 0 - 6i\sqrt{2}$$

$$r_1 = \sqrt{(0)^2 + \left(-6\sqrt{2}\right)^2}$$

$$r_1 = \sqrt{0 + 72}$$

$$r_1 = \sqrt{72}$$

$$r_1 = \sqrt{36 \cdot 2}$$

$$r_1 = 6\sqrt{2}$$

$$\theta_1 = tan^{-1}\left|\frac{-6\sqrt{2}}{0}\right|$$

$$\theta_1 = tan^{-1}(undefined)$$

$$\theta_1 = 270°$$

$$z_1 = 6\sqrt{2}cis270°$$

Trigonometric Form

$$z_2 = \frac{3\sqrt{2}}{2} + \frac{3\sqrt{6}}{2}i$$

$$r_2 = \sqrt{\left(\frac{3\sqrt{2}}{2}\right)^2 + \left(\frac{3\sqrt{6}}{2}\right)^2}$$

$$r_2 = \sqrt{\frac{18}{4} + \frac{54}{4}}$$

$$r_2 = \sqrt{\frac{72}{4}}$$

$$r_2 = \sqrt{18}$$

$$r_2 = 3\sqrt{2}$$

$$\theta_2 = tan^{-1}\left|\frac{\frac{3\sqrt{6}}{2}}{\frac{3\sqrt{2}}{2}}\right|$$

$$\theta_2 = tan^{-1}\left|\frac{3\sqrt{6}}{2} \cdot \frac{2}{3\sqrt{2}}\right|$$

$$\theta_2 = tan^{-1}\left|\sqrt{3}\right|$$

$$\theta_2 = 60°$$

$$z_2 = 3\sqrt{2}cis60°$$

Trigonometric Form

Continued on the next page.

▶ **EXAMPLE 5 (Continued)**

Find $z_1 \cdot z_2$ and $\dfrac{z_1}{z_2}$ given $z_1 = -6i\sqrt{2}$ and $z_2 = \dfrac{3\sqrt{2}}{2} + \dfrac{3\sqrt{6}}{2}i$. Write results in both rectangular form and trigonometric form. Give exact values (no decimals).

▶▶ **Solution**

$$z_1 = 0 - 6i\sqrt{2} \qquad\qquad\qquad z_2 = \frac{3\sqrt{2}}{2} + \frac{3\sqrt{6}}{2}i$$

$$z_1 = 6\sqrt{2}\,cis270° \qquad\qquad\qquad z_2 = 3\sqrt{2}\,cis60°$$

$$z_1 z_2 = r_1 r_2[\cos(\theta_1 + \theta_2) + i\sin(\theta_1 + \theta_2)]$$

$$z_1 z_2 = (6\sqrt{2})(3\sqrt{2})[\cos(270° + 60°) + i\sin(270° + 60°)]$$

$$z_1 z_2 = 36[\cos(330°) + i\sin(330°)]$$

$$z_1 z_2 = 36[cis(330°)]$$

— Trigonometric Forms

$$z_1 z_2 = 36[\cos(330°) + i\sin(330°)]$$

$$z_1 z_2 = 36\left[\frac{\sqrt{3}}{2} + \left(-\frac{1}{2}\right)i\right]$$

$$z_1 z_2 = 18\sqrt{3} - 18i \qquad \text{Rectangular Form}$$

$$\frac{z_1}{z_2} = \frac{r_1}{r_2}[\cos(\theta_1 - \theta_2) + i\sin(\theta_1 - \theta_2)]$$

$$\frac{z_1}{z_2} = \frac{6\sqrt{2}}{3\sqrt{2}}[\cos(270° - 60°) + i\sin(270° - 60°)]$$

$$\frac{z_1}{z_2} = 2[\cos(210°) + i\sin(210°)]$$

$$\frac{z_1}{z_2} = 2cis210°$$

— Trigonometric Forms

$$\frac{z_1}{z_2} = 2[\cos(210°) + i\sin(210°)]$$

$$\frac{z_1}{z_2} = 2\left[-\frac{\sqrt{3}}{2} + \left(-\frac{1}{2}\right)i\right]$$

$$\frac{z_1}{z_2} = -\sqrt{3} - i \qquad \text{Rectangular Form}$$

▶EXAMPLE 6

Find $z_1 \cdot z_2$ and $\dfrac{z_1}{z_2}$ given $z_1 = 7cis120°$ and $z_2 = 2cis300°$. Write results in both rectangular form and trigonometric form. Give exact values (no decimals).

▶▶Solution

$$z_1 = 7cis120° \qquad z_2 = 2cis300°$$

$$z_1 z_2 = r_1 r_2 [cis(\theta_1 + \theta_2)]$$

$$z_1 z_2 = (7)(2)[cis(120° + 300°)]$$

$$z_1 z_2 = 14[cis(420°)] \qquad \textit{The argument needs to be in the interval } [0, 2\pi) \textit{ or } [0°, 360°).$$
$$\textit{420° − 360° = 60°}$$

$$z_1 z_2 = 14[cis(60°)]$$

$$z_1 z_2 = 14[cos(60°) + isin(60°)] \quad \text{Trigonometric Forms}$$

$$z_1 z_2 = 14\left[\frac{1}{2} + \frac{\sqrt{3}}{2}i\right]$$

$$z_1 z_2 = 7 + 7\sqrt{3}i \qquad \text{Rectangular Form}$$

$$\frac{z_1}{z_2} = \frac{r_1}{r_2}[cis(\theta_1 - \theta_2)]$$

$$\frac{z_1}{z_2} = \frac{7}{2}[cis(120° - 300°)]$$

$$\frac{z_1}{z_2} = \frac{7}{2}[cis(-180°)] \quad \textit{The argument needs to be in the interval } [0, 2\pi) \textit{ or } [0°, 360°).$$
$$\textit{−180° + 360° = 180°}$$

$$\frac{z_1}{z_2} = \frac{7}{2}[cis(180°)]$$

$$\frac{z_1}{z_2} = \frac{7}{2}[cos(180°) + isin(180°)] \quad \text{Trigonometric Forms}$$

$$\frac{z_1}{z_2} = \frac{7}{2}[(-1) + (0)i]$$

$$\frac{z_1}{z_2} = -\frac{7}{2} \qquad \text{Rectangular Form}$$

► EXAMPLE 7

Find $z_1 \cdot z_2$ and $\dfrac{z_1}{z_2}$ given $z_1 = \sqrt{3}cis\dfrac{7\pi}{6}$ and $z_2 = \sqrt{2}cis\dfrac{5\pi}{3}$. Write results in both rectangular form and trigonometric form. Give exact values (no decimals).

►►Solution

$$z_1 z_2 = r_1 r_2 [cis(\theta_1 + \theta_2)]$$

$$z_1 z_2 = (\sqrt{3})(\sqrt{2})\left[cis\left(\frac{7\pi}{6} + \frac{5\pi}{3}\right)\right]$$

$$z_1 z_2 = \sqrt{6}\left[cis\left(\frac{7\pi}{6} + \frac{10\pi}{6}\right)\right]$$

$$z_1 z_2 = \sqrt{6}cis\frac{17\pi}{6} \qquad \textit{The argument needs to be in the interval } [0, 2\pi) \textit{ or } [0°, 360°).$$

$$z_1 z_2 = \sqrt{6}cis\frac{5\pi}{6} \left.\begin{array}{l} \\ \\ \\ \\ \end{array}\right\}$$

$$z_1 z_2 = \sqrt{6}\left(\cos\frac{5\pi}{6} + isin\frac{5\pi}{6}\right) \quad\rule[0.5ex]{0.5em}{0.4pt}\textbf{ Trigonometric Forms}$$

$$\frac{17\pi}{6} - 2\pi = \frac{17\pi}{6} - \frac{12\pi}{6} = \frac{5\pi}{6}$$

$$z_1 z_2 = \sqrt{6}\left(-\frac{\sqrt{3}}{2} + \frac{1}{2}i\right)$$

$$z_1 z_2 = -\frac{\sqrt{18}}{2} + \frac{\sqrt{6}}{2}i$$

$$z_1 z_2 = -\frac{3\sqrt{2}}{2} + \frac{\sqrt{6}}{2}i \quad \textbf{Rectangular Form}$$

$$\frac{z_1}{z_2} = \frac{r_1}{r_2}[cis(\theta_1 - \theta_2)]$$

$$\frac{z_1}{z_2} = \frac{\sqrt{3}}{\sqrt{2}}\left[cis\left(\frac{7\pi}{6} - \frac{5\pi}{3}\right)\right]$$

$$\frac{z_1}{z_2} = \frac{\sqrt{3}}{\sqrt{2}} \cdot \frac{\sqrt{2}}{\sqrt{2}}\left[cis\left(\frac{7\pi}{6} - \frac{10\pi}{6}\right)\right]$$

$$\frac{z_1}{z_2} = \frac{\sqrt{6}}{2}\left[cis\left(-\frac{3\pi}{6}\right)\right]$$

$$\frac{z_1}{z_2} = \frac{\sqrt{6}}{2}\left[cis\left(-\frac{\pi}{2}\right)\right]$$

The argument needs to be in the interval $[0, 2\pi)$ *or* $[0°, 360°)$.

$$-\frac{\pi}{2} + 2\pi = -\frac{\pi}{2} + \frac{4\pi}{2} = \frac{3\pi}{2}$$

Trigonometric Forms $\left\{\begin{array}{l} \\ \\ \\ \\ \end{array}\right.$

$$\frac{z_1}{z_2} = \frac{\sqrt{6}}{2}cis\frac{3\pi}{2}$$

$$\frac{z_1}{z_2} = \frac{\sqrt{6}}{2}\left(\cos\frac{3\pi}{2} + isin\frac{3\pi}{2}\right)$$

$$\frac{z_1}{z_2} = \frac{\sqrt{6}}{2}(0 + i(-1))$$

Rectangular Form $\quad \dfrac{z_1}{z_2} = -\dfrac{\sqrt{6}}{2}i$

► EXAMPLE 8

Perform the indicated operation: $\frac{1+i}{1-i}$. Write the results in trigonometric form and rectangular form. Round to the nearest tenth if necessary.

►►**Solution**

We need to convert the numerator and the denominator to trigonometric form.

$$1+i$$

$$r = \sqrt{(1)^2 + (1)^2}$$

$$r = \sqrt{1+1}$$

$$r = \sqrt{2}$$

$$\theta_r = tan^{-1}\left|\frac{1}{1}\right|$$

$$\theta_r = tan^{-1}|1|$$

$$\theta_r = 45°$$

$$\theta = 45°$$

$$\sqrt{2}cis45°$$

$$1-i$$

$$r = \sqrt{(1)^2 + (-1)^2}$$

$$r = \sqrt{1+1}$$

$$r = \sqrt{2}$$

$$\theta_r = tan^{-1}\left|\frac{-1}{1}\right|$$

$$\theta_r = tan^{-1}|1|$$

$$\theta_r = 45°$$

$$\theta = 315°$$

$$\sqrt{2}cis315°$$

$$\frac{1+i}{1-i} = \frac{\sqrt{2}cis45°}{\sqrt{2}cis315°} = \frac{\sqrt{2}}{\sqrt{2}}cis(45° - 315°)$$

$$= 1cis(-270°)$$

$$= 1cis(-270° + 360°)$$

$$= cis90°$$

$$= cos90° + isin90°$$ ⎤ Trigonometric Forms

$$= 0 + 1i$$

$$= i \quad \text{Rectangular Form}$$

▶EXAMPLE 9

Perform the indicated operation: $\frac{8}{\sqrt{3}+i}$**. Write the results in trigonometric form and rectangular form. Give exact values (no decimals).**

▶▶Solution

We need to convert the numerator and the denominator to trigonometric form.

$$8 + 0i$$

$$r = \sqrt{(8)^2 + (0)^2}$$

$$r = \sqrt{64 + 0}$$

$$r = \sqrt{64}$$

$$r = 8$$

$$\sqrt{3} + i$$

$$r = \sqrt{\left(\sqrt{3}\right)^2 + (1)^2}$$

$$r = \sqrt{3 + 1}$$

$$r = \sqrt{4}$$

$$r = 2$$

$$\theta_r = tan^{-1}\left|\frac{0}{8}\right|$$

$$\theta_r = tan^{-1}|0|$$

$$\theta_r = 0°$$

$$\theta = 0°$$

$$8cis0°$$

$$\theta_r = tan^{-1}\left|\frac{1}{\sqrt{3}}\right|$$

$$\theta_r = tan^{-1}\left(\frac{1}{\sqrt{3}}\right)$$

$$\theta_r = 30°$$

$$\theta = 30°$$

$$2cis30°$$

$$\frac{8}{\sqrt{3} + i} = \frac{8cis0°}{2cis30°} = \frac{8}{2}cis(0° - 30°)$$

$$= 4cis(-30°)$$

$$= 4cis(-30° + 360°)$$

$$= 4cis330°$$

$$= 4(cos330° + isin330°) \left.\right\}\ \text{Trigonometric Forms}$$

$$= 4\left(\frac{\sqrt{3}}{2} - \frac{1}{2}i\right)$$

$$= 2\sqrt{3} - 2i \quad \text{Rectangular Form}$$

De Moivre's Theorem

Multiplying a complex number $z = r(\cos\theta + i\sin\theta)$ by itself gives us

$$z \cdot z = [r(\cos\theta + i\sin\theta)] \cdot [r(\cos\theta + i\sin\theta)]$$

$$z^2 = r^2(\cos^2\theta + i\cos\theta\sin\theta + i\cos\theta\sin\theta + i^2\sin^2\theta)$$

$$z^2 = r^2[(\cos^2\theta + 2i\cos\theta\sin\theta - \sin^2\theta)]$$

$$z^2 = r^2[(\cos^2\theta - \sin^2\theta + 2i\cos\theta\sin\theta)]$$

$$z^2 = r^2[\cos(2\theta) + i\sin(2\theta)] \quad \longleftarrow$$

Multiply by $z = r(\cos\theta + i\sin\theta)$ again:

$$z^3 = z^2 \cdot z = r^2[(\cos2\theta + i\sin2\theta)] \cdot [r(\cos\theta + i\sin\theta)]$$

$$z^3 = r^3[\cos2\theta\cos\theta + i\cos2\theta\sin\theta + i\cos\theta\sin2\theta + i^2\sin2\theta\sin\theta]$$

$$z^3 = r^3[\cos2\theta\cos\theta + i\cos2\theta\sin\theta + i\cos\theta\sin2\theta - \sin2\theta\sin\theta]$$

$$z^3 = r^3[\cos2\theta\cos\theta - \sin2\theta\sin\theta + i(\cos2\theta\sin\theta + \cos\theta\sin2\theta)]$$

$$z^3 = r^3[\cos(2\theta + \theta) + i\sin(\theta + 2\theta)]$$

$$z^3 = r^3[\cos(3\theta) + i\sin(3\theta)] \quad \longleftarrow$$

If we continue the pattern we can conclude $z^n = r^n[\cos(n\theta) + i\sin(n\theta)]$. This is called De Moivre's Theorem which is named after Abraham De Moivre (1667-1754) a French mathematician.

De Moivre's Theorem

If $z = r(\cos\theta + i\sin\theta)$ is a complex number and n is a positive integer, then

$$z^n = [r(\cos\theta + i\sin\theta)]^n = r^n[\cos(n\theta) + i\sin(n\theta)].$$

► EXAMPLE 10

Use De Moivre's Theorem to find $(-2 + 2i)^6$. Write the result in both rectangular form and trigonometric form. Give exact values (no decimals).

►►Solution

We need to convert $z = -2 + 2i$ from rectangular form to trigonometric form.

$$z = -2 + 2i$$

$$r = \sqrt{(-2)^2 + (2)^2}$$ $$\theta_r = \tan^{-1}\left|\frac{2}{-2}\right|$$

$$r = \sqrt{4+4}$$ $$\theta_r = \tan^{-1}1$$

$$r = \sqrt{8}$$ $$\theta_r = 45°$$

$$r = \sqrt{4 \cdot 2}$$ $$\theta = 135°$$

$$r = 2\sqrt{2}$$

180°-45°=135°

45°

$$z = 2\sqrt{2}\,cis135°$$

$$z^6 = \left(2\sqrt{2}\,cis135°\right)^6$$

$$z^6 = \left(2\sqrt{2}\right)^6 cis(6 \cdot 135°)$$

$$z^6 = 512cis(810°)$$ *The argument needs to be in the interval $[0, 2\pi)$ or $[0°, 360°)$.*

$$z^6 = 512cis90°$$

$$z^6 = 512(cos90° + isin90°)$$ — Trigonometric Forms

$$810° - 360° = 450°$$
$$450° - 360° = 90°$$

$$z^6 = 512(0 + i)$$

$$z^6 = 512i$$ **Rectangular Form**

► EXAMPLE 11

Use De Moivre's Theorem to find $\left[3\left(cos\frac{2\pi}{3} + isin\frac{2\pi}{3}\right)\right]^6$. **Write the result in both rectangular form and trigonometric form. Give exact values (no decimals).**

►► Solution

$$\left[3\left(cos\frac{2\pi}{3} + isin\frac{2\pi}{3}\right)\right]^6$$

$$3^6\left[cos\left(6\cdot\frac{2\pi}{3}\right) + isin\left(6\cdot\frac{2\pi}{3}\right)\right]$$

$729[cos(4\pi) + isin(4\pi)]$ *The argument needs to be in the interval* $[0, 2\pi)$ *or* $[0°, 360°)$.

$\left.\begin{array}{l} 729[cos0 + isin0] \\ 729cis0 \end{array}\right\}$ Trigonometric Forms
$$\begin{array}{l} 4\pi - 2\pi = 2\pi \\ 2\pi - 2\pi = 0 \end{array}$$

$729[1 + i(0)]$

729 Rectangular Form

Roots of Complex Numbers

Recall that by the Fundamental Theorem of Algebra, a polynomial of degree n has n solutions. We define the nth root of a complex number in the following chart.

---

nth Roots of Complex Numbers

For $z = r(cos\theta + isin\theta)$ or $z = rcis\theta$ a positive integer n, and r is a real number, z has exactly n distinct nth roots determined by

$$\sqrt[n]{z} = \sqrt[n]{r}\left[cos\left(\frac{\theta + 360°k}{n}\right) + isin\left(\frac{\theta + 360°k}{n}\right)\right]$$

$$\sqrt[n]{z} = \sqrt[n]{r}\left[cis\left(\frac{\theta + 360°k}{n}\right)\right]$$

$$\sqrt[n]{z} = \sqrt[n]{r}\left[cos\left(\frac{\theta + 2\pi k}{n}\right) + isin\left(\frac{\theta + 2\pi k}{n}\right)\right]$$

$$\sqrt[n]{z} = \sqrt[n]{r}\left[cis\left(\frac{\theta + 2\pi k}{n}\right)\right]$$

for k=0, 1, 2, 3, . . ., n-1

► EXAMPLE 12

Solve $x^3 + 64i = 0$ using the nth roots theorem. Write answers in trigonometric form.

Give exact values (no decimals).

►► Solution

Solve for x^3, $x^3 = -64i$. Write $-64i$ in trigonometric form.

$r = \sqrt{(0)^2 + (-64)^2}$

$r = \sqrt{0 + 4096}$

$r = \sqrt{4096}$

$r = 64$

$\theta_r = \tan^{-1}\left|\dfrac{-64}{0}\right|$

$\theta_r = \tan^{-1}(\text{undefined})$

$\theta = 270°$

$-64i = 64cis270°$

$$\sqrt[n]{z} = \sqrt[n]{r}\left[cis\left(\frac{\theta + 360°k}{n}\right)\right] \text{ for } k=0, 1, 2, 3, \ldots, n\text{-}1$$

We are finding the cube roots, so $n = 3$. The $\sqrt[n]{r} = \sqrt[3]{64} = 4$ and for $k = 0,1,2,\ldots,n-1$ we have $k = 0,1,2$.

$k = 0$ $4\left[cis\left(\frac{270° + 360°(0)}{3}\right)\right] = 4cis\left(\frac{270°}{3}\right) = 4cis90°$

$k = 1$ $4\left[cis\left(\frac{270° + 360°(1)}{3}\right)\right] = 4cis\left(\frac{270° + 360°}{3}\right) = 4cis\left(\frac{630°}{3}\right) = 4cis210°$

$k = 2$ $4\left[cis\left(\frac{270° + 360°(2)}{3}\right)\right] = 4cis\left(\frac{270° + 720°}{3}\right) = 4cis\left(\frac{990°}{3}\right) = 4cis330°$

The three roots are $4cis90°$, $4cis210°$, and $4cis330°$.

Notice the difference between the arguments is $120°$. The difference between the arguments of the roots can be found by $\dfrac{360°}{n}$ or $\dfrac{2\pi}{n}$ where n is the root. The arguments of the solutions should be in the interval $[0°, 360°)$ or $[0, 2\pi)$.

▶EXAMPLE 13

Solve $x^5 - 1 + i\sqrt{3} = 0$ **using the *n*th roots theorem. Write answers in trigonometric form. Give exact values (no decimals).**

▶▶**Solution**

Solve for x^5, $x^5 = 1 - i\sqrt{3}$. Write $1 - i\sqrt{3}$ in trigonometric form.

$$r = \sqrt{(1)^2 + \left(-\sqrt{3}\right)^2}$$

$$r = \sqrt{1+3}$$

$$r = \sqrt{4}$$

$$r = 2$$

$$1 - i\sqrt{3} = 2cis300°$$

$$\theta_r = \tan^{-1}\left|\frac{-\sqrt{3}}{1}\right|$$

$$\theta_r = \tan^{-1}\sqrt{3}$$

$$\theta_r = 60°$$

$$\theta = 300°$$

$$\sqrt[n]{z} = \sqrt[n]{r}\left[cis\left(\frac{\theta + 360°k}{n}\right)\right] \text{ for } k=0, 1, 2, 3, \ldots, n\text{-}1$$

We are finding the fifth roots, so $n = 5$. The $\sqrt[n]{r} = \sqrt[5]{2}$ and for $k = 0,1,2,\ldots,n-1$ we have $k = 0,1,2,3,4$. The difference between the arguments is $\frac{360°}{5} = 72°$.

$$k = 0 \quad \sqrt[5]{2}\left[cis\left(\frac{300°+360°(0)}{5}\right)\right] = \sqrt[5]{2}cis\left(\frac{300°}{5}\right) = \sqrt[5]{2}cis60°$$

$$k = 1 \quad \sqrt[5]{2}\left[cis\left(\frac{300°+360°(1)}{5}\right)\right] = \sqrt[5]{2}cis\left(\frac{300°+360°}{5}\right) = \sqrt[5]{2}cis\left(\frac{660°}{5}\right) = \sqrt[5]{2}cis132°$$

$$k = 2 \quad \sqrt[5]{2}\left[cis\left(\frac{300°+360°(2)}{5}\right)\right] = \sqrt[5]{2}cis\left(\frac{300°+720°}{5}\right) = \sqrt[5]{2}cis\left(\frac{1020°}{5}\right) = \sqrt[5]{2}cis204°$$

$$k = 3 \quad \sqrt[5]{2}\left[cis\left(\frac{300°+360°(3)}{5}\right)\right] = \sqrt[5]{2}cis\left(\frac{300°+1080°}{5}\right) = \sqrt[5]{2}cis\left(\frac{1380°}{5}\right) = \sqrt[5]{2}cis276°$$

$$k = 4 \quad \sqrt[5]{2}\left[cis\left(\frac{300°+360°(4)}{5}\right)\right] = \sqrt[5]{2}cis\left(\frac{300°+1440°}{5}\right) = \sqrt[5]{2}cis\left(\frac{1740°}{5}\right) = \sqrt[5]{2}cis348°$$

The five roots are $\sqrt[5]{2}cis60°$, $\sqrt[5]{2}cis132°$, $\sqrt[5]{2}cis204°$, $\sqrt[5]{2}cis276°$ and $\sqrt[5]{2}cis348°$.

► **EXAMPLE 14**

Find the fourth roots of $z = -8\sqrt{3} - 8i$ using the *n*th roots theorem. Write answers in trigonometric form.

►►**Solution**

Write $z = -8\sqrt{3} - 8i$ in trigonometric form.

$$r = \sqrt{\left(-8\sqrt{3}\right)^2 + \left(-8\right)^2}$$

$$r = \sqrt{192 + 64}$$

$$r = \sqrt{256}$$

$$r = 16$$

$$\theta_r = \tan^{-1}\left|\frac{-8}{-8\sqrt{3}}\right|$$

$$\theta_r = \tan^{-1}\frac{1}{\sqrt{3}}$$

$$\theta_r = 30°$$

$$\theta = 210°$$

$$z = -8\sqrt{3} - 8i = 16cis210°$$

$$\sqrt[n]{z} = \sqrt[n]{r}\left[cis\left(\frac{\theta + 360°k}{n}\right)\right] \text{ for } k=0, 1, 2, 3, \ldots, n\text{-}1$$

We are finding the fifth roots, so $n = 4$. The $\sqrt[n]{r} = \sqrt[4]{16} = 2$ and for $k = 0,1,2,\ldots,n-1$ we have $k = 0,1,2,3$. The difference between the arguments is $\frac{360°}{4} = 90°$.

$$k = 0 \quad 2\left[cis\left(\frac{210°+360°(0)}{4}\right)\right] = 2cis\left(\frac{210°}{4}\right) = 2cis52.5°$$

$$k = 1 \quad 2\left[cis\left(\frac{210°+360°(1)}{4}\right)\right] = 2cis\left(\frac{210°+360°}{4}\right) = 2cis\left(\frac{570°}{4}\right) = 2cis142.5°$$

$$k = 2 \quad 2\left[cis\left(\frac{210°+360°(2)}{4}\right)\right] = 2cis\left(\frac{210°+720°}{4}\right) = 2cis\left(\frac{930°}{4}\right) = 2cis232.5°$$

$$k = 3 \quad 2\left[cis\left(\frac{210°+360°(3)}{4}\right)\right] = 2cis\left(\frac{210°+1080°}{4}\right) = 2cis\left(\frac{1290°}{4}\right) = 2cis322.5°$$

The four roots are $2cis52.5°$, $2cis142.5°$, $2cis232.5°$ and $2cis322.5°$.

HOMEWORK

Objective 1

Graph the complex numbers.

See Example 1.

1. $4 - 3i$

2. $-2 - 4i$

3. $-4 + i$

4. $5 + 3i$

Objectives 2 and 3

Write the complex number in trigonometric form. Round to the nearest tenth if necessary.

See Examples 2 and 3.

5. $-2 - 2i$

6. $2 - 2\sqrt{3}i$

7. $-4\sqrt{3} + 4i$

8. $-5\sqrt{2} + 5\sqrt{2}i$

9. $-8 + 15i$

10. $9 - 12i$

11. $-5 - 12i$

12. $6 + 8i$

13. $6 - 10i$

Objective 4

Write the complex number in rectangular form.

See Example 4.

14. $2cis120°$

15. $8cis\frac{4\pi}{3}$

16. $4\sqrt{3}cis315°$

17. $10[cos210° + isin210°]$

18. $10[cos135° + isin135°]$

19. $6\left[cos\frac{5\pi}{6} + isin\frac{5\pi}{6}\right]$

20. $4\left[cos\frac{11\pi}{6} + isin\frac{11\pi}{6}\right]$

21. $2[cos300° + isin300°]$

22. $6\left[cos\frac{\pi}{3} + isin\frac{\pi}{3}\right]$

Objective 4

Find $z_1 \cdot z_2$ and $\frac{z_1}{z_2}$. Write answers in trigonometric form. Give exact values (no decimals).

See Examples 5, 6 and 7.

23. $z_1 = 10cis60°, \ z_2 = 4cis30°$

24. $z_1 = 7cis120°, \ z_2 = 2cis300°$

25. $z_1 = 5\sqrt{2}[cos30° + isin30°], \ z_2 = 2\sqrt{2}[cos210° + isin210°]$

26. $z_1 = 6\sqrt{3}\left[cos\frac{\pi}{6} + isin\frac{\pi}{6}\right], \ z_2 = 4\sqrt{3}\left[cos\frac{\pi}{3} + isin\frac{\pi}{3}\right]$

27. $z_1 = 12cis\frac{5\pi}{3}, \ z_2 = 4cis\frac{4\pi}{3}$

28. $z_1 = \sqrt{6}cis\frac{5\pi}{6}, \ z_2 = \sqrt{2}cis\frac{\pi}{6}$

29. $z_1 = 8\left[cos\frac{\pi}{4} + isin\frac{\pi}{4}\right], \ z_2 = 2\left[cos\frac{7\pi}{4} + isin\frac{7\pi}{4}\right]$

30. $z_1 = 5\sqrt{3} + 5i, \ z_2 = \frac{5}{2} + \frac{5\sqrt{3}}{2}i$

31. $z_1 = -4 - 4\sqrt{3}i, \ z_2 = -2\sqrt{3} + 2i$

Objective 4

Find $z_1 \cdot z_2$ and $\frac{z_1}{z_2}$. Write answers in rectangular form. Give exact values (no decimals).

See Examples 5, 6, 7, 8 and 9.

32. $z_1 = 5cis90°, \ z_2 = 10cis30°$

33. $z_1 = 12cis150°, \ z_2 = 2cis300°$

34. $z_1 = 8\sqrt{2}[cos60° + isin60°], \ z_2 = 2\sqrt{2}[cos240° + isin240°]$

35. $z_1 = 6\sqrt{6}[cos150° + isin150°], \ z_2 = 2\sqrt{2}[cos30° + isin30°]$

36. $z_1 = 8cis\frac{\pi}{3}, \ z_2 = 4cis\frac{4\pi}{3}$

37. $z_1 = \sqrt{6}cis\frac{11\pi}{6}, \ z_2 = \sqrt{3}cis\frac{7\pi}{6}$

38. $z_1 = 16\left[cos\frac{3\pi}{4} + isin\frac{3\pi}{4}\right], \ z_2 = 2\left[cos\frac{7\pi}{4} + isin\frac{7\pi}{4}\right]$

39. $z_1 = -\sqrt{3} + i, \ z_2 = -\frac{3}{2} + \frac{3\sqrt{3}}{2}i$

40. $z_1 = \frac{9}{2} + \frac{9\sqrt{3}}{2}i, \ z_2 = \frac{3\sqrt{3}}{2} + \frac{3}{2}i$

Objective 4

Perform the indicated operation. Write answers in trigonometric form and rectangular form. Give exact values (no decimals).

See Examples 5, 6, 7, 8 and 9.

41. $\dfrac{2i}{-1-i\sqrt{3}}$

42. $\dfrac{10(\cos 50°+i\sin 50°)}{5(\cos 230°+i\sin 230°)}$

43. $[4(cis112°)][6(cis38°)]$

44. $\left[2\left(cis\dfrac{5\pi}{3}\right)\right]\left[3\left(cis\dfrac{\pi}{6}\right)\right]$

45. $\dfrac{2-2i}{1-i}$

Objective 5

Use De Moivre's theorem. Write the result in both rectangular form and trigonometric form. Give exact values (no decimals).

See Examples 10 and 11.

46. $\left(\sqrt{3}+i\right)^5$

47. $\left(\dfrac{\sqrt{2}}{2}-\dfrac{\sqrt{2}}{2}i\right)^8$

48. $(-1-i)^6$

49. $\left(\dfrac{\sqrt{3}}{2}-\dfrac{1}{2}i\right)^4$

50. $\left(-1-\sqrt{3}i\right)^4$

51. $(-2+2i)^3$

52. $\left(\dfrac{\sqrt{2}}{2}cis135°\right)^6$

53. $(4cis150°)^4$

54. $(3cis330°)^3$

55. $\left(\sqrt{2}cis60°\right)^8$

Objective 6

Solve each equation using the nth roots theorem. Write answers in trigonometric form. Give exact values (no decimals).

See Examples 12, 13 and 14.

56. $x^5-32i=0$

57. $x^3-27=0$

58. $x^3+64i=0$

59. $x^5+1-\sqrt{3}i=0$

60. $x^5+243=0$

61. $x^4-16=0$

62. $x^4+81i=0$

63. $x^6-64i=0$

Objective 6

Find the nth roots indicated. Write answers in trigonometric form. Give exact values (no decimals).

See Examples 12, 13 and 14.

64.   Find the fourth roots of $-16i$.

65.   Find the cube roots of $-1 + \sqrt{3}i$.

66.   Find the cube roots of 8.

67.   Find the fourth roots of $4 + 4\sqrt{3}i$.

68.   Find the fifth roots of i.

ANSWERS

1.

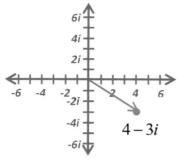

$4 - 3i$

2.

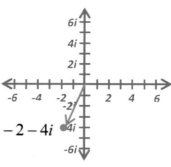

$-2 - 4i$

3.

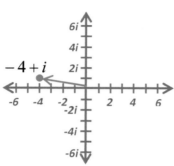

$-4 + i$

4.

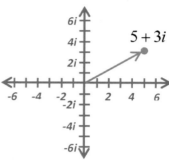

$5 + 3i$

5. $2.8cis225°$	6. $4cis300°$	7. $8cis150°$
8. $10cis135°$	9. $17cis118.1°$	10. $15cis306.9°$
11. $13cis247.4°$	12. $10cis53.1°$	13. $11.7cis301.0°$
14. $-1 + \sqrt{3}i$	15. $-4 - 4\sqrt{3}i$	16. $2\sqrt{6} - 2\sqrt{6}i$
17. $-5\sqrt{3} - 5i$	18. $-5\sqrt{2} + 5\sqrt{2}i$	19. $-3\sqrt{3} + 3i$
20. $2\sqrt{3} - 2i$	21. $1 - \sqrt{3}i$	22. $3 + 3\sqrt{3}i$

23. $40cis90°, \frac{5}{2}cis30°$

24. $14cis60°, \frac{7}{2}cis180°$

25. $20cis240°, \frac{5}{2}cis180°$

26. $72cis\frac{\pi}{2}, \frac{3}{2}cis\frac{11\pi}{6}$

27. $48cis\pi, 3cis\frac{\pi}{3}$

28. $2\sqrt{3}cis\pi, \sqrt{3}cis\frac{2\pi}{3}$

29. $16cis0, 4cis\frac{\pi}{2}$

30. $50cis90°, 2cis330°$

31. $32cis30°, 2cis90°$

32. $-25 + 25\sqrt{3}i, \frac{1}{4} + \frac{\sqrt{3}}{4}i$

33. $24i, -3\sqrt{3} - 3i$

34. $16 - 16\sqrt{3}i, -4$

35. $-24\sqrt{3}, -\frac{3\sqrt{3}}{2} + \frac{9}{2}i$

36. $16 - 16\sqrt{3}i, -2$

37. $-3\sqrt{2}, -\frac{\sqrt{2}}{2} + \frac{\sqrt{6}}{2}i$

38. $32i, -8$

39. $-6i, \frac{\sqrt{3}}{3} + \frac{1}{3}i$

40. $27i, \frac{3\sqrt{3}}{2} + \frac{3}{2}i$

41. $cis210°, -\frac{\sqrt{3}}{2} - \frac{1}{2}i$

42. $2cis180°, -2$

43. $24cis150°, -12\sqrt{3} + 12i$

44. $6cis\frac{11\pi}{6}, 3\sqrt{3} - 3i$

45. $2cis0°, 2$

46. $32cis150°, -16\sqrt{3} + 16i$

47. $cis0°, 1$

48. $8cis270°, -8i$

49. $cis240°, -\frac{1}{2} - \frac{\sqrt{3}}{2}i$

50. $16cis240°, -8 - 8\sqrt{3}i$

51. $16\sqrt{2}cis45°, 16 + 16i$

52. $\frac{1}{8}cis90°, \frac{1}{8}i$

53. $256cis240°, -128 - 128\sqrt{3}i$

54. $27cis270°, -27i$

55. $16cis120°, -8 + 8\sqrt{3}i$

56. $2cis18°, 2cis90°, 2cis162°, 2cis234°, 2cis306°$

57. $3cis0°, 3cis120°, 3cis240°$

58. $4cis90°, 4cis210°, 4cis330°$

59. $\sqrt[5]{2}cis24°, \sqrt[5]{2}cis96°, \sqrt[5]{2}cis168°, \sqrt[5]{2}cis240°, \sqrt[5]{2}cis312°$

60. $3cis36°, 3cis108°, 3cis180°, 3cis252°, 2cis324°$

61. $2cis0°, 2cis90°, 2cis180°, 2cis270°$

62. $3cis67.5°, 3cis157.5°, 3cis247.5°, 3cis337.5°$

63. $2cis15°, 2cis75°, 2cis135°, 2cis195°, 2cis255°, 2cis315°$

64. $2cis67.5°, 2cis157.5°, 2cis247.5°, 2cis337.5°$

65. $\sqrt[3]{2}cis40°, \sqrt[3]{2}cis160°, \sqrt[3]{2}cis280°$

66. $2cis0°, 2cis120°, 2cis240°$

67. $\sqrt[4]{8}cis15°, \sqrt[4]{8}cis105°, \sqrt[4]{8}cis195°, \sqrt[4]{8}cis285°$

68. $cis18°, cis90°, cis162°, cis234°, cis306°$

Section 22:   Polar Equations

Learning Outcomes:

- The student will illustrate and examine the graphs of: the six trigonometric functions and their inverses, parametric equations, and polar equations.
- The student will correctly memorize and apply trigonometric formulas, definitions, identities, and properties.

Objectives: At the conclusion of this lesson you should be able to:

1. Graph polar points.
2. Convert from rectangular coordinates to polar coordinates.
3. Convert from polar coordinates to rectangular coordinates.
4. Convert from rectangular equations to polar equations.
5. Convert from polar equations to rectangular equations.
6. Test for symmetry with respect about the polar axis, about the pole and about the line $\theta = \frac{\pi}{2}$.
7. Graph polar equations.

Polar Coordinate System

The polar coordinate system uses angles and directed distances to specify the location of a point in the plane. To form the polar coordinate system in the plane, we choose a fixed point O, called the pole (or origin), and draw from O a ray (half line) in the direction of the positive x-axis called the polar axis. See **Figure 22.1**. We can assign each point P in the plane a polar coordinate (r, θ) where r is the directed distance from O to P and θ is the directed angle, counterclockwise from the polar axis to the line segment $\overline{OP}$, as shown in **Figure 22.2**.

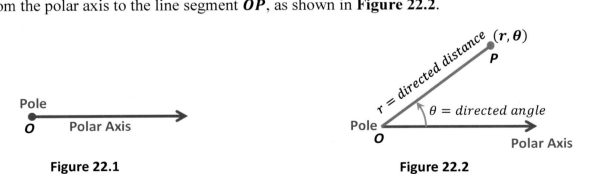

Pole
O   Polar Axis

Figure 22.1

Figure 22.2

When we plot a polar coordinate $P(r, \theta)$ in the polar coordinate plane, we go a distance of $|r|$ at $0°$ then move $\theta°$ counterclockwise along a circle of radius r. If $r > 0$, the point P lies on the terminal side of θ and if $r < 0$, the point P lies on the circle of same radius but $180°$ in the opposite direction as shown in **Figure 22.3**. Consider the polar coordinates $P(2, 60°)$ and

$Q(-2,60°)$, as shown in **Figure 22.4.** They are graphed by starting at the pole and going a distance of 2 and moving counterclockwise along the circle to 60°. The polar coordinate $P(2,60°)$ has $r > 0$, plot the point P. Since the polar coordinate $Q(-2,60°)$ has $r < 0$, we go 180° in the opposite direction (notice this puts at 240°) and with the same r value of 2 we plot the point Q.

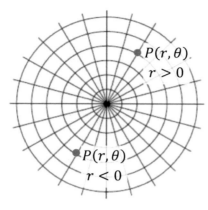

Figure 22.3

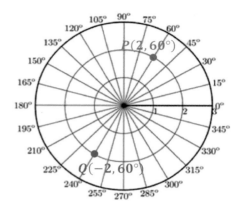

Figure 22.4

► EXAMPLE 1

Plot $A(3,135°)$, $B(-2,30°)$, $C(1,-45°)$, **and** $D(-3,-90°)$.

►►Solution

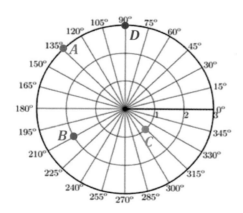

$A(3,135°)$ -- Go 3 units at 0°, then rotate 135° counterclockwise.

$B(-2,30°)$-- Go 2 units at 0°, then rotate 30° counterclockwise. Since $r < 0$, go 180° in the opposite direction.

$C(1,-45°)$-- Go 1 unit at 0°, then rotate 45° clockwise.

$D(-3,-90°)$--Go 3 units at 0°, then rotate 90° clockwise. Since $r < 0$, go 180° in the opposite direction.

► EXAMPLE 2

Plot $A\left(4, \frac{2\pi}{3}\right)$, $B\left(-2, \frac{5\pi}{6}\right)$, $C\left(-3, -\frac{3\pi}{4}\right)$, **and** $D\left(-5, -\frac{5\pi}{3}\right)$.

►►Solution

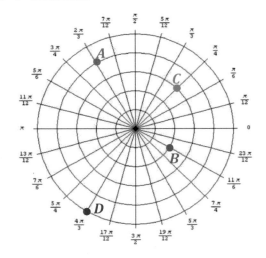

$A\left(4, \frac{2\pi}{3}\right)$--Go 4 units at 0, then rotate $\frac{2\pi}{3}$ counterclockwise.

$B\left(-2, \frac{5\pi}{6}\right)$--Go 2 units at 0, then rotate $\frac{5\pi}{6}$ counterclockwise. Since $r < 0$, go π in the opposite direction.

$C\left(-3, -\frac{3\pi}{4}\right)$--Go 3 units at 0, then rotate $\frac{3\pi}{4}$ clockwise.

$D\left(-5, -\frac{5\pi}{3}\right)$--Go 5 units at 0, then rotate $\frac{5\pi}{3}$ clockwise. Since $r < 0$, go π in the opposite direction.

► EXAMPLE 3

Plot $(2, 150°)$ and find three additional polar representations of this point using $-360° < \theta < 360°$.

►►Solution

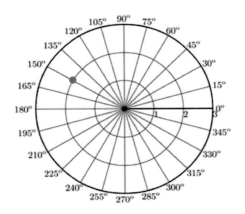

$(2, 150°)$--Go 2 units at 0°, then rotate 150° counterclockwise.

$(2, -210°)$--Subtract 360° from θ.

$(-2, 330°)$--Replace r with $-r$ and add 180° to θ.

$(-2, -30°)$--Replace r with $-r$ and subtract 180° from θ.

If we place the pole at the origin of a rectangular coordinate system so that the polar axis coincides with the positive x-axis, we have what is shown in **Figure 22.5**. Graphing the point $P(r, \theta)$ gives us the graph shown in **Figure 22.6**.

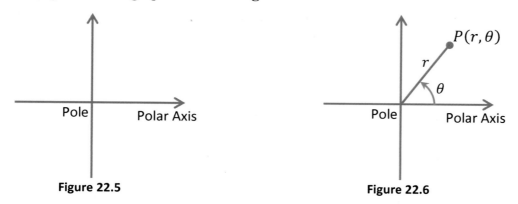

Figure 22.5

Figure 22.6

Converting from Rectangular Coordinates to Polar Coordinates

The polar form of a rectangular coordinate can be found by dropping the perpendicular from the point $P(x, y)$ to the x-axis, as shown in **Figure 22.7**. The horizontal distance is x and the vertical distance is y. The directed distance, r, is found using the Pythagorean Theorem $r = \sqrt{x^2 + y^2}$. The directed angle θ formed with the x-axis is found using $\theta = \tan^{-1}\left|\dfrac{y}{x}\right|$.

$$r = \sqrt{x^2 + y^2}$$

$$\theta = \tan^{-1}\left|\frac{y}{x}\right|$$

Figure 22.7

▶ EXAMPLE 4

Convert $\left(6, -6\sqrt{3}\right)$ to polar coordinates, where $\theta \in [0°, 360°)$.

▶▶ Solution

$$r = \sqrt{x^2 + y^2}$$

$$r = \sqrt{(6)^2 + \left(-6\sqrt{3}\right)^2}$$

$$r = \sqrt{36 + 108}$$

$$r = \sqrt{144}$$

$$r = 12$$

$$\theta = \tan^{-1}\left|\frac{y}{x}\right|$$

$$\theta = \tan^{-1}\left|\frac{-6\sqrt{3}}{6}\right|$$

$$\theta = \tan^{-1}\left|\sqrt{3}\right|$$

$$\theta = 60° \quad \textit{Quadrant I}$$

$$\theta = 300° \quad \textit{Quadrant IV}$$

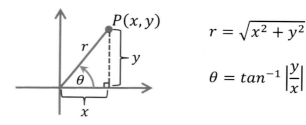

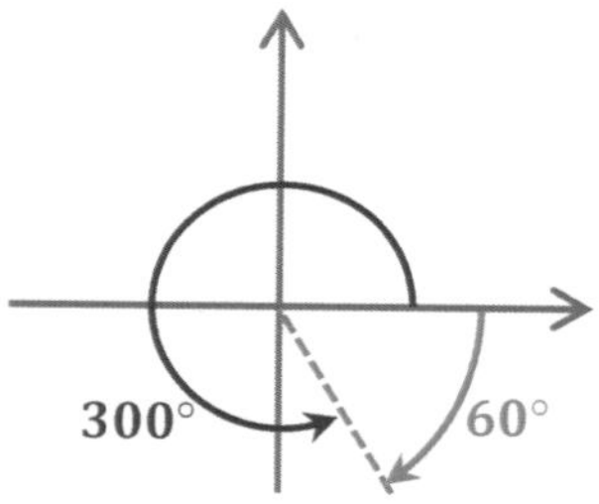

$$360° - 60° = 300°$$

Converting from Polar Coordinates to Rectangular Coordinates

To convert from a polar coordinate to a rectangular coordinate, we drop the perpendicular from the point $P(r, \theta)$ to the x-axis, as shown in **Figure 22.8**. The horizontal distance x can be found by using cosine and the vertical distance y can be found by using sine.

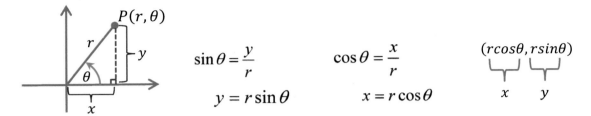

$$\sin \theta = \frac{y}{r}$$

$$y = r \sin \theta$$

$$\cos \theta = \frac{x}{r}$$

$$x = r \cos \theta$$

$$(r\cos\theta, r\sin\theta)$$

$$x \qquad y$$

Figure 22.8

▶EXAMPLE 5

Convert $\left(-8, \frac{2\pi}{3}\right)$ to rectangular coordinates. Write answer with exact values (no decimals).

▶▶Solution

$$(r\cos\theta, r\sin\theta)$$

$$\left(-8\cos\frac{2\pi}{3}, -8\sin\frac{2\pi}{3}\right)$$

$$\left(-8\left(-\frac{1}{2}\right), -8\left(\frac{\sqrt{3}}{2}\right)\right)$$

$$\left(4, -4\sqrt{3}\right)$$

Relationship between Polar and Rectangular Coordinates

1. To change from rectangular to polar coordinates, use

$$r = \sqrt{x^2 + y^2} \text{ and } \theta = tan^{-1}\left|\frac{y}{x}\right|$$

2. To change from polar to rectangular coordinates, use

$$x = rcos\theta \text{ and } y = rsin\theta$$

Equation Conversion

An equation that contains the variables x and y is a rectangular (or Cartesian) equation and an equation that contains the variables r and θ is a polar equation. We can convert a rectangular equation to a polar equation by recalling that $x = r\cos\theta$ and $y = r\sin\theta$. To convert from a polar equation to a rectangular equation we recall that $r = \sqrt{x^2 + y^2}$ which implies the form $r^2 = x^2 + y^2$ and $\theta = tan^{-1}\left|\frac{y}{x}\right|$ which implies $tan\theta = \frac{y}{x}$.

$$\boxed{\begin{array}{c} \textit{Equation Conversion} \\[2mm] x = r\cos\theta \quad\quad y = r\sin\theta \\[2mm] r = \sqrt{x^2 + y^2} \ \text{ or } \ r^2 = x^2 + y^2 \\[2mm] \theta = tan^{-1}\left|\frac{y}{x}\right| \ \text{ or } \ tan\theta = \frac{y}{x} \end{array}}$$

▶EXAMPLE 6

Convert the rectangular equation $2x - 3y + 4 = 0$ to a polar equation.

▶▶Solution

$$2x - 3y + 4 = 0 \quad \text{Rectangular Equation}$$

$$2(r\cos\theta) - 3(r\sin\theta) + 4 = 0 \quad \text{Replace } x \text{ with } r\cos\theta \text{ and } y \text{ with } r\sin\theta.$$

$$2r\cos\theta - 3r\sin\theta = -4$$

$$r(2\cos\theta - 3\sin\theta) = -4 \quad \text{Factor out the } r.$$

$$\frac{r(2\cos\theta - 3\sin\theta)}{2\cos\theta - 3\sin\theta} = \frac{-4}{2\cos\theta - 3\sin\theta} \quad \text{Solve for } r.$$

$$r = \frac{-4}{2\cos\theta - 3\sin\theta} \quad \text{Polar Equation}$$

▶EXAMPLE 7

Convert the rectangular equation $x^2 + y^2 = 16$ to a polar equation.

▶▶Solution

$$x^2 + y^2 = 16 \quad \text{Rectangular Equation}$$

$$r^2 = 16 \quad \text{Replace } x^2 + y^2 \text{ with } r^2.$$

$$\sqrt{r^2} = \pm\sqrt{16} \quad \text{Take the square root of both}$$

$$r = \pm 4 \quad \text{Polar Equation}$$

► EXAMPLE 8

Convert the rectangular equation $2xy = 3$ to a polar equation.

►►Solution

$$2xy = 3 \quad \text{Rectangular Equation}$$

$$2(r\cos\theta)(r\sin\theta) = 3 \quad \text{Replace } x \text{ with } r\cos\theta \text{ and } y \text{ with } r\sin\theta.$$

$$2r^2\cos\theta\sin\theta = 3$$

$$r^2(2\cos\theta\sin\theta) = 3 \quad \text{Factor out the } r^2.$$

$$r^2(\sin2\theta) = 3 \quad \text{Replace } 2\cos\theta\sin\theta \text{ with the double angle identity } \sin2\theta.$$

$$\frac{r^2(\sin2\theta)}{\sin2\theta} = \frac{3}{\sin2\theta} \quad \text{Solve for } r^2.$$

$$r^2 = \frac{3}{\sin2\theta}$$

$$r^2 = 3\csc2\theta \quad \text{Polar Equation}$$

► EXAMPLE 9

Convert the rectangular equation $y = x$ to a polar equation.

►►Solution

$$y = x \quad \text{Rectangular Equation}$$

$$\frac{y}{x} = 1 \quad \text{Recall that } \tan\theta = \frac{y}{x}, \text{ divide both sides by } x.$$

$$\tan\theta = 1 \quad \text{Replace } \frac{y}{x} \text{ with } \tan\theta. \text{ Where does tangent equal 1?}$$

$$\theta = 45°$$
$$\left.\begin{array}{c}\\ \theta = \frac{\pi}{4}\end{array}\right\} \text{ Polar Equation}$$

► EXAMPLE 10

Convert the polar equation $r = 4\cos\theta$ to a rectangular equation.

►►Solution

$$r = 4\cos\theta \quad \text{Polar Equation}$$

$$r \cdot r = r \cdot 4\cos\theta \quad \text{Since the } \cos\theta \text{ does not have the } r \text{ in front, we multiply both sides by } r.$$

$$r^2 = 4r\cos\theta$$

$$x^2 + y^2 = 4x \quad \text{Substitute } x^2 + y^2 \text{ for } r^2 \text{ and } x \text{ for } r\cos\theta \text{ and solve for } y^2.$$

$$y^2 = 4x - x^2 \quad \text{Rectangular Equation}$$

►EXAMPLE 11

Convert the polar equation $r = \dfrac{3}{4+sin\theta}$ **to a rectangular equation.**

►►Solution

$$r = \frac{3}{4 + sin\theta} \quad \text{Polar Equation}$$

$$r(4 + sin\theta) = 3(1) \quad \text{Cross multiply.}$$

$$4r + rsin\theta = 3 \quad \text{Replace } r \text{ with } \sqrt{x^2 + y^2} \text{ and } rsin\theta \text{ with}$$

$$4\sqrt{x^2 + y^2} + y = 3 \quad \text{Isolate } \sqrt{x^2 + y^2}.$$

$$\sqrt{x^2 + y^2} = \frac{3 - y}{4}$$

$$\left(\sqrt{x^2 + y^2}\right)^2 = \left(\frac{3 - y}{4}\right)^2 \quad \text{Square both sides.}$$

$$x^2 + y^2 = \left(\frac{3 - y}{4}\right)^2 \quad \text{Solve for } x^2.$$

$$x^2 = \left(\frac{3 - y}{4}\right)^2 - y^2 \quad \text{Rectangular Equation}$$

►EXAMPLE 12

Convert the polar equation $r = 2csc\theta$ **to a rectangular equation.**

►►Solution

$$r = 2csc\theta \quad \text{Polar Equation}$$

$$r = \frac{2}{sin\theta} \quad \text{Rewrite cosecant using a reciprocal identity and cross multiply.}$$

$$rsin\theta = 2 \quad \text{Replace } rsin\theta \text{ with } y.$$

$$y = 2 \quad \text{Rectangular Equation}$$

►EXAMPLE 13

Convert the polar equation $r = 4$ **to a rectangular equation.**

►►Solution

$$r = 4 \quad \text{Polar Equation}$$

$$\sqrt{x^2 + y^2} = 4 \quad \text{Replace } r \text{ with } \sqrt{x^2 + y^2}.$$

$$\left(\sqrt{x^2 + y^2}\right)^2 = (4)^2 \quad \text{Square both sides.}$$

$$x^2 + y^2 = 16 \quad \text{Rectangular Equation}$$

Symmetry

When graphing polar equations, it his helpful to use symmetry. Polar equations can have symmetry about the polar axis, symmetry about the pole, and symmetry about the line $\theta = \frac{\pi}{2}$ as shown in **Figure 22.9**.

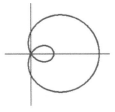

Symmetry about
the polar axis

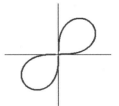

Symmetry about
the pole

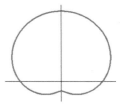

Symmetry about
the line $\theta = \frac{\pi}{2}$

Tests for Symmetry

Symmetry about the Polar Axis

1.   Replace θ with $-\theta$ or r with $-r$ and θ with $\pi - \theta$.

2.   No change, symmetry about the polar axis.

Symmetry about the Pole

1.   Replace r with $-r$ or θ with $\theta + \pi$.

2.   No change, symmetry about the pole.

Symmetry about the line $\theta = \frac{\pi}{2}$

1.   Replace θ with $\pi - \theta$ or θ with $-\theta$ and r with $-r$.

2.   No change, symmetry about the line $\theta = \frac{\pi}{2}$.

►**EXAMPLE 14**

Test $r = 2 + 2cos\theta$ for symmetry about the polar axis, about the pole and about the line $\theta = \frac{\pi}{2}$.

►►**Solution**

Symmetry about the Polar Axis

$r = 2 + 2cos\theta$

$r = 2 + 2cos(-\theta)$ Replace θ with $-\theta$. Cosine is an even function, so $cos(-\theta) = cos\theta$.

$r = 2 + 2cos\theta$ No change, symmetry about the polar axis

Symmetry about the Pole

$r = 2 + 2cos\theta$

$r = 2 + 2cos(\theta + \pi)$ Replace θ with $\theta + \pi$ and apply the cosine sum identity.

$r = 2 + 2[cos\theta cos\pi - sin\theta sin\pi]$

$r = 2 + 2[cos\theta(-1) - sin\theta(0)]$

$r = 2 + 2[-cos\theta]$

$r = 2 - 2cos\theta$ There is a change, no symmetry about the pole.

$r = 2 + 2cos\theta$ Replace r with $-r$.

$-r = 2 + 2cos\theta$ There is a change, no symmetry about the pole.

Symmetry about the line $\theta = \frac{\pi}{2}$

$r = 2 + 2cos\theta$

$r = 2 + 2cos(\pi - \theta)$ Replace θ with $\pi - \theta$ and apply the cosine sum identity.

$r = 2 + 2[cos\theta cos\pi + sin\theta sin\pi]$

$r = 2 + 2[cos\theta(-1) - sin\theta(0)]$

$r = 2 + 2[-cos\theta]$

$r = 2 - 2cos\theta$ There is a change, no symmetry about the pole.

Symmetry with respect to the polar axis.

▶EXAMPLE 15

Test $r^2 = cos2\theta$ for symmetry about the polar axis, about the pole and about the line $\theta = \frac{\pi}{2}$.

▶▶Solution

Symmetry about the Polar Axis

$$r^2 = cos2\theta$$

$r^2 = cos2(-\theta)$ Replace θ with $-\theta$. Cosine is an even function, so $cos(-\theta) = cos\theta$.

$r^2 = cos2\theta$ No change, symmetry about the polar axis

Symmetry about the Pole

$$r^2 = cos2\theta$$

$(-r)^2 = cos2\theta$ Replace r with $-r$.

$r^2 = cos2\theta$ No change, symmetry about the pole.

Symmetry about the line $\theta = \frac{\pi}{2}$

$$r^2 = cos2\theta$$

$r^2 = cos2(\pi - \theta)$ Replace θ with $\pi - \theta$ and apply the cosine sum identity.

$r^2 = cos(2\pi - 2\theta)$ Apply the cosine sum identity.

$r^2 = cos2\pi cos2\theta - sin2\pi sin2\theta$

$r^2 = (1)cos2\pi - (0)sin2\theta$

$r^2 = cos2\theta$ No change, symmetry about the pole.

Symmetry with respect about the polar axis, about the pole and about the line $\theta = \frac{\pi}{2}$.

Graphing Polar Equations

The graph of a polar equation $r = f(\theta)$ consists of coordinates (r, θ) that satisfy the equation. The following are some common polar curves.

Circles and Spiral

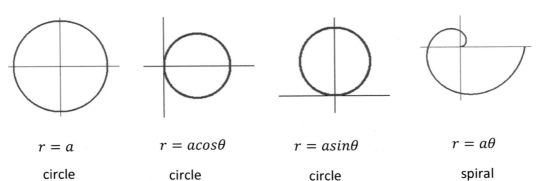

| $r = a$ | $r = acos\theta$ | $r = asin\theta$ | $r = a\theta$ |
| circle | circle | circle | spiral |

Limacons

$$r = a \pm b\sin\theta \text{ or } r = a \pm b\cos\theta \text{ where } a > 0 \text{ and } b > 0$$

Orientation depends on the trigonometric function and the sign of b.

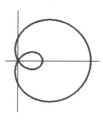

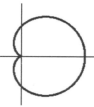

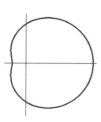

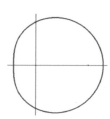

$a < b$	$a = b$	$a > b$	$a \geq 2b$
limacon with inner loop	Cardiod	dimpled limacon	convex limacon

Roses

$$r = a\sin(n\theta) \text{ or } r = a\cos(n\theta)$$

If n is odd, n-leaved.

If n is even, 2n-leaved.

$r = a\cos2\theta$	$r = a\cos3\theta$	$r = a\cos4\theta$	$r = a\cos5\theta$
4-leaved rose	3-leaved rose	8-leaved rose	5-leaved rose

Lemniscates

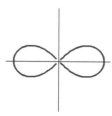

$r^2 = a^2\cos2\theta$	$r^2 = a^2\sin2\theta$
lemniscate	lemniscate

► EXAMPLE 16

Graph the polar equation, $r = 2 + 2cos\theta$

►►Solution

θ°	r
0°	$2 + 2cos0° = 4$
15°	$2 + 2cos15° = 3.9$
30°	$2 + 2cos30° = 3.7$
45°	$2 + 2cos45° = 3.4$
60°	$2 + 2cos60° = 3$
75°	$2 + 2cos75° = 2.5$
90°	$2 + 2cos90° = 2$
105°	$2 + 2cos105° = 1.5$
120°	$2 + 2cos120° = 1$
135°	$2 + 2cos135° = 0.6$
150°	$2 + 2cos150° = 0.3$
165°	$2 + 2cos165° = 0.1$
180°	$2 + 2cos180° = 0$
195°	$2 + 2cos195° = 0.1$
210°	$2 + 2cos210° = 0.3$
225°	$2 + 2cos225° = 0.6$
240°	$2 + 2cos240° = 1$
255°	$2 + 2cos255° = 1.5$
270°	$2 + 2cos270° = 2$
285°	$2 + 2cos285° = 2.5$
300°	$2 + 2cos300° = 3$
315°	$2 + 2cos315° = 3.4$
330°	$2 + 2cos330° = 3.7$
345°	$2 + 2cos345° = 3.9$
360°	$2 + 2cos360° = 4$

1. Recognize that the polar equation will graph as a cardioid.

2. Symmetry about the polar axis (see Example 14).

3. Complete the table.

4. Plot the points.

5. Draw the curve.

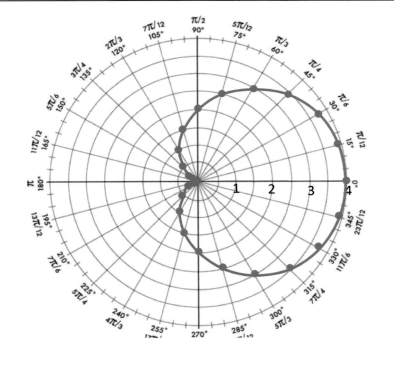

► EXAMPLE 17

Graph the polar equation, $r = 5sin3\theta$.

►►Solution

θ°	r
0°	$5sin3(0°) = 0$
15°	$5sin3(15°) = 3.5$
30°	$5sin3(30°) = 5$
45°	$5sin3(45°) = 3.5$
60°	$5sin3(60°) = 0$
75°	$5sin3(75°) = -3.5$
90°	$5sin3(90°) = -5$
105°	$5sin3(105°) = -3.5$
120°	$5sin3(120°) = 0$
135°	$5sin3(135°) = 3.5$
150°	$5sin3(150°) = 5$
165°	$5sin3(165°) = 3.5$
180°	$5sin3(180°) = 0$
195°	$5sin3(195°) = -3.5$
210°	$5sin3(210°) = -5$
225°	$5sin3(225°) = -3.5$
240°	$5sin3(240°) = 0$
255°	$5sin3(255°) = 3.5$
270°	$5sin3(270°) = 5$
285°	$5sin3(285°) = 3.5$
300°	$5sin3(300°) = 0$
315°	$5sin3(315°) = -3.5$
330°	$5sin3(330°) = -5$
345°	$5sin3(345°) = -3.5$
360°	$5sin3(360°) = 0$

1. Recognize that the polar equation will graph as a 3-leaved rose.

2. Symmetry to the line $\theta = \frac{\pi}{2}$.

3. Complete the table.

4. Plot the points.

5. Draw the curve.

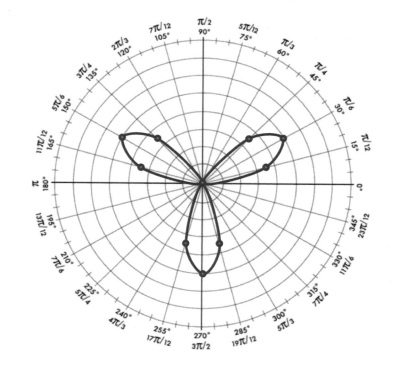

▶ EXAMPLE 18

Graph the polar equation, $r = 3 + 5sin\theta$.

▶▶ Solution

θ°	r
0°	$3 + 5sin\theta = 3$
15°	$3 + 5sin\theta = 4.3$
30°	$3 + 5sin\theta = 5.5$
45°	$3 + 5sin\theta = 6.5$
60°	$3 + 5sin\theta = 7.3$
75°	$3 + 5sin\theta = 7.8$
90°	$3 + 5sin\theta = 8$
105°	$3 + 5sin\theta = 7.8$
120°	$3 + 5sin\theta = 7.3$
135°	$3 + 5sin\theta = 6.5$
150°	$3 + 5sin\theta = 5.5$
165°	$3 + 5sin\theta = 4.3$
180°	$3 + 5sin\theta = 3$
195°	$3 + 5sin\theta = 1.7$
210°	$3 + 5sin\theta = 0.5$
225°	$3 + 5sin\theta = -0.5$
240°	$3 + 5sin\theta = -1.3$
255°	$3 + 5sin\theta = -1.8$
270°	$3 + 5sin\theta = -2$
285°	$3 + 5sin\theta = -1.8$
300°	$3 + 5sin\theta = -1.3$
315°	$3 + 5sin\theta = -0.5$
330°	$3 + 5sin\theta = 0.5$
345°	$3 + 5sin\theta = 1.7$
360°	$3 + 5sin\theta = 3$

1. Recognize that the polar equation will graph as a limacon with inner loop.

2. Symmetry to the line $\theta = \frac{\pi}{2}$.

3. Complete the table.

4. Plot the points.

5. Draw the curve.

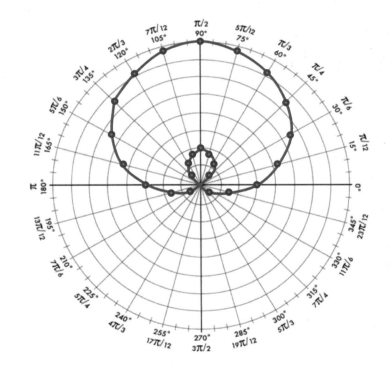

HOMEWORK

Objective 1

Plot the polar coordinates.

See Example 1.

1. $(4, 120°)$ 2. $(-3, 210°)$ 3. $\left(-2, \frac{4\pi}{3}\right)$

4. $\left(4, \frac{5\pi}{6}\right)$ 5. $(-4, -60°)$ 6. $\left(3, -\frac{3\pi}{4}\right)$

Objective 2

Convert the rectangular coordinates to polar coordinates, where $[0°, 360°)$. Give exact values (no decimals) for r and round θ values to the nearest tenth when necessary.

See Example 4.

7. $(3, 0)$ 8. $(\sqrt{3}, -1)$ 9. $(4, 4)$

10. $(-1, -\sqrt{3})$ 11. $(0, -5)$ 12. $(6, -8)$

13. $(-3, -4)$ 14. $(0, 7)$ 15. $(6, -6)$

Objective 3

Convert the polar coordinates to rectangular coordinates. Give exact values (no decimals).

See Example 5.

16. $(4, 60°)$ 17. $(8, 150°)$ 18. $(4, -45°)$

19. $\left(-5, \frac{5\pi}{3}\right)$ 20. $\left(5, \frac{2\pi}{3}\right)$ 21. $\left(-8, \frac{5\pi}{4}\right)$

22. $\left(-4, \frac{11\pi}{6}\right)$ 23. $(10, 135°)$ 24. $(3, 270°)$

Objective 4

Convert the rectangular equation to a polar equation.

See Examples 6, 7, 8 and 9.

25. $x^2 + y^2 = 36$ 26. $x^2 + y^2 = 49$ 27. $3x - 2y - 4 = 0$

28. $3x + y + 2 = 0$ 29. $3xy = 4$ 30. $y = 2x$

31. $x = 6$ 32. $y = 2x - 1$ 33. $y^2 = 2x$

34. $x^2 = 2y$ 35. $2xy = 1$ 36. $x^2 + y^2 = 3x$

Objective 5

Convert the polar equation to a rectangular equation.

See Examples 10, 11, 12 and 13.

37. $r = 3$

38. $3r = 5$

39. $\theta = \dfrac{5\pi}{4}$

40. $\theta = \dfrac{\pi}{3}$

41. $r + r\sin\theta = 5$

42. $r = 3\sin\theta$

43. $r = \dfrac{4}{\cos\theta}$

44. $r = \dfrac{-2}{\sin\theta}$

45. $r = 3\sec\theta$

46. $r = 4\csc\theta$

47. $r = \dfrac{2}{1+\sin\theta}$

48. $r = \dfrac{3}{2-\cos\theta}$

Objective 6

Test for symmetry with respect about the polar axis, about the pole and about the line $\theta =$ $\dfrac{\pi}{2}$.

See Examples 14 and 15.

49. $r = 2 + 2\sin\theta$

50. $r = 2\sin3\theta$

51. $r = 1 + 2\cos\theta$

52. $r = 3 + \cos\theta$

Objective 7

Graph each polar equation.

See Examples 16, 17 and 18.

53. $r = 3\cos3\theta$

54. $r = 4 - 4\cos\theta$

55. $r^2 = 4\cos2\theta$

56. $r = 2\sin2\theta$

57. $r = 4 + \cos\theta$

58. $r = 3 - 2\sin\theta$

59. $r = 1 + 2\cos\theta$

60. $r^2 = 4\sin2\theta$

61. $r = 3 + 2\cos\theta$

62. $r = 2 + 2\sin\theta$

ANSWERS

1.

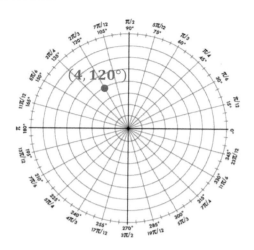

2.

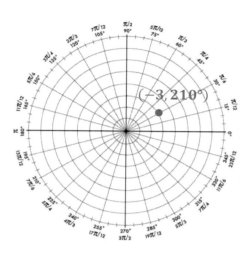

3.

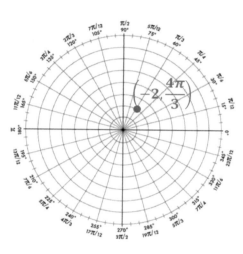

4.

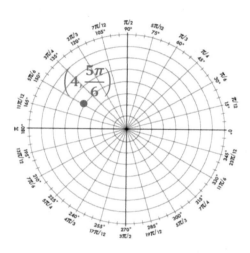

5.

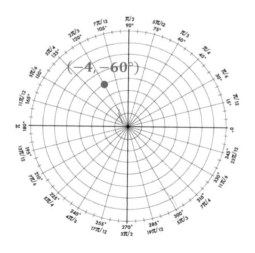

6.

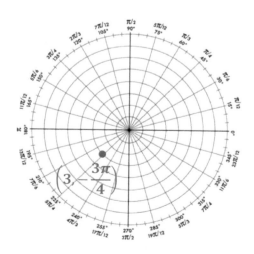

7. $(3, 0°)$

8. $(2, 330°)$

9. $\left(4\sqrt{2}, 45°\right)$

10. $(2, 240°)$

11. $(5, 270°)$

12. $(10, 306.9°)$

13. $(5, 233.1°)$

14. $(7, 90°)$

15. $\left(6\sqrt{2}, 315°\right)$

16. $\left(2, 2\sqrt{3}\right)$

17. $\left(-4\sqrt{3}, 4\right)$

18. $\left(2\sqrt{2}, -2\sqrt{2}\right)$

19. $\left(-\frac{5}{2}, \frac{5\sqrt{3}}{2}\right)$

20. $\left(-\frac{5}{2}, \frac{5\sqrt{3}}{2}\right)$

21. $\left(4\sqrt{2}, 4\sqrt{2}\right)$

22. $\left(-2\sqrt{3}, 2\right)$

23. $\left(-5\sqrt{2}, 5\sqrt{2}\right)$

24. $(0, -3)$

25. $r = 6$

26. $r = 7$

27. $r = \dfrac{4}{3\cos\theta - 2\sin\theta}$

28. $r = \dfrac{-2}{3\cos\theta + \sin\theta}$

29. $r^2 = \dfrac{4}{3\cos\theta\sin\theta}$

30. $\tan\theta = 2$

31. $r = 6\sec\theta$

32. $r = \dfrac{-1}{\sin\theta - 2\cos\theta}$

33. $r = 2\cot\theta\csc\theta$

34. $r = 2\tan\theta\sec\theta$

35. $r^2 = \csc 2\theta$

36. $r^2 = 3r\cos\theta$

37. $x^2 + y^2 = 9$

38. $x^2 + y^2 = \dfrac{25}{9}$

39. $y = x$

40. $y = \sqrt{3}x$

41. $x^2 + y^2 = (5 - y)^2$

42. $x^2 + y^2 = 3y$

43. $x = 4$

44. $y = -2$

45. $x = 3$

46. $y = 4$

47. $x^2 + y^2 = (2 - y)^2$

48. $x^2 + y^2 = \left(\dfrac{x+3}{2}\right)^2$

49. $\theta = \dfrac{\pi}{2}$

50. $\theta = \dfrac{\pi}{2}$

51. Polar axis

52. Polar axis

53.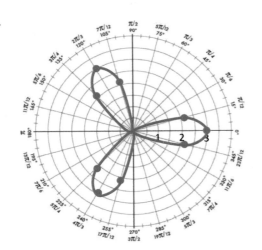

54.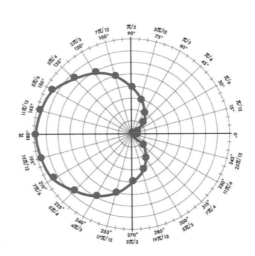

55.

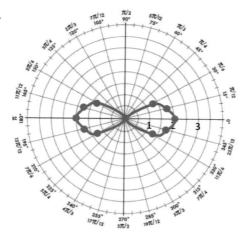

56.

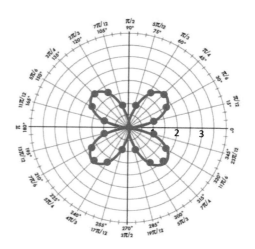

57.

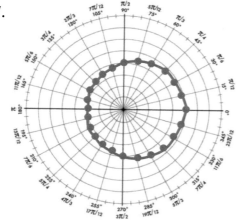

58.

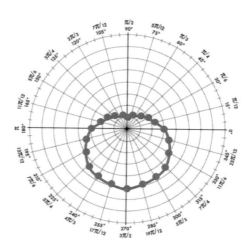

59.

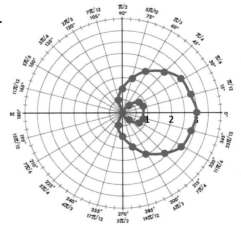

60.

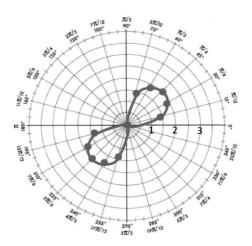

61.

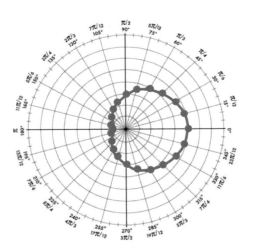

62.

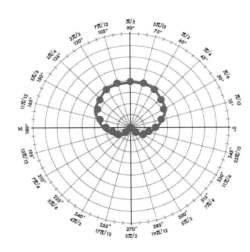

Section 23:   Parametric Equations

Learning Outcomes:

- The student will illustrate and examine the graphs of: the six trigonometric functions and their inverses, parametric equations, and polar equations.
- The student will correctly memorize and apply trigonometric formulas, definitions, identities, and properties.

Objectives: At the conclusion of this lesson you should be able to:

1.   Graph parametric equations.

Given a set of points $P(x, y)$ where $x = f(t)$ and $y = g(t)$ with f and g both defined over a given interval, the equations $x = f(t)$ and $y = g(t)$ are called parametric equations and the parameter is t. In other words, both x and y change, usually depending on time, t. Not all parametric equations have a parameter that represents time. Sometimes an angle θ is the parameter.

When graphing the curve of a pair of parametric equations, we plot points in the xy-plane. Each ordered pair (x, y) is determined from a chosen value of the parameter, t or θ. Plotting the resulting ordered pairs in the order of increasing values of the parameter, t or θ, traces the curve in a specific direction. This is called the orientation of the curve.

Steps for Graphing a Parametric Equation

Step 1: Eliminate the parameter and write as a single equation.

Step 2: Apply the restrictions on the parameter and get ordered pairs to graph.

Step 3: Plot the ordered pairs.

Step 4: Draw the curve and indicate the orientation (direction) of the graph.

►EXAMPLE 1

Use the four steps and sketch the curve $x = t + 3$, $y = 1 + t^2$, for $t \in [-3,3]$.

►►Solution

Step 1: Eliminate the parameter and write as a single equation.

$x = t + 3$ *Solve for the parameter, t.* $y = 1 + t^2$ $y = 1 + (x - 3)^2$

$x - 3 = t$

Substitute for t.

Single equation without
the parameter.

Step 2: Apply the restrictions on the parameter and get ordered pairs to graph.

t	$x = t + 3$	$y = 1 + t^2$
-3	$-3 + 3 = 0$	$1 + (-3)^2 = 10$
-2	$-2 + 3 = 1$	$1 + (-2)^2 = 5$
-1	$-1 + 3 = 2$	$1 + (-1)^2 = 2$
0	$0 + 3 = 3$	$1 + (0)^2 = 1$
1	$1 + 3 = 4$	$1 + (1)^2 = 2$
2	$2 + 3 = 5$	$1 + (2)^2 = 5$
3	$3 + 3 = 6$	$1 + (3)^2 = 10$

Step 3: Plot the ordered pairs.

Step 4: Draw the curve and indicate the
orientation (direction) of the graph.

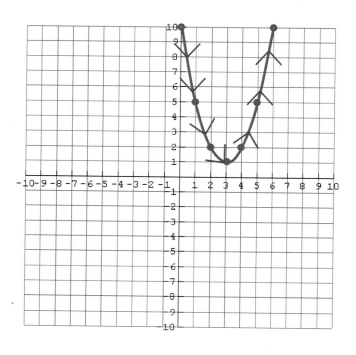

► EXAMPLE 2

Use the four steps and sketch the curve $x = t - 1$, $y = 2 - t^2$, for $t \in [-3,3]$.

►► Solution

Step 1: Eliminate the parameter and write as a single equation.

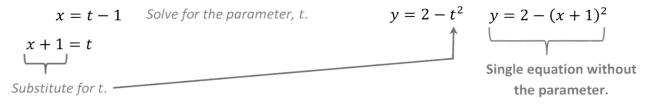

$x = t - 1$ *Solve for the parameter, t.*

$x + 1 = t$

Substitute for t.

$y = 2 - t^2$ $y = 2 - (x + 1)^2$

Single equation without
the parameter.

Step 2: Apply the restrictions on the parameter and get ordered pairs to graph.

t	$x = t - 1$	$y = 2 - t^2$
-3	$-3 - 1 = -4$	$2 - (-3)^2 = -7$
-2	$-2 - 1 = -3$	$2 - (-2)^2 = -2$
-1	$-1 - 1 = -2$	$2 - (-1)^2 = 1$
0	$0 - 1 = -1$	$2 - (0)^2 = 2$
1	$1 - 1 = 0$	$2 - (1)^2 = 1$
2	$2 - 1 = 1$	$2 - (2)^2 = -2$
3	$3 - 1 = 2$	$2 - (3)^2 = -7$

Step 3: Plot the ordered pairs.

Step 4: Draw the curve and indicate the
orientation (direction) of the graph.

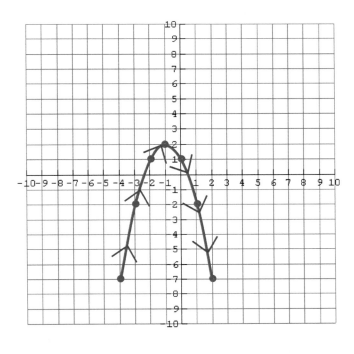

► EXAMPLE 3

Use the four steps and sketch the curve $x = 2sin\theta$, $y = -3cos\theta$, for $\theta \in [0, 2\pi)$.

►► Solution

Step 1: Eliminate the parameter and write as a single equation.

$$x = 2sin\theta \qquad y = -3cos\theta \qquad\qquad cos^2\theta + sin^2\theta = 1$$

$$\frac{x}{2} = sin\theta \qquad \frac{y}{-3} = cos\theta \qquad\qquad \left(\frac{x}{2}\right)^2 + \left(-\frac{y}{3}\right)^2 = 1$$

$$\frac{x^2}{4} + \frac{y^2}{9} = 1 \qquad \text{Single equation without the parameter.}$$

Step 2: Apply the restrictions on the parameter and get ordered pairs to graph.

t	$x = 2sin\theta$	$y = -3cos\theta$
0	$2sin0 = 0$	$-3cos0 = -3$
$\frac{\pi}{4}$	$2sin\frac{\pi}{4} = 1.4$	$-3cos\frac{\pi}{4} = -2.1$
$\frac{\pi}{2}$	$2sin\frac{\pi}{2} = 2$	$-3cos\frac{\pi}{2} = 0$
$\frac{3\pi}{4}$	$2sin\frac{3\pi}{4} = 1.4$	$-3cos\frac{3\pi}{4} = 2.1$
π	$2sin\pi = 0$	$-3cos\pi = 3$
$\frac{5\pi}{4}$	$2sin\frac{5\pi}{4} = -1.4$	$-3cos\frac{5\pi}{4} = 2.1$
$\frac{3\pi}{2}$	$2sin\frac{3\pi}{2} = -2$	$-3cos\frac{3\pi}{2} = 0$
$\frac{7\pi}{4}$	$2sin\frac{7\pi}{4} = -1.4$	$-3cos\frac{7\pi}{4} = -2.1$
2π	$2sin2\pi = 0$	$-3cos2\pi = -3$

Step 3: Plot the ordered pairs.

Step 4: Draw the curve and indicate the orientation (direction) of the graph.

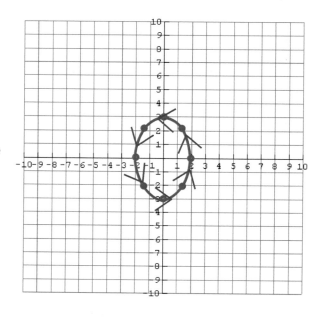

► **EXAMPLE 4**

Use the four steps and sketch the curve $x = \tan^2\theta$, $y = 3\cos\theta$, **for** $\theta \in [0, \pi]$, $\theta \neq \frac{\pi}{2}$.

►►**Solution**

Step 1: Eliminate the parameter and write as a single equation.

$$x = \tan^2\theta \qquad y = 3\cos\theta \qquad\qquad 1 + \tan^2\theta = \sec^2\theta$$

$$\frac{y}{3} = \cos\theta \qquad\qquad 1 + x = \left(\frac{3}{y}\right)^2$$

$$\frac{3}{y} = \sec\theta \qquad\qquad 1 + x = \frac{9}{y^2}$$

$$y^2 = \frac{9}{1+x} \qquad \text{Single equation without the parameter.}$$

Step 2: Apply the restrictions on the parameter and get ordered pairs to graph.

t	$x = \tan^2\theta$	$y = 3\cos\theta$
0	$\tan^2 0 = 0$	$3\cos 0 = 3$
$\frac{\pi}{6}$	$\tan^2\frac{\pi}{6} = 0.3$	$3\cos\frac{\pi}{6} = 2.6$
$\frac{\pi}{4}$	$\tan^2\frac{\pi}{4} = 1$	$3\cos\frac{\pi}{4} = 2.1$
$\frac{\pi}{3}$	$\tan^2\frac{\pi}{3} = 3$	$3\cos\frac{\pi}{4} = 1.5$
$\frac{\pi}{2}$	$undefined$	
$\frac{2\pi}{3}$	$\tan^2\frac{2\pi}{3} = 3$	$3\cos\frac{2\pi}{3} = -1.5$
$\frac{3\pi}{4}$	$\tan^2\frac{3\pi}{4} = 1$	$3\cos\frac{3\pi}{4} = -2.1$
$\frac{5\pi}{6}$	$\tan^2\frac{5\pi}{6} = 0.3$	$3\cos\frac{5\pi}{6} = -2.6$
π	$\tan^2\pi = 0$	$3\cos\pi = -3$

Step 3: Plot the ordered pairs.

Step 4: Draw the curve and indicate the orientation (direction) of the graph.

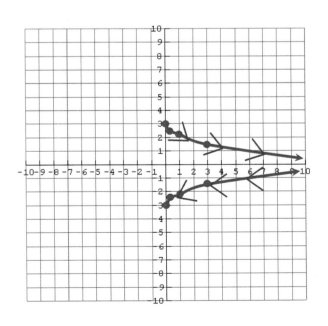

HOMEWORK

Objective 1

Use the four steps and sketch the curve of the parametric equation.

See Examples 1-4.

1. $x = t + 2$, $y = t^2 - 1$, for $t \in [-3,3]$

2. $x = 2t$, $y = t + 1$, for $t \in [-2,3]$

3. $x = \sqrt{t}$, $y = 3t - 4$, for $t \in [0,4]$

4. $x = \sqrt{t}$, $y = t^2$, for $t \in [0,2]$

5. $x = (t + 2)^2$, $y = t^2 - 1$, for $t \in [-3,1]$

6. $x = t - 2$, $y = \sqrt{t - 1}$, for $t \in [1,10]$

7. $x = 4\cos\theta$, $y = 3\sin\theta$, for $\theta \in [0, 2\pi)$

8. $x = -3\sin\theta$, $y = 2\cos\theta$, for $\theta \in [0, 2\pi)$

9. $x = 2\cos\theta$, $y = \sin\theta$, for $\theta \in [0, 2\pi)$

10. $x = \cos2\theta$, $y = \cos\theta$, for $\theta \in [0, \pi)$

11. $x = \tan^2\theta$, $y = 4\cos\theta$, for $\theta \in [0, \pi]$, $\theta \neq \dfrac{\pi}{2}$

12. $x = \sec^2\theta$, $y = 5\cos\theta$, for $\theta \in [0, \pi]$, $\theta \neq \dfrac{\pi}{2}$

ANSWERS

1. $y = (x - 2)^2 - 1$

t	x	y
-3	-1	8
-2	0	3
-1	1	0
0	2	-1
1	3	0
2	4	3
3	5	8

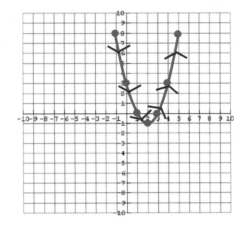

2. $y = \frac{1}{2}x + 1$

t	x	y
-2	-4	-1
-1	-2	0
0	0	1
1	2	2
2	4	3
3	6	4

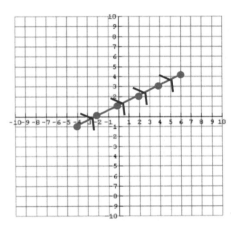

3. $y = x^2 - 4$

t	x	y
0	0	-4
1	1	-1
2	1.4	2
3	1.7	5
4	2	8

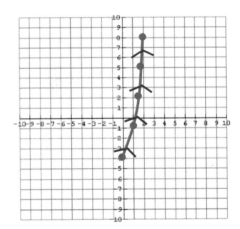

4. $y = x^4$

t	x	y
0	0	0
1	1	1
2	1.4	4

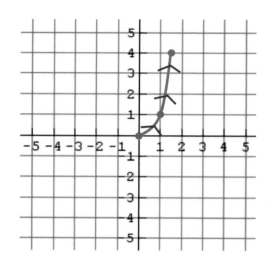

5. $y = \left(\sqrt{x} - 2\right)^2 - 1$

t	x	y
-3	1	8
-2	0	3
-1	1	0
0	4	-1
1	9	0

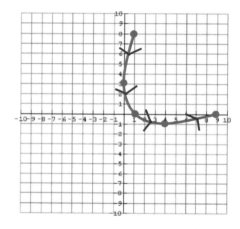

6. $y = \sqrt{x} + 1$

t	x	y
1	-1	0
2	0	1
3	1	1.4
4	2	1.7
5	3	2
6	4	2.2
7	5	2.4
8	6	2.6
9	7	2.8
10	8	3

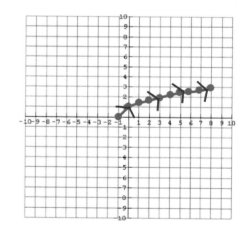

7. $\dfrac{x^2}{16} + \dfrac{y^2}{9} = 1$

t	x	y
0	4	0
$\dfrac{\pi}{4}$	2.8	2.1
$\dfrac{\pi}{2}$	0	3
$\dfrac{3\pi}{4}$	-2.8	2.1
π	-4	0
$\dfrac{5\pi}{4}$	-2.8	-2.1
$\dfrac{3\pi}{2}$	0	-3
$\dfrac{7\pi}{4}$	2.8	-2.1
2π	4	0

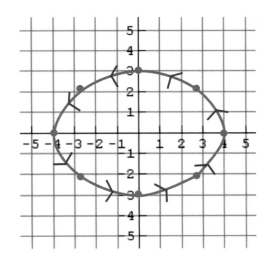

8. $\dfrac{x^2}{9} + \dfrac{y^2}{4} = 1$

t	x	y
0	0	2
$\dfrac{\pi}{4}$	-2.1	1.4
$\dfrac{\pi}{2}$	-3	0
$\dfrac{3\pi}{4}$	-2.1	-1.4
π	0	-2
$\dfrac{5\pi}{4}$	2.1	-1.4
$\dfrac{3\pi}{2}$	3	0
$\dfrac{7\pi}{4}$	2.1	1.4
2π	0	2

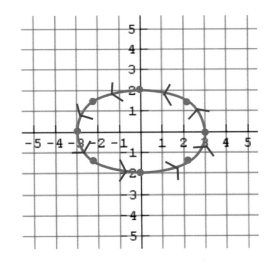

9. $\dfrac{x^2}{4} + y^2 = 1$

t	x	y
0	2	0
$\dfrac{\pi}{4}$	1.4	0.7
$\dfrac{\pi}{2}$	0	1
$\dfrac{3\pi}{4}$	-1.4	0.7
π	-2	0
$\dfrac{5\pi}{4}$	-1.4	-0.7
$\dfrac{3\pi}{2}$	0	-1
$\dfrac{7\pi}{4}$	1.4	-0.7
2π	2	0

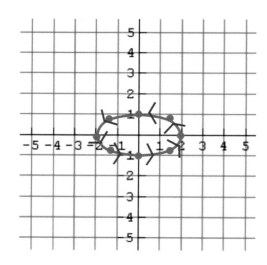

10. $x = 2y^2 - 1$

t	x	y
0	1	1
$\dfrac{\pi}{4}$	0	0.7
$\dfrac{\pi}{2}$	-1	0
$\dfrac{3\pi}{4}$	0	-0.7
π	1	-1

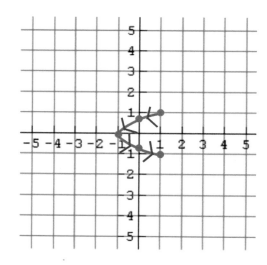

11. $x = \dfrac{16}{y^2} - 1$

t	x	y
0	0	4
$\dfrac{\pi}{6}$	0.3	3.4
$\dfrac{\pi}{4}$	1	2.8
$\dfrac{\pi}{3}$	3	2
$\dfrac{3\pi}{8}$	5.8	1.5
$\dfrac{\pi}{2}$	undefined	
$\dfrac{5\pi}{8}$	5.8	-1.5
$\dfrac{2\pi}{3}$	3	-2
$\dfrac{3\pi}{4}$	1	-2.8
$\dfrac{5\pi}{6}$	0.3	-3.4
π	0	-4

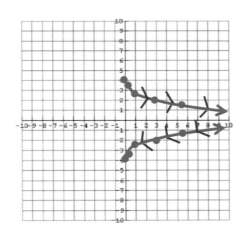

12. $x = \dfrac{25}{y^2}$

t	x	y
0	1	5
$\dfrac{\pi}{6}$	1.3	4.3
$\dfrac{\pi}{4}$	2	3.5
$\dfrac{\pi}{3}$	4	2.5
$\dfrac{3\pi}{8}$	6.8	1.9
$\dfrac{\pi}{2}$	undefined	
$\dfrac{5\pi}{8}$	6.8	-1.9
$\dfrac{2\pi}{3}$	4	-2.5
$\dfrac{3\pi}{4}$	2	-3.5
$\dfrac{5\pi}{6}$	1.3	-4.3
π	1	-5

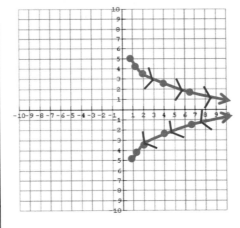

Index

Plane Trigonometry

Made in the USA
Columbia, SC
01 January 2019